Fuel Production with Heterogeneous Catalysis

Fuel Production with Heterogeneous Catalysis

Edited by Jacinto Sá

CRC Press
Taylor & Francis Group
Boca Raton London New York

CRC Press is an imprint of the
Taylor & Francis Group, an **informa** business

CRC Press
Taylor & Francis Group
6000 Broken Sound Parkway NW, Suite 300
Boca Raton, FL 33487-2742

First issued in paperback 2017

© 2015 by Taylor & Francis Group, LLC
CRC Press is an imprint of Taylor & Francis Group, an Informa business

No claim to original U.S. Government works

ISBN-13: 978-1-4822-0371-4 (hbk)
ISBN-13: 978-1-138-07719-5 (pbk)

Library of Congress Cataloging-in-Publication Data

Fuel production with heterogeneous catalysis / edited by Jacinto Sá.
    pages cm
   Includes bibliographical references and index.
   ISBN 978-1-4822-0371-4
   1. Fuel. 2. Heterogeneous catalysis. I. Sá, Jacinto, editor.

TP318.F835 2015
338.4'76626--dc23   2014022169

Visit the Taylor & Francis Web site at
http://www.taylorandfrancis.com

and the CRC Press Web site at
http://www.crcpress.com

Contents

Preface...vii
Editor ix
Contributors xi

Chapter 1 Solar Water Splitting Using Semiconductor Systems 1
          Lorenzo Rovelli and K. Ravindranathan Thampi

Chapter 2 Photocatalytic Hydrogen Evolution......................................63
          Yusuke Yamada and Shunichi Fukuzumi

Chapter 3 CO$_2$ to Fuels 93
          Atsushi Urakawa and Jacinto Sá

Chapter 4 Methane Activation and Transformation over Nanocatalysts 123
          Rajaram Bal and Ankur Bordoloi

Chapter 5 Fischer–Tropsch: Fuel Production with Cobalt Catalysts 147
          Cristina Paun, Jacinto Sá, and Kalala Jalama

Chapter 6 Syngas to Methanol and Ethanol......................................169
          Martin Muhler and Stefan Kaluza

Chapter 7 Steam Reforming 193
          Karin Föttinger

Chapter 8 Biomass to Liquid Biofuels via Heterogeneous Catalysis................213
          Michael Stöcker and Roman Tschentscher

Chapter 9 Catalytic Pyrolysis of Lignocellulosic Biomass..............................253
          K. Seshan

Chapter 10 Recent Trends in the Purification of H$_2$ Streams by Water–Gas
            Shift and PROX 281

            A. Sepúlveda-Escribano and J. Silvestre-Albero

Index 303

Preface

This book describes the importance of catalysis for the sustainable production of fuels, focused primarily on the state-of-the-art catalysts and catalytic processes anticipated to play a pivotal role in the production of fuels. The growth in energy prices, environmental trepidations, and population commanded the development of new and/or improved processes to attain sustainability and energy stability, and ultimately mankind's current way of life.

Heterogeneous catalysis plays an essential role in improving industrial processes, including a water–gas shift, and coal combustion in respect to greenhouse gas emissions and profitability. Novel technologies, including solar energy harvesting and conversion, provide an attractive means toward carbon independence; however, they are not yet economical, thus requiring further research.

Herein, we compile the latest developments on fuel production processes with heterogeneous catalysis, including reaction mechanism schemes, engineering solutions, and perspectives for the field. The book is aimed for undergraduate and graduate students as well as scientists working in the area. It should be mentioned that the goal of the book is to provide an overview of the field. The examples were chosen to cover a larger range of scientific experiments as concisely as possible. Therefore, and on behalf of the authors, we would like to apologize for any work that has not been referenced or mentioned. The omission was decided simply on the basis of chapter concision. Finally, the book was written in a language that we consider accessible to most science undergraduate students. Technical terms were mentioned only when strictly necessary.

On a personal note, I thank the chapter authors for their contribution and all the scientific works that made the execution of this book possible.

Jacinto Sá
Polish Academy of Sciences

Editor

Dr. Jacinto Sá (PhD in physical chemistry) is the Modern Heterogeneous Catalysis (MoHCa) group leader at the Institute of Physical Chemistry, Polish Academy of Sciences, Warsaw, Poland. He earned an MSc in analytical chemistry at the Universidade de Aveiro, Portugal, and did his research project at the Vienna University of Technology, Vienna, Austria. He earned a PhD degree at the University of Aberdeen, Scotland, in the field of catalysis and surface science. In 2007, he moved to the CenTACat group at Queen's University Belfast, Belfast, Northern Ireland, to start his first postdoctoral fellowship under the guidance of Professors Robbie Burch and Chris Hardacre. During his stay, he was awarded an R&D100 for his involvement in the development of SpaciMS equipment (Hiden Analytical, Warrington, UK). In 2010, he moved to Switzerland to start his second postdoctoral fellowship at ETH Zurich and the Paul Scherrer Institute. His research efforts were focused on the adaptation of high-resolution x-ray techniques to the study of catalysts and nanomaterials under working conditions. In 2013, he joined the Laboratory of Ultrafast Spectroscopy, the École Polytechnique Fédérale de Lausanne (EPFL), Switzerland, to expand the use of high-resolution x-ray techniques into the ultrafast domain and to take advantage of the newly developed XFEL facilities.

Currently, Dr. Sá's research efforts are focused on understanding the elemental steps of catalysis, in particular those taking place in artificial photosynthesis and nanocatalytic systems used in the production of fine chemicals and pharmaceuticals. His experience in using accelerator-based light sources to diagnose the mechanisms by which important catalytic processes proceed, and more recently conventional ultrafast laser sources, makes him one of the most experienced researchers in the world in this area. He has more than 70 publications in international scientific journals and more than 20 oral and 50 poster presentations at scientific congresses.

Sá is married to Cristina Paun and is expecting his first child, a boy to be named Lucca V. Sá. He is a member of Portuguese think tank O Contraditorio, which is part of the English volunteer group of the Red Cross Zurich, and a part-time DJ (DJ Sound It). He enjoys fine art, in particular impressionism and surrealism, traveling, music, cinema, and fine dining.

Contributors

Rajaram Bal
CSIR-Indian Institute of Petroleum
Dehradun, India

Ankur Bordoloi
CSIR-Indian Institute of Petroleum
Dehradun, India

Karin Föttinger
Vienna University of Technology
Vienna, Austria

Shunichi Fukuzumi
Graduate School of Engineering
Osaka University, ALCA, Japan Science
    and Technology Agency (JST)
Osaka, Japan

and

Ewha Womans University
Seoul, Korea

Kalala Jalama
University of Johannesburg
Johannesburg, Republic
    of South Africa

Stefan Kaluza
Fraunhofer Institute for Environmental,
    Safety, and Energy Technology
    (UMSICHT)
Oberhausen, Germany

Martin Muhler
Ruhr-University Bochum
Bochum, Germany

Cristina Paun
ETH Zurich
Zurich, Switzerland

Lorenzo Rovelli
SB-SCGC, École Polytechnique
    Fédérale de Lausanne
Lausanne, Switzerland

Jacinto Sá
Institute of Physical Chemistry
Polish Academy of Sciences
Warsaw, Poland

A. Sepúlveda-Escribano
Universidad de Alicante
Alicante, Spain

K. Seshan
University of Twente
Enschede, The Netherlands

J. Silvestre-Albero
Universidad de Alicante
Alicante, Spain

Michael Stöcker
SINTEF Materials and Chemistry
Oslo, Norway

K. Ravindranathan Thampi
University College Dublin
Dublin, Ireland

Roman Tschentscher
SINTEF Materials and Chemistry
Oslo, Norway

Atsushi Urakawa
Institute of Chemical Research
  of Catalonia (ICIQ)
Tarragona, Spain

Yusuke Yamada
Osaka University
Osaka, Japan

1 Solar Water Splitting Using Semiconductor Systems

Lorenzo Rovelli and K. Ravindranathan Thampi

CONTENTS

1.1 Introduction on Solar Water Splitting..2
    1.1.1 General Introduction...2
    1.1.2 Historical Background on Solar Water Splitting 3
    1.1.3 Theoretical Principles of Semiconductor-Based Solar
          Water Splitting 5
          1.1.3.1 Key Properties of Semiconductor Materials.........................5
          1.1.3.2 Semiconductor–Liquid Junction and Semiconductor
                Photoelectrodes 6
    1.1.4 An Overview of the Main Approaches toward Solar
          Water Splitting..7
1.2 Materials for Solar Water Splitting: Requirements and Current Trends 8
    1.2.1 Semiconductor Materials for Solar Water Splitting: General
          Requirements 8
          1.2.1.1 Photoactivity of Semiconductor Materials 9
          1.2.1.2 Stability of Semiconductor Materials 10
    1.2.2 Modification and Nanostructuring of Materials for Solar Water
          Splitting..12
          1.2.2.1 Surface Modifications for Enhanced Light Harvesting 13
          1.2.2.2 Bulk Modifications for Enhanced Light Harvesting............14
          1.2.2.3 Nanostructuring of Semiconductor Materials 16
    1.2.3 Materials for Water Oxidation and Water Reduction Catalysis 19
          1.2.3.1 Water Reduction Catalysts...22
          1.2.3.2 Water Oxidation Catalysts 25
1.3 Progress on Semiconductor-Based Solar Water Splitting 27
    1.3.1 Photocatalytic Water Splitting..27
          1.3.1.1 Single-Step Photosystems...29
          1.3.1.2 Dual-Step Z-Schemes..30
    1.3.2 Photoelectrochemical Water Splitting 31
          1.3.2.1 Photoanode Materials 32
          1.3.2.2 Photocathode Materials 34

1.3.3   Tandem Photoelectrochemical Systems 35
       1.3.3.1   Photoanode–Photocathode Tandem Cells 36
       1.3.3.2   Photoelectrode–Photovoltaic Hybrid Tandem Cells 37
   1.3.4   Photovoltaic-Based Water Splitting 38
   1.3.5   Dye-Sensitized Water Splitting 42
1.4   Conclusions and Outlook 44
References 47

This chapter deals with sustainable hydrogen production by water splitting under sunlight using semiconductor materials. We will start with a broad overview of the field, briefly dealing with its history and evolution, and thereafter present the state of the art in a logical manner by highlighting the current and future research trends, and finally conclude with a critical perspective.

1.1   INTRODUCTION ON SOLAR WATER SPLITTING

1.1.1   General Introduction

It is often said that artificial photosynthesis is the "Holy Grail" of chemistry. After almost two decades since the first introduction of this metaphor by A. J. Bard and M. A. Fox [1], it is still pretty common to find similar statements among the scientific community. And in some ways this is not surprising; indeed, energy is often considered to be the most important problem for mankind in the coming years [2]. In particular, the need for clean and sustainable energy sources is of high priority for at least two reasons. First, it is well known that the fossil fuel resources are limited and the extraction of such fuels will require continuously growing financial investments. Secondly, there is general agreement that threats like global warming and climate change in general are associated to the significantly huge and still growing anthropogenic emissions of greenhouse gases to the atmosphere [3]. It is also widely recognized that, being completely renewable and plentiful, solar energy is one of the best candidates to feed the energy needs of humanity. However, this is obviously an intermittent energy source and therefore needs to be associated with an efficient energy storage system. Hydrogen produced from nonfossil sources appears to be the best fuel that qualifies as a clean energy vector [4]; the ideal energy system would then associate solar light as the energy source and hydrogen produced from water as an energy storage means. Conceptually, two general configurations can be adopted to achieve such a system; the most obvious "brute-force" approach simply involves using a photovoltaic module (PV) to power a conventional dark electrolyzer. A most elegant, yet most challenging approach is the direct conversion of sunlight into hydrogen using a stand-alone integrated device, a solar water-splitting cell. Despite the several underlying challenges, as will be pointed out in this chapter, this second approach has several advantages over the conventional "brute-force" approach, in particular for decentralized energy production, still deserving the title of "Holy Grail" of chemistry.

1.1.2 HISTORICAL BACKGROUND ON SOLAR WATER SPLITTING

The origins of the concept of photoelectrochemical energy conversion date back to the middle of the nineteenth century, as Antoine C. Becquerel together with his son Alexandre-Edmond Becquerel discovered the photoelectrochemical effect (which at the time was called Becquerel effect, in their honour) in 1839 [5]. In their device, a voltage and an electric current were generated upon illumination of a silver chloride photoelectrode immersed in a liquid electrolyte and connected to a metallic electrode. This setup represented the first photovoltaic device ever reported. A similar scheme, involving illumination of silver halides, was by the way utilized in the same period for photographic applications, although the underlying mechanism was not understood until about one century later [6]. Interestingly, the development of film photography and photoelectrochemical cells turned out to be tightly correlated in subsequent years, as the two technologies rely on similar photochemical mechanisms; however, extensive practical applications and widespread commercial interest turned out to be restricted only to the field of photography. The role that photochemical techniques could potentially play for power generation was however highlighted in 1912 by Giacomo L. Ciamician in his brilliant contribution in *Science*. In this paper, he already realized that the fundamental problem is "how to fix the solar energy through suitable photochemical reactions" and that a potential key to the problem would be the "artificial reproduction of [...] the assimilating processes of plants" [7].

However, significant advances, in both understanding of the photoelectrochemistry and its applications, did not occur until the 1950s, when extensive research was carried out on the optical, electronic, and photochemical properties of single-crystal semiconductors. In parallel, biochemists have made also fundamental discoveries about how natural photosynthesis occurs, notably the roles of Photosystems I and II, as well as that of the thylakoid membrane, and the unique architecture of light-harvesting chloroplasts in plants. Theoretical physicists and chemists have continuously looked for clues in natural photosynthesis in order to be able to replicate it and—ideally—improve it. Excellent reviews are available on the mechanisms underlying the plant photosynthesis [8]. Basically, the natural photosynthetic machinery involves two functional units: a light-harvesting center and a reaction center. The first unit is responsible for absorbing sunlight, converting its energy into photogenerated charges through electronic excitation, and effectively mediating their separation over large distances to the site of the second functional unit. This second unit is then responsible for the conversion of the net excitation energy into an electrochemical potential difference, which ultimately provides the driving force for all biological energetics [9]. The integration of knowledge arising from these various fields has allowed researchers to bring out path breaking progress in artificial photosynthetic systems research and exploitation.

The early advances in artificial photosynthesis are in particular due to extensive research carried out in the mid-1950s at the Bell Laboratories, where the electrochemical properties of germanium and of silicon were first investigated in aqueous solutions [10]. While the same Bell Laboratories developed the first practical photovoltaic device in the same years [11], the first actual photoelectrochemical devices were reported only many years later, and research in the field of photoelectrochemistry was until the 1970s essentially of fundamental nature. In particular, the photocatalytic properties of some metal oxide semiconductors (zinc oxide, titanium oxide, and antimony trioxide) were first studied by Sister Clare Markham in 1955 [12]. Although the most well-studied materials were TiO_2 [13] and ZnO [14–16], these fundamental investigations were extended in the 1960s to other semiconductor materials such as CdS [17] and GaP [18], and allowed the development of a theoretical model to describe the junction forming at the interface between a metal or semiconductor and a liquid electrolyte. The main contributor to the understanding of such solid–liquid junction was the German chemist Heinz Gerischer. Despite devoting most of his efforts to understanding the underlying mechanisms of the photochemical processes occurring at the solid–liquid junction [19–21], Gerischer later recognized that the phenomena observed could actually lead to the realization of photoelectrochemical solar cells, for which he gave some design principles and some examples of materials, such as CdS, CdSe, and GaP [22] as well as ZnO [23]. For an exhaustive collection of the work carried out by H. Gerischer in the field of photoelectrochemistry, the interested reader is referred to one of his several book chapters [24].

The most significant advance for applied photoelectrochemistry occurred, however, in 1972 as the Japanese chemists Akira Fujishima and Kenichi Honda first reported in *Nature* the demonstration of electrochemical photolysis of water into hydrogen and oxygen using an illuminated TiO_2 photoelectrode and a dark Pt electrode [25]. This experiment represented the first successful use of a semiconductor electrode for light-driven water splitting. However, more than 10 years before this, an organic photosensitive material had already been effectively investigated for the same purpose by H. Kallmann and H. Pope, who reported in a 1960 *Nature* article successful photodecomposition of water into hydrogen and oxygen using an illuminated anthracene crystal [26]. After the groundbreaking article by Honda and Fujishima, the interest for metal oxides and in particular TiO_2 for photoelectrochemical and photocatalytic applications grew dramatically, especially in the 1970s and the 1980s. Simultaneously, a new class of semiconductor materials emerged in the fields of solar energy conversion: III–V semiconductors. The photoelectrochemical properties of these materials, in particular GaAs, InP, and GaP, were thoroughly investigated in the late 1970s and early 1980s in particular by the group of Allen J. Bard at the University of Texas at Austin [27–30] as well as several other groups [31–34]. The most significant breakthrough obtained till date with this class of materials was in 1998 as John A. Turner and Oscar Khaselev obtained a 12.4% record efficiency for a water-splitting device using $GaInP_2$, described in Section 3.3; unfortunately, the stability of the device was limited only to few hours [35].

While the earlier studies regarded single-crystal and later polycrystalline materials, the tendency of the last two decades is to employ nanostructured materials. This was largely due to the breakthrough obtained by Michael Grätzel and Brian O'Regan in 1991, as they showed that by nanostructuring a semiconducting electrode it is possible to increase the device solar energy conversion efficiency by over an order of magnitude [36]. Although this work involved a regenerative device (i.e., a photovoltaic cell), the same concept was extended to photoelectrochemical cells for water splitting. Photoelectrodes involving the use of surface modifications and of composite materials are another direction that was allowed in the last two decades by the advances in nanotechnology; these involve in particular dye- and quantum-dots-sensitized photoelectrodes, as well as deposition of protective layers, catalytic materials, or plasmonic metal nanoparticles. In the last decade, research was also focused on finding a suitable catalyst for oxygen evolution and hydrogen evolution. This indeed represents a very old and significant problem, since the only effective catalysts traditionally involve noble metals. An important advance in this field was enabled by the discovery in 2008 by the group of Daniel G. Nocera at Massachusetts Institute of Technology of a new noble-metal-free oxygen evolution catalyst composed of inexpensive cobalt, now widely known as Nocera's catalyst [37].

1.1.3 THEORETICAL PRINCIPLES OF SEMICONDUCTOR-BASED SOLAR WATER SPLITTING

In some of the solar water-splitting device types considered here, the presence of a solid–liquid junction between a semiconductor and an electrolyte plays a fundamental role. Therefore, some general considerations about semiconductors and a short description of the phenomena occurring at the solid–liquid junction between a semiconductor and an electrolyte are imperative. However, this section will be limited to a brief description of the most important concepts; for a comprehensive discussion of semiconductor properties and semiconductor–liquid junctions (SCLJs), the interested reader is referred to the available literature covering the field [38–40].

1.1.3.1 Key Properties of Semiconductor Materials

Unlike metals, semiconductors are materials characterized by a gap in the density of electronic states. This gap however is sufficiently small so that electronic conduction is possible under certain conditions. At room temperature, the concentrations of electrons in the conduction band and holes in the valence band are generally very small. This is in accordance with the Arrhenius-like relation (Equation 1.1), where n and p are the concentrations of electrons and holes in the conduction band and the valence band, respectively; N_C and P_V are the corresponding density of states in the conduction and the valence band, respectively; E_g is the bandgap of the material; k is the Boltzmann constant; and T is the absolute temperature.

$$n \cdot p = N_C \cdot P_V \cdot e^{-E_g/kT} \tag{1.1}$$

However, this situation can be drastically modified by heating or illuminating the material. In the case of intrinsic semiconductors, the amount of negative charges in the conduction band always equals the amount of positive charges in the valence band. For doped semiconductors, this is no longer the case: considering the case of p-type semiconductors, the amount of holes in the valence band is significantly higher than the amount of electrons in the conduction band. In such a material, electrons and holes are therefore referred to as the minority and majority carriers, respectively.

$$E_F = \frac{E_V + E_C}{2} + \frac{kT}{2} \cdot \ln\left(\frac{n \cdot P_V}{N_C \cdot p}\right) \tag{1.2}$$

According to the above equation, where E_F is the Fermi level, and E_V and E_C are the energies of the top edge of the valence band and the bottom edge of the conduction band, respectively, this implies that the Fermi level of a p-type semiconductor is not in the middle of the bandgap (as is the case for intrinsic semiconductors, neglecting the contribution of the reduced mass of the electron and of the hole), but rather is shifted toward the top edge of the valence band. How close this is to the valence band depends on the doping level: the higher the p-doping level, the closer the Fermi level to the valence band. In some cases, the doping level can be so substantial that the Fermi level lies within the valence (for p-doping) or the conduction band (for n-doping). Transparent conductive oxides, such as Al-doped zinc oxide, are important examples of this family of semiconductors, which are said to be degenerate.

1.1.3.2 Semiconductor–Liquid Junction and Semiconductor Photoelectrodes

It is well known that a photovoltaic device can be obtained upon contact between a p-doped and an n-doped semiconductor. Considering the simple case of a homojunction, that is, when the two sides of the junction are composed of the same material (e.g., n-Si and p-Si), it is well known that a band bending will establish at the interface upon equilibration of the Fermi levels of the two materials. An analogous situation originates upon contact between a doped semiconductor and an electrolyte. In the context of solar water splitting, this interfacial band bending and the electrical field associated can be exploited for the formation of photoanodes and photocathodes for water oxidation and water reduction, respectively. The case of a p-type semiconductor material in contact with an aqueous electrolyte (typically constituting a photocathode where water reduction can occur) will be considered here in detail, while the case of n-type photoanodes for water oxidation is entirely analogous. In commonly used photocathode systems, the Fermi level of the p-type semiconductor (and thus the electrochemical potential of the electron in the solid) turns out to be lower in energy than the redox potential of the electrolyte. This implies that upon equilibration of the electrochemical potentials for the electron, a net flow of negative charges will occur from the electrolyte to the p-type semiconductor. This will cause an increase in the energy of the conduction and the valence bands of the semiconductor; however, the energy of these bands turns out to be fixed right at the interface, due to the presence of adsorbed ions (called potential-determining ions) within the Helmholtz layer. Therefore, the energy of the bands will increase only in the bulk of the semiconductor; as a consequence, a band bending situation is encountered, where the bands bend "downward" from the bulk of the semiconductor toward the interface. This is exactly the situation that is needed to drive electrons toward the interface (and thus react with any adsorbed species having a suitable redox potential) and holes toward the bulk of the material (and thus be collected at the back-contact), and is therefore the desired situation for a photocathode.

However, one should note that this charge separation is only effective for electron–hole pairs that are generated sufficiently close to the interface: in principle, only electron–hole pairs that are generated within the space-charge layer are readily separated by migration within the electrical field (band bending). In fact, it turns out that pairs that are generated close enough to the space-charge layer can be separated as well, provided that the diffusion of the electrons is faster than the competing recombination of the electron–hole pair. Therefore, one would ideally look for a space-charge layer as large as possible; unfortunately, a larger space-charge layer implies a lower doping level, as shown by Equation 1.3, where W_{SC} is the width of the space-charge layer, N_D is the concentration of dopants, ε_0 is the permittivity of vacuum, ε_r is the dielectric constant of the material, V_a is the applied potential, V_{fb} is the flat-band potential, and e is the elementary charge.

$$W_{SC} = \sqrt{\frac{2 \cdot \varepsilon_0 \cdot \varepsilon_r \left| V_a - V_{fb} \right|}{e \cdot N_D}} \tag{1.3}$$

There is therefore an evident trade-off between electrical conductivity on the one side and the width of the space-charge layer on the other. Experimentally, the concentration of dopants (doping level), and therefore the width of the space-charge layer, can be found by performing a Mott–Schottky analysis from capacitance measurements. According to the Mott–Schottky Equation 1.4, where C is the interfacial capacitance and A is the interfacial area, the plot of the inverse square of the capacitance against the applied potential allows establishing the flat-band potential of the semiconductor as well, by simple extrapolation on the potential axis.

$$\frac{1}{C^2} = \frac{2}{\varepsilon_0 \cdot \varepsilon_r \cdot e \cdot N_D \cdot A^2} \left(\left| V_a - V_{fb} \right| - \frac{kT}{e} \right) \tag{1.4}$$

One has to note that the above equation only applies under depletion conditions, that is, it only holds for the range of applied potential leading to a depletion of majority charge carriers within the space-charge region. Furthermore, the Mott–Schottky analysis is based on the assumption that the capacitance of the electrochemical double layer is several orders of magnitude higher than the capacitance of the space-charge layer, and its contribution to the total (i.e., measured) capacitance is therefore negligible. The flat-band potential is an important characteristic of a semiconductor, as it reflects the position of its Fermi level before contact with the electrolyte. In the case of p-type semiconductors, the flat-band potential therefore gives an indication of the energy of the valence band edge. The energy of the conduction band edge can then be estimated by adding the energy of the valence band edge to the value of the bandgap, which can in turn be determined experimentally by transmittance or reflectance spectroscopy measurements.

1.1.4 AN OVERVIEW OF THE MAIN APPROACHES TOWARD SOLAR WATER SPLITTING

Several different approaches have been developed toward solar water splitting using semiconductor materials. An important distinction among these approaches is the one between systems using suspended semiconductor particles and systems featuring semiconductor electrodes. The first approach is commonly and broadly referred as photocatalytic systems (described in Section 1.3.1) and is arguably the simplest conceivable type of solar water-splitting device. The second approach, on the other hand, involves both photoelectrochemical cells (presented in Section 1.3.2) and photovoltaic-based cells (described in Section 1.3.4). In all these three approaches semiconductor materials play an essential role, as they are responsible, among the other things, for harvesting the incident sunlight. However, there is a fourth approach, namely dye-sensitized solar water splitting, where this does not happen to be the case. In this concept (presented in Section 1.3.5), the semiconductor material acts instead as a porous scaffold and as a charge transfer mediator, while the light harvesting is carried out by a dye (or another sensitizer) attached to its surface. An important common point between dye-sensitized and photocatalytic systems in that owing to the size of the particles involved, their mechanism does not necessarily rely on an SCLJ, but rather on the kinetic competition between the various charge transfer processes involved in the system.

The opposite situation is encountered with photoelectrochemical and photovoltaic-based water-splitting cells, as in both cases the semiconductors form a junction, which plays a fundamental role toward effective separation of the photogenerated charges. This can be either a solid–solid junction between two semiconductors (as in most photovoltaic-driven cells considered here) or a solid–liquid junction forming at the interface between the semiconductor and the aqueous electrolyte. There exists a fifth class of water-splitting cells, where two of these junctions are combined in the same device, known as water-splitting tandem cells. Broadly speaking, two different classes of tandem cells can be conceived, which are known as photoanode–photocathode cells and photoelectrode–photovoltaic hybrid cells, respectively (as described in Section 1.3.3).

A concept related to solar water splitting is solar-driven hydrogen evolution from wastewater and biomass; within this type of devices, hydrogen generation is coupled to degradation of organic molecules, such as dyes, sugars, and alcohols. This concept is extremely interesting from a practical point of view, as the goal of hydrogen production is coupled to decontamination of industrial effluents or efficient utilization of energy-rich biomass. Furthermore, this concept shares many aspects with solar water splitting from a mechanistic point of view. However, as the scope of this chapter is confined to overall water-splitting systems, this concept will not be discussed in detail here; the interested reader is instead referred to Chapter 2.

1.2   MATERIALS FOR SOLAR WATER SPLITTING: REQUIREMENTS AND CURRENT TRENDS

1.2.1   Semiconductor Materials for Solar Water Splitting: General Requirements

In practice, a suitable semiconductor material for solar water splitting must satisfy a number of conditions. Foremost, the material has to yield a reasonable photoactivity toward water splitting; in other words, a reasonable solar-to-hydrogen conversion efficiency. Generally, the accepted energy conversion threshold for a commercial solar water-splitting system would be 10%, although the vast majority of the materials investigated so far are not capable of yielding such high performances. To achieve this high activity, the semiconductor material must be capable of absorbing sunlight over a wide range of wavelengths with photonic energy being larger than the minimum energy required to split water and to subsequently perform an efficient conversion into chemical energy. The second most critical criterion for the choice of a suitable semiconductor material for a practical solar water-splitting device is its stability. This implies that the photoactivity (and therefore the conversion efficiency or in other words the amount of hydrogen generated) has to be sustained for a sufficiently long period of time, typically several years to be practically useful. Typically, a system is considered to achieve satisfying stability if its performances do not change significantly over the course of a 1000-hour test under simulated solar light illumination. In addition, the device should ideally be composed of nontoxic, inexpensive, and readily available materials. Aspects that are even more important for a commercially viable device are that the materials and manufacturing processes involved must be easily upscalable, reproducible, environmentally safe, and relatively inexpensive.

1.2.1.1 Photoactivity of Semiconductor Materials

First and foremost, photoactive semiconductor materials for solar energy conversion must be able to absorb as much light as possible from the solar spectrum. Translated in terms of electronic properties, this implies that their electronic bandgap should be sufficiently small with respect to the photon energy of the incident sunlight. Absorption of solar light allows the generation of electron–hole pairs within the semiconductor absorber that will subsequently need to be separated at its surface; in practical systems, that generally occurs at the interface with another material or another phase. But even before this can occur, however, the photogenerated charges must be able to reach the surface of the material before recombining; this implies good electrical conductivity within the material and ideally the absence of any trap states. Typically, this in turn requires the use of high purity and defect-free materials, as impurities and defects are among the major factors responsible for reduced charge mobility within semiconductors. In addition, to ensure that the semiconductor material is capable of overall water splitting, its conduction and valence band edges must straddle the relevant redox potentials for the water-splitting reaction: the conduction band bottom edge of the semiconductor must be higher in energy than the redox potential of the H^+/H_2 couple, while its valence band top edge must be lower in energy than the redox potential of the O_2/OH^- couple. As the standard potential of the O_2/OH^- redox couple is 1.23 V/RHE, this condition imposes a thermodynamic minimum value of 1.23 eV for the bandgap of the material, which corresponds to photons having maximum wavelength of 1008 nm being absorbed. However, in addition to thermodynamic considerations, kinetic losses have to be taken into account in any practical system. In accordance to the Marcus theory for heterogeneous charge transfer, the energy difference between the band edges and the potential of the redox couples have to be sufficiently high to ensure enough driving force for the electron (or hole) transfer to happen at a reasonable rate [41]. Due to these ubiquitous kinetic losses, as well as to resistive Ohmic losses which are present within any realistic system, the optimum bandgap for a semiconductor absorber material turns out to be substantially increased, to over 2 eV [42,43]. Obviously, there is therefore a fundamental trade-off, as a large bandgap material is desirable to obtain the sufficient driving force for unassisted water splitting, while on the other hand a small bandgap material would harvest a higher amount of photons from sunlight, thus yielding more photogenerated charges. This is clearly exemplified by the wide-bandgap material anatase TiO_2, which has a bandgap of 3.2 eV (thus absorbing only the UV portion of the solar spectrum), for which the maximum theoretical solar-to-hydrogen (STH) efficiency is only 1%, while hematite (α-Fe$_2$O$_3$)—which has a bandgap of 2.2 eV (thus absorbing UV and a portion of the visible wavelengths)—features a maximal theoretical STH efficiency of 15% [44]. Since the overall efficiency of the water-splitting process is dictated by both the amount of charges generated and their driving force toward water splitting, this trade-off leads in practice to limited efficiencies when only one semiconductor material is used to harvest sunlight.

In practice, various strategies have been proposed and investigated in order to relax the various material constraints and in particular to overcome some of the fundamental trade-offs. The two most common approaches in the literature are the use of multiple absorber materials (in what are known as photoelectrochemical tandem cells and photocatalytic Z-schemes, discussed in Sections 1.3.1.2 and 1.3.3, respectively) and the sensitization of a wide-bandgap semiconductor with sunlight-harvesting dye molecules. A further possible approach is of course to simply connect the photocatalytic or photoelectrochemical cell to any external bias, in order to generate the required voltage for overall water splitting; such a configuration, however, cannot be classified as unassisted solar water splitting, unless the external bias is generated by a photovoltaic cell. In some experiments, like the Fujishima and Honda's seminal work, a wide difference in chemical potentials between the liquid medium (aqueous electrolyte) present in a two-compartment cell was maintained in order to take advantage of the fact that for every 1 pH unit about 59 mV of potential difference could be generated between the two electrodes [25]. Again, this is akin to applying an external chemical potential and involves the use of bases and acids. Finally, the use of sacrificial agents is another possible option to relax the constraints in terms of material photoactivity. In this situation the difference between two relevant redox potentials involved in the overall reaction turns out to be smaller than 1.23 eV, which allows for the use of lower bandgap semiconductors. In these systems, however, only one of the two water-splitting half-reactions is carried out by the photogenerated charges – the other half-reaction arising from decomposition of the sacrificial agent – and overall water splitting, is therefore not achieved.

1.2.1.2   Stability of Semiconductor Materials

All of the aforementioned conditions are necessary to ensure a reasonable photoelectrochemical or photocatalytic activity of the material, translating into a high efficiency of the solar water-splitting cell. However, activity is far from being the only important parameter for a commercially viable device; the second most important factor being the stability of the materials, implying that the efficiency of the device should not significantly decrease with time. As pointed out by Butler and Ginley, a candidate semiconductor material for solar water splitting has to be stable on various different levels [45]; for photoelectrochemical energy conversion there are in fact three main factors which can lead to material instability. These are the nature of the electrolyte with which the material is in contact, the electrical potential applied, and the effect of light illumination. This means that the semiconductor material has to be stable against chemical dissolution, electrochemical corrosion, and photocorrosion.

The third factor, photocorrosion, is unique to photosystems and represents is in fact a self-inflicted damage caused by kinetic factors besides thermodynamic reasons and arising from the photogenerated holes consuming the semiconductor itself. In fact, for a given combination of pH value and potential, a material can be stable in the dark according to its Pourbaix diagram but unstable under light illumination; this is because of the photogenerated holes and electrons within the material, which are strongly oxidizing and reducing species, respectively, and can thus lead to oxidative or reductive decomposition of the material. From a *thermodynamic* point of view, a semiconductor is said to be unstable against oxidative (or anodic) decomposition if the redox potential for its anodic decomposition reaction lies above the valence band edge, while it is classified as unstable against reductive (or cathodic) decomposition if the redox potential for its cathodic decomposition reaction lies below the conduction band edge. The vast majority of the materials reported turn out to be thermodynamically unstable and susceptible of photodecomposition; however, several materials appear to be relatively stable when tested in practice.

This apparent contradiction is related to favorable *kinetic* competition between the decomposition reactions and the reactions of interest (water oxidation and water reduction). In fact, what is critical from a practical point of view is the relative position of the redox potential of the decomposition reactions with respect to the water reduction and water oxidation potentials. In accordance to the Marcus theory for electron transfers, the rate at which a charge transfer occurs scales to the square of the driving force. Therefore, for a system within the normal Marcus region, if the driving force for water reduction into hydrogen is sufficiently larger than the driving force for reductive decomposition of the semiconductor material, the latter will not occur at an appreciable rate in practice. Analogously, oxidative decomposition of the material can be avoided if the driving force for water oxidation into oxygen is sufficiently larger than the driving force for reductive decomposition. The latter situation can be exemplified by TiO_2, which is considered to be among the most stable and robust semiconductor materials, despite actually turning out to be unstable against photodecomposition from a thermodynamic point of view. On the other hand, CdS is perhaps the prototypical example of a material, which is unstable against photodecomposition both from a thermodynamic and a kinetic point of view. It is interesting to note here that CdS, unlike TiO_2, can absorb a considerable portion of the visible spectrum and hence would have been a suitable material to improve the spectral absorption range and thus attain higher energy conversion efficiencies in water-splitting reactions. It appears that a few general empirical rules can be drawn after decades of experience: in particular, it has been observed that oxide materials typically offer superior electrochemical stability when compared to other semiconductors and that wide-bandgap semiconductors are generally more stable than low bandgap materials. Although these rules bear some rationale associated to the electronic structure of the materials, they are not to be taken for granted as several exceptions exist; for instance, ZnO, despite being a wide-bandgap oxide, is particularly unstable when in contact with an aqueous electrolyte under light illumination. Unfortunately, ZnO is an excellent photocatalyst except for its precarious stability issues arising from its amphoteric nature.

With such unstable materials (such as ZnO or CdS) there still exist some possible strategies to overcome this problem. The most straightforward, which does not involve any modification of the semiconductor material, is to modify the electrolyte by introducing a redox couple, which owing to its appropriate redox potential can favorably compete with the decomposition reaction, kinetically. These redox couples are known as *sacrificial agents* as they are not regenerated after their oxidation or reduction. A sacrificial electron donor is added to the electrolyte to efficiently compete against oxidative decomposition, while a sacrificial electron acceptor is introduced in the system to prevent reductive decomposition of the semiconductor. Typical examples of sacrificial donors include simple organic compounds like alcohols or inorganic chalcogenide salts or multivalent ions such as $Ce^{4+/3+}$. This solution however—which does not represent overall water splitting—presents two major drawbacks: besides the increased overall cost due to the utilization of the sacrificial agent, the fact that a species has to be continuously fed in the system introduces some additional complications from a technical point of view.

Two other major approaches have been pursued to overcome the stability issue, which both involve modification of the surface of the semiconductor itself: an unstable material can be made more stable by deposition of a suitable catalyst or by deposition of a stable passivating material. Owing to its chemical stability, TiO_2 is a commonly employed passivating material, while prototypical examples of catalysts employed are Pt for water reduction and RuO_2 for water oxidation (although a wide range of alternative catalyst materials has emerged in the last decade, described in Section 1.2.3). The two approaches can be seen as fundamentally different, as the first is designed to remove highly reactive photogenerated charge carriers as quick as possible from the semiconductor material, whereas the second is mainly designed to provide a physical barrier between the unstable semiconductor material and the electrolyte. Contrary to a catalyst, which is essentially designed to prevent photocorrosion and electrochemical corrosion of the semiconductor material, a stable passivating material will therefore also be effective to prevent physical dissolution of the semiconductor in the electrolyte. Ideally, a passivating material will therefore coat the entire surface of the semiconductor, while on the other hand the presence of a catalyst will be characterized by a small surface coverage. In practice, however, these two strategies are often used in combination and in some systems a material is introduced that can behave as both a catalyst and a passivating agent; even from a mechanistic point of view, it is sometimes not straightforward to determine whether a material rather acts as a catalyst or merely as a physical barrier.

1.2.2 MODIFICATION AND NANOSTRUCTURING OF MATERIALS FOR SOLAR WATER SPLITTING

In the previous section, it has been pointed out how difficult it is to develop semiconductor materials, which are both highly photoactive toward water reduction and/or oxidation and sufficiently stable in contact with an aqueous electrolyte and under light illumination. In fact, in practical water-splitting systems, surface modification of the photoactive material with one or more other materials, whether passivating or catalytic, turns out to be virtually always necessary to obtain a reasonably efficient and stable system. While surface passivation and deposition of a suitable catalyst are two major strategies toward stable photoactive architectures, various other approaches have been proposed and successfully demonstrated, in particular to improve the photoactivity of semiconductor materials and therefore the overall solar-to-hydrogen efficiency of water-splitting cells. These strategies can be classified into three major approaches: surface modification, bulk modification, and nanostructuring.

1.2.2.1 Surface Modifications for Enhanced Light Harvesting

Regarding *surface modification*, besides the two aforementioned approaches involving passivation and catalysis, two additional strategies have been pursued in the last decade: surface sensitization, through anchoring of sensitizer molecules on the surface of a wide-bandgap semiconductor, and—more recently—plasmonics, through deposition of plasmonic metallic nanoparticles. These approaches are somewhat similar in that they both have as the main objective an increase of light absorption by the system in the visible spectral range, ultimately leading to an increase in the amount of photo-generated charge carriers and thus an enhancement in the rate of hydrogen production.

The idea of using visible-light absorbing molecules to sensitize wide-bandgap semiconductors is nowadays well known because of the successful development of the dye-sensitized solar cell (DSSC), a photovoltaic device which can currently reach efficiencies of over 12% [46]. Surprisingly perhaps, this concept has been around for well over a century, as it had been first used in the field of color photography in 1873 [47] and had been applied to the field of energy conversion devices already in 1887 [48]. In the 1960s dye-sensitization has been "rediscovered," in particular thanks to the work carried out by Heinz Gerischer and Helmut Tributsch [23], but it was not until the late 1980s and early 1990s, with the development of fractal and then the mesoporous films of TiO_2 and other semiconductor materials that dye-sensitization became a credible strategy for effective solar energy conversion [36].

Understandably, the development of dye-sensitized water-splitting cells turned out to be significantly more challenging than their regenerative counterpart (i.e., dye-sensitized photovoltaic cells) and significant progress has only been carried out in the last decade. This is essentially due to the different electron transfer kinetics involved: while in dye-sensitized regenerative cells one can choose a suitable redox couple in order to optimize the energetics and the kinetics of the system, the redox potentials of the electrolyte are obviously fixed in the case of a water-splitting cell. In a very similar way to DSSCs, this class of water-splitting cells are typically based on the absorption of visible light by ruthenium-based sensitizers, which are anchored onto the surface of a mesoporous TiO_2 nanocrystalline film and attached to a water oxidation catalyst (WOC), such as iridium oxide nanoparticles [49,50]; these devices are a subclass of photoelectrochemical water-splitting cells. In fact, dye-sensitized photocatalytic water-splitting systems, involving a suspension of TiO_2 nanoparticles, have been investigated as well. These devices are based on various scaffold materials and architectures, ranging from simple TiO_2 nanoparticles [51] to more complex semiconductor/zeolite composites [52] and layered metal oxides semiconductors [53], and will be described in more detail in Section 1.3.5.

The second major class of surface modification involves deposition of plasmonic metallic nanoparticles on the surface of a semiconductor. Despite the effects of metallic nanoparticles first being reported and investigated in the 1970s, their use for water-splitting purposes emerged only in the last decade and is still limited to few materials. Plasmonic effects arising from metallic nanoparticles deposited on semiconducting materials are manifold and they generally all lead to increased light absorption and thus increased amount of photogenerated charge carriers. The two most important mechanisms involve hot carrier transfer and plasmon resonance energy transfer (PRET) from the metal to the semiconductor, although purely optical effects may come into play as well. The so-called hot carrier transfer refers to the transfer of a photogenerated electron from the conduction band of the confined metal to the conduction band of the semiconductor. This is in many ways analogous to dye sensitization, with the role of the dye molecule played here by a metallic nanoparticle, and allows for visible-light sensitization of wide-bandgap semiconductors, such as TiO_2. On the other hand, PRET does not involve the transfer of photogenerated charge carriers, but rather of excitation energy. Analogous to the phenomenon of Förster resonant energy transfer (FRET), this phenomenon requires overlap of the emission spectrum of the metallic nanoparticles with the absorption spectrum of the underlying semiconductor. Therefore, this mechanism is mainly exploited in combination with lower bandgap semiconductors, such as Fe_2O_3 or WO_3. Further details on the various mechanism involved in plasmonic-enhanced photocatalysis can be found in the various reviews recently published in this field [54–56].

A significant milestone on the way to plasmonic-enhanced water splitting has been achieved in 2004, as a TiO_2 film immersed in an aqueous electrolyte had been reported to be photoactive in the visible range upon deposition of gold and silver nanoparticles, which was attributed to the hot carrier transfer mechanism [57]. The beginning of the 2010s showed a dramatic increase in the interest in the field, and several examples of plasmonic-enhanced water splitting have been reported. A significant number of studies have been devoted to the deposition of Au or Ag nanoparticles on Fe_2O_3 [58–60] and N-doped [61,62] and undoped [63,64] TiO_2 photoelectrodes. In the case of Fe_2O_3 and N-doped TiO_2, the enhancement in the photoactivity is usually attributed to PRET because of the spectral overlap between the plasmonic nanoparticles and the semiconductor [55]. While gold and silver appear to be the best plasmonic metals so far, the approach has been extended to other semiconductor materials besides TiO_2. A heterogeneous Au/CeO_2 composite showed enhanced water oxidation upon visible-light illumination in the presence of Ag^+ as a sacrificial electron donor [65]. The activity of CdS nanoparticles toward photocatalytic hydrogen evolution also appeared to be significantly enhanced when these were connected to Au nanoparticles. In order to achieve an optimum distance between the Au and CdS nanoparticles and thus avoid detrimental charge transfer from the semiconductor to the metal, both CdS and Au nanoparticles were coated with a thin insulating layer of SiO_2 [66]. Enhanced water oxidation has also been reported on Ag/WO_3 nanocomposite photoanodes; in this study, however, the enhanced performances of the composite appeared to be mostly due to optical factors, as the plasmonic nanoparticles can also lead to enhanced light scattering and reduced reflection of light [67]. This constitutes a simple light management system, through which the photons shining on the composite remain "trapped" within the underlying semiconductor and have thus a higher probability to be absorbed by the latter.

1.2.2.2   Bulk Modifications for Enhanced Light Harvesting

Besides surface modification strategies, *bulk modifications* of semiconductor materials offer an additional approach to enhance the photocatalytic activity of semiconductors and essentially involve changing the composition and/or the structure of the material. Traditionally, most modifications of semiconductor materials generally involved a change in the composition of the material; however, in the last few years, structural modification of semiconductor materials has developed as a novel approach to enhance their light-harvesting capabilities and thus their photocatalytic activity. This approach, however, can be ascribed to nanostructuring and therefore it will be presented in the next section. The most popular approach toward modifications of the composition of a semiconductor material is by doping, that is, by purposely introducing selected impurities (known as dopants) in the host material. The dopants can be either metals or nonmetals and can be introduced into the host material through various chemical and physical techniques. Traditionally, doping of semiconductors has been used to enhance the conductivity of materials; however, more recently this approach has been widely used to extend the absorption spectrum of wide-bandgap material into the visible range. This effect is generally obtained through the introduction of intra-band states, which are additional energy states lying between the valence and the conduction band of the semiconductor. Upon light illumination, these will provide electronic states, which were forbidden in the untreated material, from or to which electronic excitation can occur. This in turn leads to the presence of additional bands in the absorption spectrum of the material, characterized by lower energies than the band arising from bandgap excitation.

This approach can be easily illustrated for the case of a polar metal oxide, such as TiO_2. For this kind of materials, a suitable nonmetal will be generally used as dopant to replace oxygen atoms in the lattice and thus create new energy states lying slightly above the valence band, by virtue of its high electronegativity. On the other hand, a suitable metal will be used to replace the cation in the lattice, thus introducing new electronic states lying slightly below the conduction band. ZnO and TiO_2 are the two materials for which this approach has been used most extensively, owing to their lack of absorption in the visible spectrum. A comprehensive description of the various metallic and nonmetallic dopants reported for these materials can be found in several reviews [68–70]. N-doping of TiO_2 is the most well-studied example of anion-doping reported for TiO_2, followed by C- and S-doping. A seminal report was published by Asahi et al. in 2001 [71], showing that N-doped TiO_2 was acting as visible-light photocatalyst. One year later, visible-light absorbing C-doped TiO_2 was reported to split water with application of a small external bias, with an astonishing efficiency of over 8.3% (and very good stability). However, several research groups later pointed out that the reported efficiency was probably highly overestimated and the real efficiency was more likely to be in the range of 1%–3% [72]. C-doped TiO_2 photocatalysts have been later prepared by several groups, although a substantial increase in the photocatalytic performances was never achieved [73,74].

It is well known in the field of semiconductor physics that impurities within a material may act as recombination centers thus reducing the mobility of charge carriers. This situation, which is typically encountered with energy states located far from the bandgap edges—so-called deep defects, can indeed lead to reduced photocatalytic activities. This problem appears to be particularly acute when metal cations are used as dopants [68], while on the other hand anions appear to be less likely to generate recombination centers [75]. To mitigate the effects of recombination introduced by doping, a general approach is to confine the impurities to the surface of the host material in such a way that charge carriers trapped in the recombination centers are more likely to reach the surface of the material [75]. Co-doping is another strategy to reduce dopant-induced recombination, which consists in introducing two different types of dopants in the lattice of the host. Boron–nitrogen co-doping of TiO_2 has been shown to reduce dopant-associated recombination, by removing the Ti^{3+} recombination centers due to charge compensation effects [76]. Density functional calculations have been used to predict the most suitable combinations of co-dopants, unveiling the potential of combinations of co-dopants involving ions of Mo, W, Nb, and Ta as electron donors, while C and N appear as the two most promising acceptor dopants [77].

Besides dopant-induced recombination, which is believed to be the main reason for the overall low efficiency of visible-light absorbing materials such as $N-TiO_2$ [78], doping of wide-bandgap semiconductors to introduce intra-band states has a second major intrinsic limitation, namely, that the charges generated at the dopant levels may not have enough driving force to reduce and/or oxidize water. This has been suggested in particular for the $N-TiO_2$ system, where holes generated by visible-light absorption appear not to have enough driving force to oxidize water into oxygen [78]. Nonetheless, the overall approach of doping wide-bandgap semiconductors to achieve visible-light sensitization is extremely promising and has been shown to be applicable to other semiconductors besides TiO_2. Recently, N-doped ZnO has been demonstrated for the first time to evolve oxygen from water in a photocatalytic system without external bias under visible-light illumination [79].

1.2.2.3 Nanostructuring of Semiconductor Materials

As already pointed out, that it is extremely difficult to find simple materials that are suitable for solar water-splitting cells. Chemical surface and bulk modifications of pristine materials are very important routes through which complex assemblies showing suitable photoactivity and stability can be obtained. Another fundamental route to achieve this target is offered by *nanostructuring*. In the last decades there has been a dramatic increase in the number of available synthetic routes toward nanostructured material, allowing for increasingly higher control over the morphology, the shape, the size, and ultimately the various properties of nanomaterials. On the other hand, the continuous development of new analytical tools, accompanied by the increased performances of computational chemistry, has allowed a more detailed understanding of the relations between structure and properties at the nanoscale. In the field of solar energy conversion, DSSCs are the most striking and successful example of how nanostructured materials can dramatically improve the properties of a system. In this system, the use of a nanostructured mesoporous TiO_2 electrode produced a 3- to 4-order of magnitude improvement in both the incident photon-to-current efficiency (IPCE) and the photocurrent of the device, when compared to a flat single-crystalline electrode [80]. Nanostructured materials can improve the efficiency of solar water-splitting cells in several different ways. The most obvious effect is an increase in the roughness factor of the material, leading to a dramatically enhanced surface area; this can in turn offer several advantages, in terms of enhanced light absorption, increased catalyst loading, and increased number of catalytically active sites.

An increasingly popular device architecture is represented by ordered arrays of nanowires or nanotubes. The potential for the improvement in the catalytic activity offered by nanowire electrodes has been estimated by Liu et al. using a simple model [81], showing that these architectures offer much reduced overpotentials for solar-to-fuel conversion over their planar counterpart. In addition, these architectures are generally characterized by a very high aspect ratio, and therefore offer the additional combined advantages of long optical paths, leading to efficient light absorption and reduced charge diffusion distances, ultimately leading to reduced charge recombination. The distance that photogenerated charges have to travel before reaching an active interface is a critical factor for the separation of photogenerated carriers and thus the photoactivity; this is particularly the case for materials characterized by short charge diffusion lengths, where recombination occurs relatively fast and/or the mobility of charges is typically moderate. By exploiting these effects, significant improvements in the photoactivity of several materials have been obtained in the recent years by using arrays of various high-aspect ratio architectures, such as TiO_2 nanotubes [82] and nanowires [83,84], Si microwires [85], Si/TiO_2 composite nanowires [81,86,87], ZnO/Si nanowires [88], Ti–Fe–O nanotubes [89], Ta_3N_5 nanotubes [90,91], Fe_2O_3 nanorods [92], and other Fe_2O_3 high-aspect ratio architectures [93,94].

Another avenue to improve the device performance through nanotechnology is the controlled synthesis of nanostructured materials to create complex morphologies and nanocomposites, in order to spatially separate the various components and thus decouple the various functionalities within the system. The prototypical example of this approach is represented by core–shell nanocomposites, although lamellar assemblies and other host–guest nanoarchitectures have been extensively investigated as well. The core/shell approach is extensively used in photocatalytic systems, where the inner and the outer surface of the shell can be functionalized with water reduction and WOCs, respectively. In addition to spatially separate the two water-splitting half-reactions and thus decrease the reverse reaction, this architecture has been designed to facilitate the separation of the charge carriers and thus reduce electron–hole recombination. These two effects combined can lead to significantly enhanced overall water-splitting activity as it has been recently shown, for example, with SiO_2/Ta_3N_5 core/shell photocatalysts [95].

The use of lamellar assemblies is another approach to spatially separate components for effective charge separation and reduced recombination in photocatalytic systems. This route has been pursued in particular by the group of Thomas E. Mallouk at the Pennsylvania State University. The different active components of these multilayer assemblies, such as sensitizers, redox-active electron-relays, and catalysts, are confined to individual layers and separated by the presence of inorganic spacers, typically metal-organic phosphates, or layered metal oxides semiconductors, such as niobates [96,97]. Naturally, the need for compartmentalization and spatial separation of the various components of the systems is particularly stringent for photocatalytic systems, where there is generally no electrical field that can drive the separation of charge carriers. This approach, however, can be useful in photoelectrochemical cells as well, where it is also generally introduced in order to facilitate charge separation, and thus reduce the degree of charge recombination. An important example is represented by host–guest nanoarchitectures, composed of a host scaffold and a guest absorber; this strategy has recently been successfully implemented with Fe_2O_3 photoanodes. Photocurrents using this material, which is characterized by poor light absorption and a small hole diffusion length, have indeed been significantly increased using high–surface area hosts (consisting of various oxides, such as WO_3 [98], Ga_2O_3 [99], or Nb-SnO_2 [100]) as a substrate for the subsequent deposition of a very thin layer of hematite. A quite different approach to achieve improved charge separation in semiconductor systems is gradient doping. This concept, very simple from a conceptual point of view, though only recently rediscovered, consists in introducing a gradient dopant concentration in a semiconductor, in order to establish an additional band bending that is not confined to the interface but can also span through the bulk of the semiconductor. Using this concept of gradient doping, the group of R. van de Krol at Delft University of Technology recently achieved record-breaking conversion efficiencies using a W-doped $BiVO_4$ photoanode in a hybrid tandem cell [101].

Finally, nanostructuring techniques also turned out to be particularly beneficial toward enhanced light harvesting of semiconductor materials, through bandgap engineering, spatial structuring, and other related concepts. It is well known that the bandgap of a nanosized material can be easily tuned according to the size and the shape of the particles; this phenomenon can be useful to enhance the light harvesting of photocatalytic systems by synthesizing particles having a suitable absorption spectrum, ideally matching the solar emission spectrum. A more subtle method recently developed to tune the bandgap of semiconductor material is through introduction of disorders in the lattice. This approach has been recently developed with hydrogenated TiO_2, which has been shown to be significantly more active than pristine TiO_2 for photocatalytic water reduction due to the resulting absorption spectrum being extended in the visible and infrared range [102]. This system was shown to perform photocatalytic hydrogen generation from water at a high rate of 10 mmol/h per gram of photocatalyst, although using methanol as a sacrificial hole scavenger. Nowadays, bandgap engineering is a very common way for optimized light harvesting through nanotechnology; besides the aforementioned approaches of doping and disorder engineering, it should be mentioned that solid solutions represent a long-known yet still extremely valuable route toward the preparation of materials with an optimized bandgap; this has been demonstrated by the advances with photocatalytic water splitting using materials such as GaN:ZnO [103].

A more recent approach toward enhanced light harvesting in nanomaterials consists in resonant light trapping using photonic crystals. This novel strategy for light management, which has been classified as "spatial structuring" of nanosized materials, exploits the fact that photonic crystals represent resonant optical cavities, where light of selected wavelengths is reflected with very low losses and has thus an increased probability to interact with a light absorbing medium, owing to increased optical path and contact time with the latter [68,104]. This approach, initially dealing with three-dimensional periodic lattices such as zeolites [104], has been recently extended to thin films, where the resonant light trapping arises from the light being reflected from a metallic "back-reflector" substrate back into the absorbing film. By depositing an ultrathin layer of a hematite photoanode with a controlled thickness on selected metallic substrates, it has been shown that an optical gain of up to a factor of 5 can be achieved [105]. This very simple approach leads to a double advantage: not only the light absorption by the hematite electrode gets increased by the above factor, but also only ultrathin films (approximately 20–30 nm) of photoactive material are required. Clearly, such reduced film thicknesses are extremely beneficial for materials where charge transport is typically very slow (of which hematite is a prototypical example). Finally, the approaches of bandgap engineering and light trapping using photonic crystals can be combined within the same device, to further enhance the photocatalytic properties. In particular, TiO_2 inverse opal nanostructures have been fabricated in such a way to match the electronic bandgap of the material with its photonic bandgap [106]. This system has been employed as a photocatalyst in combination with gold photonic nanoparticles, showing remarkable photocatalytic activity toward hydrogen evolution from water under sacrificial conditions. Interestingly, the enhanced photocatalytic activity does not arise in this system from enhanced light harvesting, but rather from reduced electron–hole recombination: in such systems, where the photonic bandgap of a material overlaps with its electronic absorption band, spontaneous emission is indeed forbidden, which in turn suppresses a major route toward electron–hole recombination.

1.2.3 MATERIALS FOR WATER OXIDATION AND WATER REDUCTION CATALYSIS

One of the major challenges related to solar water splitting is represented by catalysis of hydrogen and oxygen evolution on the surface of the light-harvesting material. In principle, the catalytic photodissociation of water can be achieved by various redox couples, when their redox potential is larger than 1.23 V. Heidt and McMillan have demonstrated already in 1949 the evolution of hydrogen gas from water by illumination of aqueous solution of Ce^{3+} ions with UV light under acidic pH conditions [107,108]. Later, Kiwi and Grätzel succeeded in increasing the oxygen evolution yield by adding solid redox catalysts like PtO_2 and IrO_2 [109]. Further, Vonach and Getoff have reported oxygen evolution when a suspension of TiO_2 in an acidic solution of Ce^{4+} was illuminated [110]. From some of these earlier and seminal reports, it is clear that the most significant fraction of the overpotential required to split water in a solar water-splitting device is due to kinetic barriers toward interfacial charge transfers to oxidize, respectively reduce, water. Water oxidation into oxygen gas, in particular, is an extremely complex reaction, which requires accumulation and storage of four oxidative equivalents per molecule of oxygen generated; therefore, the role played by WOCs in solar water splitting is of paramount importance. Interestingly, despite water reduction into hydrogen gas being characterized by a simpler mechanism, where only two reductive equivalents are required per molecule of hydrogen evolved, water reduction catalysts (WRCs) turn out to be virtually always necessary in semiconductor-based solar water-splitting devices. This is mainly due to the fact that water reduction typically occurs at very high overpotential at the surface of semiconductors.

It is pertinent here to consider that photosplitting of water on semiconductor powders, unlike in photoelectrochemical cells, can be achieved in either vapor phase or liquid phase. In fact, reactions of water vapor over doped and undoped TiO_2 and Fe_2O_3 have been reported in the 1970s and one of the detailed earlier studies is that of Schrauzer and Guth, who noticed that besides the slow rate of H_2 and O_2 production, the system failed to produce gases beyond a few hours [111]. Kawai and Sakata investigated water splitting in the presence of TiO_2 to which RuO_2, a well-known oxygen evolution catalyst with low overpotential for O_2 evolution, was added [112,113]. Similarly, the effects of Pt on sustained H_2 production during water (vapor) splitting under illumination has been highlighted by the work of Sato and White [114]. Around the same period, Domen et al. have reported the photocatalytic decomposition of water vapor on $SrTiO_3$ loaded with NiO to steadily produce H_2 and O_2. In many of these earlier studies where liquid or gaseous water was illuminated over photocatalysts, stoichiometric production of H_2 and O_2 was not obtained. This was attributed to the presence of oxidizable impurities on the catalysts surface by which a substantial portion of O_2 would be consumed; list of suspected impurities included carbon compounds. Since the overall reactions were not sustaining over a very long period, it was not evident whether these attributions were true by running the reactions until all the oxidizable impurities could be consumed. Subsequently, in photosplitting of water in liquid water, the nonstoichiometric production of oxygen was clearly attributed to the formation of peroxides and stable surface complexes of peroxide radicals [115]. All these earlier studies during the evolutionary phase of water decomposition under light-irradiated semiconductor materials clearly illustrate the necessity of attaching active catalysts for both H_2 and O_2 evolution from water in order to run the reaction in a sustainable manner over long periods of time. This led to the standard practice of using metallized semiconductor powders for photocatalysis. In fact, the first ever report of using metallized semiconductor dispersions (Pt on TiO_2) was from Bulatov and Khidekel [116]. Subsequently, Lehn et al. reported the photolysis of water on Rh-$SrTiO_3$ [117]. Later, the concept of bifunctional catalysts containing both hydrogen evolution and oxygen evolution functions tailored to the same particle has come into the picture, however in the presence of a sensitizer [118]. When compared to the evolution of hydrogen, oxygen evolution generally tends to be sluggish due to the fact that it requires four oxidative equivalents to be successful.

The main function of a catalyst within a solar water-splitting cell is obviously to accelerate sluggish charge transfer kinetics. However, the presence of catalytically active materials (which are typically metals or semiconductors) at the surface of a semiconductor material may generally cause dramatic and nontrivial effects on the thermodynamics of the system. In the relevant case of a metallic catalyst (such as platinum, which is currently the most widely used WRC in the academic context), an additional metal–semiconductor junction will be created in the system. Ideally, this junction should be highly rectifying and possibly free of surface states; in practice, any real metal–semiconductor junction will involve the presence of additional surface states, which could lead to detrimental effects, in particular by acting as charge recombination centers. In addition, the choice of a catalyst for solar water splitting has to satisfy an additional condition: the catalytic material must not interfere with the light-harvesting abilities of the photoactive material, that is, it should not absorb or reflect the incoming light. This constraint typically limits the choice of catalytic materials to extremely thin films or islands-like clusters of nanoparticles. As pointed out in Section 2.2.1, the choice of plasmonic materials, having both remarkable catalytic and optical properties, appears to be an interesting avenue to successfully overcome this limitation [55].

Naturally, the two most important characteristics of a catalytic material are its activity and its stability. The catalytic activity of a material is typically measured in terms the amount of product generated per unit time. For photoelectrochemical systems, the photocurrent generated is a good and convenient indicator of this; however, it bears an important drawback in the case of catalysts with a nonquantitative Faradaic yield toward water reduction or oxidation, for it does not allow distinguishing between the different products of the reaction. To characterize the catalytic activity in the case of photocatalytic systems, which generally do not involve electrodes, the turnover frequency (TOF) is generally used, which is typically defined as the amount of product generated per unit time and normalized with respect to the catalyst loading. The turnover number (TON) is a related concept, used to assess the stability as well as the activity of a catalyst and defined as the absolute amount of product generated, normalized with respect to the catalyst loading. The product is in this case generally detected and quantified by gas chromatography or by other techniques such as volume displacement measurements.

In addition to being highly active toward accelerating the kinetics of the reaction, ideal catalytic materials have to be stable during the lifetime of the (solar) water-splitting device. In the same way as the photoactive semiconductor materials, the catalysts are typically exposed to harsh environments, determined by the presence of water and oxygen, by the pH of the solution, and by the electrolyte composition. Catalysts, however, are particularly sensitive functional materials, and readily suffer from poisoning: their activity can indeed dramatically decrease upon contamination of trace amounts of impurities, which could directly arise from the solution or even from the dissolution of electrode materials. In addition, it is important to point out here that within photosystems the environment is typically particularly harsh, owing to the presence of highly reducing and oxidizing species, potentially leading to photocorrosion, as already illustrated in Section 1.2.1.2. This is particularly true for the sites where water oxidation occurs, owing to the typically highly oxidizing holes generated by most metal oxide semiconductors used for water oxidation. It appears logical, then, that the majority of the most stable and active WOCs investigated till date consists of metal oxide materials, which are generally more robust than their metallic counterparts.

It has been mentioned that electrochemical measurements can give a good indication of the catalytic activity in the case of photoelectrochemical devices. Charge transfers between an electrode and an electrolyte can be generally described with the Butler–Volmer model; when the overpotential applied at the electrode is sufficiently large, the model reduces to the Tafel relation, which predicts a linear relationship between the overpotential applied and the logarithm of current. This relation is widely used to characterize the catalytic performances of water-splitting devices. In particular, performing a Tafel analysis on experimental j-V curves allows extracting two important parameters that define the catalytic activity of an electrocatalyst: the exchange current density and the Tafel slope. These two parameters are related to the two main type of effects that determine the overall activity of a catalyst: intrinsic or electronic effects, which are related to the electronic structure of the catalytic material, and geometric effects, which are related to the effective catalytically active area of the material. Broadly speaking, the Tafel slope is related to the mechanism of the charge transfer reaction and is therefore a measure of the intrinsic or electronic properties of a catalyst. The exchange current density on the other hand is related to the rate of forward and reverse reaction at equilibrium conditions (i.e., when the overpotential is nil) and is therefore mostly a measure of the geometric effects of a catalyst.

The overall activity of a catalyst generally arises from a combination of electronic and geometric effects; some catalysts mainly act through their electronic structure (by determining a decrease in the Tafel slope), while others mainly operate through geometric effects (by increasing the exchange current). By their very definition, it is clear that the exchange current mostly determines the charge transfer rate next to the equilibrium conditions (i.e., at low overpotentials), while the Tafel slope becomes increasingly important in determining the overall rate of reaction at high values of overpotential. This fact has important practical implication for the design of water-splitting devices, for a catalyst material that is suitable for cells where small current densities (and rates of reaction) are required might be less adapted for cells where large current densities are envisaged. In particular, for the relevant case of solar water-splitting, current densities are generally comparatively much lower than those required in commercial dark electrolyzers, as the maximum achievable current densities are ultimately limited by the solar spectrum. An important consequence of this, already pointed out by Walter et al. [119], is that catalysts whose performances are mostly related to geometric factors (e.g., nanostructured materials characterized by high roughness factors and high surface areas) could be comparatively more successful for solar water-splitting devices than those whose activity mainly arises from electronic effects. A final consideration, which particularly concerns WOCs and WRCs for solar water splitting, must be made regarding the limitations on their synthetic routes. Because of the photoactive species naturally present in these systems, WOCs and WRCs catalysts are required to be deposited in relatively mild conditions. Typically these are either vacuum-based or solution-based techniques, of which electrodeposition is probably the most popular, while high-temperature synthetic routes would generally lead to degradation of the underlying photoactive material.

1.2.3.1 Water Reduction Catalysts

From a mechanistic point of view, the water reduction reaction occurring on the surface of a solid material is believed to be composed of three elementary steps [120]:

$$M + H^+ + e^- \rightarrow M - H_{ads} \qquad \text{(I) Tafel step}$$

$$M - H_{ads} + H^+ + e^- \rightarrow M + H_2 \qquad \text{(IIa) Heyrovsky step}$$

$$M - H_{ads} + M - H_{ads} \rightarrow 2M + H_2 \qquad \text{(IIb) Volmer step}$$

After the initial Tafel step, the reaction may proceed according to the Heyrovsky or to the Volmer mechanism. This mechanistic model clearly illustrates that the key intermediates common to both reaction pathways are adsorbed hydrogen atoms. Therefore, in accordance with the Sabatier principle, an ideal WRC should be characterized by interactions with the adsorbed intermediate which are neither too weak, in order for the binding event to occur, nor too strong, in order to facilitate the release [121]. This principle is clearly illustrated by the well-known volcano plots, empirical relations between catalytic activity and bond energies introduced almost 50 years ago [122]. Experimental volcano plots have been obtained for the water reduction reaction occurring on metallic surfaces, showing a clear correlation between the metal—hydrogen bond strength [123]. This empirical observation has been rationalized in the 1950s by H. Gerischer [124] and R. Parsons [125], which showed that the exchange current density for the water reduction reaction is maximized for a standard free energy of adsorption equal to zero. A fundamental question is to determine the main physical and chemical properties responsible for the strength of the bond between the surface and the hydrogen adatoms, and hence for the intrinsic catalytic activity. For the case of metallic catalysts, the work function and the electronegativity of the metal have been proposed as two possible descriptors by S. Trasatti [123]. However, no general model has been formulated and the description for compound materials such as metal oxides is more complex and is complicated in particular by the fact that the hydrogen atoms can bind to both the metal and the nonmetal. As a consequence, the vast majority of known WRCs have been found experimentally; in the last decade, however, a few computational studies based on density functional theory have shown some degree of predictive power, in particular for metallic catalysts [126] and even for bimetallic catalysts [127], indicating that in the near future computational chemistry will be complementary to experiments in the quest for new WRCs.

Four broad classes of materials have been investigated as WRCs. These are pure metals, metal alloys and composites, compound materials composed of metals and nonmetals, and finally molecular inorganic compounds. In terms of catalytic activity, platinum has long been known as the best available material [126]; as a consequence, the vast majority of the reports on solar water reduction employ platinum as the WRC. Obviously, because of the scarcity and the high cost of platinum, intensive research has been devoted to finding alternative WRCs, which do not involve noble metals. Another possible strategy is to reduce the cost of platinum-based catalysts by reducing the amount of platinum required without compromising excessively the catalytic performances. This approach has been demonstrated by alloying platinum with other less costly compounds, such as with bismuth [127], or by depositing it in extremely small amounts, as recently demonstrated using atomic layer deposition [128]. Among the non-noble metals, nickel is considered the most active WRC; however, its performances are generally well behind those of platinum, with typical exchange current densities being about two orders of magnitude lower [123]. Alloying nickel with other metals, in particular left-hand-side transition metals such as iron, molybdenum, or tungsten, has shown to be a promising strategy to enhance the catalytic performances; the improved activity of these alloys has been attributed to both geometrical and electronic factors [129]. Ruthenium-based materials have recently been shown to be another potential alternative to platinum; RuO_2 is probably the most active metal oxide WRC [130] and has been successfully applied to photoelectrochemical water reduction in combination with TiO_2-protected Cu_2O photocathodes. In a recent work, it has been shown to have comparable activity to platinum and even a significantly improved stability, which could be attributed to its lower sensitivity to poisoning by trace impurities in the electrolyte [131]. Nickel and ruthenium nanoparticles have also been recently shown to be very active WRC species for photocatalytic water splitting, yet in sacrificial systems: while the activity of nickel nanoparticles was about half of that of platinum nanoparticles, ruthenium nanoparticles were essentially as active as platinum nanoparticles over a wide range of pH values [132]. Core–shell nanoparticles are another interesting class of WRCs, which has recently been investigated in the context of photocatalytic water splitting. In particular, the Rh/Cr_2O_3 core/shell system, loaded on the surface of GaN:ZnO photocatalyst, has been shown to effectively promote water reduction in an overall photocatalytic water-splitting system. In particular, the core–shell structure turned out to be particularly beneficial as water formation from hydrogen and oxygen, which is otherwise catalyzed by Rh nanoparticles, is largely suppressed by the Cr_2O_3 shell [133].

A third class of WRC materials is represented by compounds formed by metallic and nonmetallic elements. Within this class of materials, molybdenum-based compounds emerged in the last decade as important class of potentially inexpensive WRCs. Molybdenum sulfides are probably the most widely investigated species; while bulk MoS_2 is known to be a poor WRC [134], MoS_2 nanoparticles have been first anticipated by the virtue of density functional theory calculations [135] and subsequently experimentally verified to be a promising WRC material [136]. This material was successfully integrated in photocatalytic systems in combination with CdS nanoparticles. Remarkably, this system showed activities comparable or even superior to platinum, which was partly attributed to the favorable interface between CdS and MoS_2; however, these extraordinary results have not been demonstrated yet in real water-splitting systems, but have rather been limited to sacrificial systems [137,138]. More recently, amorphous molybdenum sulfide (MoS_x) electrodeposited from an aqueous solution has shown promising catalytic activities [139]; incorporation of other transition metals such as Fe, Co, and Ni was shown to further increase its catalytic activity by about one order of magnitude, which was mainly attributed to geometric effects [140]. Very recently, the potential of this class of material has been demonstrated in an actual photoelectrochemical cell, in combination with TiO_2-protected Cu_2O photocathodes, showing photocurrents comparable to those obtained using platinum as WRC, yet significantly enhanced electrode stability [141]. Molybdenum boride and molybdenum carbide are other two promising recently reported molybdenum-based compounds showing high activity as WRCs; interestingly, these two compounds turn out to compete with the catalytic activity of platinum in both acidic and basic conditions [142].

Molecular compounds have also been extensively investigated as hydrogen evolution catalysts. Generally, however, these compounds have two major disadvantages with respect to inorganic materials: the higher production cost, due to the difficulties related to their synthesis, and their limited stability in water. Among the non-noble metals, iron and nickel have historically been the two most investigated compounds; the rationale behind this trend is related to the fact that these are the elements utilized by nature for proton reduction in hydrogenase enzymes. For a comprehensive overview of these compounds, the reader is directed to the specific literature [143]. An outstanding example of a Ni-based molecular WRC has been recently reported in combination with CdSe nanoparticles for photocatalytic water reduction; although the system included a sacrificial electron acceptor, the report represents a significant advance owing to the remarkable stability of the system [144]. Cobalt is a third metal showing promising catalytic activity toward hydrogen generation in molecular systems [143,145]. Very recently, a cobaloxime-type compound has been successfully shown to catalyze water reduction in combination with a GaP photocathode in the absence of external bias or sacrificial agents [146]. An emerging class of molecular WRCs involves molybdenum as the metal; in the last few years, in particular, a molecular molybdenum sulfide complex [147] and a molybdenum-oxo compound [148] have been shown to function as active and robust hydrogen evolution catalysts from water. Remarkably, a molecular molybdenum-based WRC has even been demonstrated to function in a photoelectrochemical water-splitting cell with an activity compared to platinum. In this work, a bio-inspired cubane-like cluster based on molybdenum and sulfur (Mo_3S_4) was grafted on p-type silicon, showing very efficient water splitting in the absence of external bias or sacrificial agents [149].

1.2.3.2 Water Oxidation Catalysts

Compared to water reduction, the mechanism of water oxidation reaction is much more complex and strongly depends on the pH conditions as well as on the nature of the surface on which it occurs. The general mechanism for water oxidation occurring in alkaline solutions at the surface of metals is believed to be the following [150]:

$$M + OH^- \rightarrow M - OH_{ads} \tag{I}$$

$$M - OH_{ads} + OH^- \rightarrow M - O_{ads} + H_2O + e^- \tag{IIa}$$

$$2M - OH_{ads} \rightarrow M - O_{ads} + H_2O + M \tag{IIb}$$

$$2M - O_{ads} \rightarrow 2M + O_2 \tag{III}$$

In acidic solutions, the mechanism is believed to be analogous, with the reactive species in the first step (I) being H_2O instead of OH^-. The mechanism believed to occur at the surface of metal oxides is somewhat more complex and has been extensively discussed by S. Trasatti [151]. In a similar way to the water reduction reaction, the key intermediates involve a chemical bond between the metal and an oxygen atom. It should not be a surprise therefore that a correlation has been determined experimentally between the catalytic activity on various surfaces and the strength of the metal–oxygen bond, giving rise to characteristic volcano plots [152]. In addition to this correlation, other descriptors have been developed to explain and predict the water oxidation activity of metals and oxides. In particular, the occupancy of the 3d orbitals with an e_g symmetry of surface transition metal cations has been shown to correlate very well with the catalytic activity of metal oxides [153]. Because of the more complex nature of the water oxidation reaction, computational studies have so far shown less accuracy and predictive power than for the water reduction reaction and will probably require a few more years before bringing a substantial contribution in the quest for new WOCs.

Although platinum is among the most widely used materials in the literature for oxygen evolution from water, in particular in the context of solar water splitting, its performances as a WOC are actually moderate. In fact, it is believed that only metal oxides are expected to be sufficiently robust to withstand the highly oxidizing conditions necessary for oxygen evolution from water [154]. Until the end of 2000s, the most effective WOCs were considered to be RuO_2 and IrO_2, which are both based on quite expensive and scarce metals. Because of the highly oxidizing conditions required to oxidize water, these materials usually require to be stabilized, by combination with inert substrates such as TiO_2. A colloidal version of iridium oxide, $IrO_2 \cdot xH_2O$ has been developed and subsequently thoroughly investigated in various (mostly sacrificial) photocatalytic systems [155]. Doped $IrO_2 \cdot xH_2O$ colloids have later been shown to enhance the catalytic performances, although expensive platinum and osmium were used as dopants and the enhancement in the catalytic activity was not spectacular [156]. Iridium hydroxide [$Ir(OH)_x$] nanoparticles with enhanced catalytic performances have been recently reported by S. Fukuzumi and Y. Yamada in the context of photocatalytic water oxidation [132]. In the same report, the authors also present a non-noble-metal-based alternative, based on cobalt hydroxide [$Co(OH)_x$] nanoparticles. Interestingly, the nanoparticles are formed *in situ* from less active homogeneous complexes during the catalytic water oxidation. In addition to Ru- and Ir-based oxides, a limited number of alternative WOC materials showing moderate activity have been reported until quite recently. These were mostly spinel compounds such as $NiCo_2O_4$ and Co_3O_4 [157] and perovskite compounds, in particular doped lanthanum oxides such as $LaCoO_3$ and $LaNiO_3$ [158,159]. In this context, it is worth mentioning a recent work where the spinel compound Mn_3O_4 was used as WOC in combination with core/shell-structured Rh/Cr_2O_3 as WRC and GaN:ZnO nanoparticles as photoactive material to achieve overall photocatalytic water splitting under visible light [160]. For a comprehensive review of these earlier materials (mostly Ru and Ir dioxides and spinel compounds) and their mechanism of action as WOCs, the reader is referred to the specific literature [161].

Although perovskite WOCs were already known to perform well in alkaline environments when compared to noble metal dioxides, a breakthrough perovskite material has been identified in the early 2010s by Suntivich et al. Using a design principle based on the 3d electrons occupancy, BSCF ($Ba_{0.5}Sr_{0.5}Co_{0.8}Fe_{0.2}O_{3-\delta}$) was found to catalyze water oxidation in basic conditions with an intrinsic activity over one order of magnitude higher than IrO_2 [153]. A second recent breakthrough was represented by the development of an entirely new generation of amorphous oxide–based WOCs, deposited by solution-based techniques and composed of inexpensive first-row transition metals, which have shown to act as stable and active WOCs. This represented a fundamental advance not only from a practical side, but also from a conceptual point of view, as according to the general understanding, it was believed that highly crystalline oxide materials (and therefore high manufacturing temperatures) were required to withstand the harsh conditions at which water oxidation takes place [162]. Chronologically, the first active amorphous WOCs were reported by the group of D. Nocera in 2008. An amorphous film composed of cobalt ions and phosphate (Co–Pi catalyst) was reported to perform water oxidation at low overpotential, and benign conditions such as 1 atm, room temperature, and neutral pH [37]. The Co–Pi films can be deposited *in situ* on virtually any substrate by simple electrodeposition from an aqueous solution containing Co(II) ions and phosphate electrolyte. In addition, the self-assembled Co–Pi catalyst was later verified to possess self-healing properties, thus mimicking a further important feature of biological systems and ensuring the required robustness [163]. Remarkably, this Pi catalyst also functions in the presence of high chloride concentrations, potentially allowing for (solar) water splitting from ubiquitous seawater [164]. Despite this material being significantly less active overall than the traditional WOCs IrO_2 and RuO_2, it still enables for current densities commensurate with those allowed by the solar spectrum and appears therefore to be a suitable candidate material for solar-driven water oxidation. This has been already demonstrated in several proof-of-concept reports, in particular in combination with ZnO [165], WO_3 [166], and Fe_2O_3 photoanodes [167,168], as well as silicon photovoltaic cells [169–171].

A related amorphous material, composed of nickel and borate (Ni–Bi catalyst) was reported 2 years later by the same group [172]. The Ni–Bi catalyst operates at near-neutral conditions (pH = 9) and shares several features with the Co–Pi system, thus expanding the class of inexpensive amorphous WOCs. The last and perhaps most significant breakthrough in the field of WOC consists in a novel synthetic technique to access amorphous mixed metal oxide WOCs. This technique relies on photochemical metal-organic deposition, a simple and low-temperature process, and is broadly applicable to several metals and combinations thereof and allows for very high stoichiometric control [173]. A series of proof-of-concept amorphous mixed metal materials (the most active being $Fe_{100-y-z}Co_yNi_zO_x$) having catalytic properties comparable to those of the best WOCs has been recently demonstrated [173], indicating that several other active mixed metal oxides composed of inexpensive first-row transition metals are likely to be reported in the near future.

As for water reduction, molecular compounds have been extensively studied as WOCs; however, very limited success has been achieved till date, mainly because of the sensitivity of these compounds to the highly oxidizing conditions in which water oxidation occurs. Among the various metals investigated, cobalt- and iron-based complexes appear to be the most credible potential candidate to catalyze water oxidation with the sufficient stability and activity, although some significant advances would certainly be required [143]. An additional and emerging class of inorganic complexes which could potentially prove the required robustness and catalytic properties are polyoxometalates. Owing to their all-inorganic nature, these compounds have already proved to be significantly stable in water [174] and could therefore potentially be of high relevance for solar water splitting in the near future.

1.3 PROGRESS ON SEMICONDUCTOR-BASED SOLAR WATER SPLITTING

1.3.1 PHOTOCATALYTIC WATER SPLITTING

From a practical point of view, the most straightforward and potentially most inexpensive route for the conversion of solar energy into hydrogen fuel consists in photocatalytic water splitting. In such a system, light-absorbing semiconducting particles are suspended in an aqueous solution and the photogenerated electrons and holes are utilized to reduce and oxidize water molecules, respectively. Unfortunately, after over 40 years of intense efforts, this apparently simple goal is still elusive; when compared to schemes involving photoelectrodes and photovoltaic junctions, water splitting using photocatalytic systems achieved only very low conversion efficiencies so far. Since achieving overall water splitting with photocatalysts is particularly difficult to achieve, the overwhelming majority of the studies in this area involve either hydrogen evolution or oxygen evolution, in the presence of sacrificial compounds. These can be either sacrificial electron donors, which are much easier to oxidize than water, or sacrificial electron acceptors, which are compounds particularly easy to reduce. Examples of sacrificial donors widely utilized are sugars, simple carboxylic acids (such as lactic acid and ascorbic acid), and alcohols (most typically methanol), as well as other reducing agents such as triethanolamine, EDTA, and NADH. On the other hand, persulfate anions and salts of Ag^+ and Ce^{4+} are the most commonly employed sacrificial oxidizing agents. It must be highlighted that in systems using sacrificial agents, the overall reaction performed is generally thermodynamically downhill; these are therefore proper photocatalytic systems, contrary to water splitting which being thermodynamically uphill, should be considered to be a photosynthetic process. In view of this important distinction, it appears clear that the terminology "photocatalytic water splitting" is in itself incorrect when referring to overall water splitting in the absence of an external bias; however, this expression is so widely adopted by the scientific community that has become a sort of convention, and will therefore be adopted here as well.

In the early 1980s, a series of early studies on this type of reactions, involving $RuO_2/TiO_2/Pt$ bifunctional photocatalysts was published by Sakata and Kawai as well as Pichat's group in France [175,176]. Photochemical water oxidation by the persulfate anion in aqueous suspensions of $SrTiO_3$ containing p-type $LaCrO_3$ has been investigated by Thewissen et al. The p-type $LaCrO_3$ was thought to be responsible for the high rate of electron transfer from the conduction band of the semiconductor to the persulfate observed in this system [177]. Sacrificial systems are of fundamental interest from an academic standpoint as, by excluding one half-reaction, they facilitate the understanding of the reaction mechanisms and the study of the charge carrier dynamics. In addition, they could become very relevant from a practical point of view in systems where sacrificial agents are naturally present, such as wastewater and industrial effluents, where dyes and other organic contaminants could be oxidized and hydrogen fuel could be simultaneously generated. For example, production of hydrogen from water and carbohydrates using $RuO_2/TiO_2/Pt$ bifunctional photocatalysts will be particularly interesting for practical effluent water treatment processes in food industries [175]. The abundant literature on photocatalytic hydrogen evolution from sacrificial systems has been reviewed by several authors [178–180] and will be covered by Yusuke Yamada and Shunichi Fukuzumi in another chapter of this book, while the scope of this section will be limited to examples of systems where overall water splitting has been achieved.

Very few materials have been demonstrated to date to be capable of overall water splitting in a photocatalytic system; the difficulties encountered so far while attempting to achieve overall water splitting using photocatalytic systems can be ascribed to two fundamental reasons. First, in such systems it is extremely difficult to separate the photogenerated electrons and hole; this is because the powder utilized is generally composed of semiconductor particles that are too small to host an electric field when in contact with a liquid electrolyte. Therefore, contrary to the situation encountered using semiconducting electrodes in contact with an electrolyte, the charge separation within the particles is not sustained by an electrical field and tends therefore to decay much faster. The second major challenge with photocatalytic systems is that hydrogen evolution and oxygen evolution occur on the same particle. This constitutes a major practical disadvantage, as the oxygen and hydrogen molecules generated in close proximity tend to recombine to from water again, a strongly exothermic reaction. Besides decreasing the net solar-to-hydrogen efficiency, this also represents a safety issue and a technical challenge in the reactor design, as the two gases have to be separated as soon as possible after they are generated.

Because the electron–hole recombination is particularly sustained, moderate quantum yields for water splitting can very often be achieved only when specific catalyst (generally called cocatalysts in the context of photocatalytic systems) are deposited on the semiconductor particles. In a similar way to sacrificial agents, these materials act as electron and/or hole scavengers and therefore play a fundamental role in achieving effective charge separation. Noble-metal-based materials are known to be among the most active catalysts for water reduction and oxidation and have been traditionally widely used as cocatalysts for photocatalytic water splitting; this class of materials, however, is generally very efficient in catalyzing the reverse reaction (combination of oxygen and hydrogen to yield water) as well. Core/shell-structured nanoparticles have been put forward as an interesting approach to address this issue. A relevant example is the Rh/Cr_2O_3 core/shell cocatalyst developed by the Domen group: while the noble-metal-based core of the cocatalyst efficiently performs proton reduction, the Cr_2O_3 shell acts as a barrier for O_2 molecules (but not for protons and H_2 molecules), thus preventing chemical combination of oxygen and hydrogen as well as reduction of oxygen [181].

1.3.1.1 Single-Step Photosystems

The first studies involving semiconductor photocatalysis for water splitting were made with the wide-bandgap TiO_2 and $SrTiO_3$. In particular, TiO_2 was long studied during most of the 1980s, following the seminal report by Fujishima and Honda [25]. In the following decade, ternary oxides such as niobates and tantalates emerged as particularly attractive alternatives [182,183]. Analogous to titanates and TiO_2, these oxides are characterized by a d^0 electron configuration for the metal ion, which has been recognized to be an important descriptor for photocatalytic activity [184]. An important drawback with this first generation of materials is their relatively high bandgap, translating in an absorption edge limited to the blue-violet part of the visible range. In order to tackle this limitation, starting from the early 2000s, significant efforts were devoted to extending the absorption edge of these materials to the visible range. This led to a second generation of photocatalysts, consisting of doped secondary and ternary metal oxides.

Doped tantalates were the first class of materials for which water splitting was achieved under visible light, although with very low conversion efficiency: Ni-doped $InTaO_4$ was reported to split water with a quantum yield of 0.66% (at wavelengths close to 420 nm), using NiO_x as cocatalyst [182,185]. The efforts devoted to doping of TiO_2 lead to several novel materials, most of which show photoactivity in the visible range. However, after over a decade it can be established that in most cases this approach did not translate into the expected improvement in the overall conversion efficiency [186]. While the introduction of dopants in the lattice of TiO_2 generates the presence of visible-light active centers, their photocatalytic activity appears reduced as a consequence of the reduced driving force available to perform oxidation and reduction of water. In addition, one should bear in mind that introduction of impurities in the lattice inevitably leads to enhanced scattering of charge carriers and could therefore hamper the charge mobility.

In the last 10 years, solid solutions and non-oxide materials emerged as serious candidates for overall photocatalytic water splitting. Among the most active photocatalyst for water splitting, a material derived from titanium disilicide ($TiSi_2$) was shown to split water with a quantum yield of 3.9% with 540 nm light and without the need of noble-metal-based cocatalysts [187]. Water splitting was also achieved using nitride materials, such as Ge_3N_4 [188] and oxy-nitrides. Although the former can only function under UV light, owing to its very wide bandgap, a solid solution of zinc oxide and germanium nitride was reported to split water under visible-light irradiation, using RuO_2 as cocatalyst [189]. Synthesis of solid solutions, consisting of a mixture of solids having the same crystal structure, has emerged as a promising strategy to easily tune the bandgap—and thus the absorption edge—of semiconductor materials. The most active photocatalyst for water splitting reported to date is a solid solution of gallium nitride and zinc oxide (GaN:ZnO). While the initial quantum yield reported for this material was 2.5% at 420–440 nm [103,190], this value was later increased to almost 6% by high-temperature calcination, which was reported to reduce the concentration of defects, and upon loading of a mixed rhodium–chromium oxide catalyst [191]. This material is also characterized by a remarkable long-term stability, which makes it the most promising photocatalyst for water splitting investigated so far [192].

1.3.1.2 Dual-Step Z-Schemes

Besides sustained charge carriers recombination, the two major factors limiting the conversion efficiency in the aforementioned systems are the absorption edge still too close to the UV region and chemical recombination of oxygen and hydrogen generated in close proximity. A novel approach, which is designed to overcome or at least mitigate these limitations, is the two-step Z-scheme. Analogously to what occurs in the context of natural photosynthesis, two different photosystems are utilized to harvest as much solar light as possible and generate the sufficient driving force to split water. A reversible redox couple is needed to shuttle charges across the different particles, thus electrically connecting the two photosystems, mimicking the role of the electron transport chain between Photosystem I and Photosystem II in natural photosynthesis. The Z-scheme concept has the potential to lead to enhanced light harvesting because—contrary to the single-step photocatalytic scheme—two small bandgap materials can be employed here. In addition, by depositing the catalysts for water oxidation and water reduction on different particles, the product gases are not evolved in the same point, which facilitates separation and reduces chemical recombination. This concept has been widely investigated in the last decade and various overall water-splitting systems have been constructed; for an exhaustive review, the interested reader is directed to the available literature [179,180,193]. So far, the conversion efficiencies obtained using Z-schemes have already proven to be higher than those obtained using a single-step photosystem. However, the quantum yields obtained are not significantly higher so far, as the state of the art is currently represented by a system composed of Pt-loaded ZrO_2/TaON as hydrogen evolution photocatalyst, connected through a (IO^{3-}/I^-) redox shuttle to Pt-loaded WO_3 as oxygen evolution photocatalyst, showing a quantum yield of 6.3% at 420 nm [194].

An important drawback generally associated with Z-schemes is the sustained recombination between photogenerated carriers and the redox shuttle. This limitation can be avoided in redox-shuttle-free dual-step systems, where the two photosystems are in intimate contact, allowing for interparticle charge transfer to occur between them. This concept has been demonstrated using a suspension of Rh-doped $Ru/SrTiO_3$ particles for hydrogen evolution together with $BiVO_4$ particles for oxygen evolution, which is capable of splitting water under illumination (even under simulated AM1.5G sunlight) without any redox shuttle [195].

Despite the advances with bandgap engineering through solid solutions and with dual-step Z-schemes, quantum yields achieved so far with photocatalytic systems are still far from the tentative target for commercial applications. Indeed, it has been suggested that a quantum yield of 30% at wavelengths of around 600 nm, roughly corresponding to 5% solar-to-hydrogen conversion efficiency, could be a reasonable starting point for practical applications. In fact, while a conversion efficiency of 10% is generally believed to be required for more costly architectures, such as photoelectrochemical and photovoltaic-driven cells, it has been pointed out that an efficiency of 5% could be sufficient in the case of photocatalytic cells, given the simpler and more scalable design [184]. Clearly, in order to achieve this goal, the absorption edge of the absorber materials must be significantly red-shifted, which points toward the development of materials having a bandgap of around 2 eV. In addition, the quantum yield must be increased by almost a factor of 5 with respect to the current state of the art. An important avenue indicated by K. Domen consists in achieving a better control over the defects in the absorber materials, which are strongly contributing to the sustained charge recombination and the low quantum yield [181]. Several other novel approaches that are emerging as promising strategies to enhance the performances of current photocatalysts, such as plasmonics-enhanced water splitting and multiphoton excitation, have been recently described in an excellent review by N. Serpone and A. V. Emeline [186]. Another technological challenge associated with photocatalytic water-splitting cells is the separation of oxygen and hydrogen; it is important to highlight that this factor—which is particularly important in the single-step systems, where the two gases evolve from the same particles—may have a significant influence on the final cost of the water-splitting device.

1.3.2 PHOTOELECTROCHEMICAL WATER SPLITTING

Despite approximately four decades of intensive research, only a very limited number of materials have been reported to perform full photoelectrochemical water splitting under solar light illumination without any external bias. This is because very few materials have a conduction band sufficiently negative with respect to the redox potential for water reduction and simultaneously a valence band edge sufficiently positive with respect to the redox potential for water oxidation. As already pointed out in Section 1.2.1, while from a thermodynamic point of view a bandgap of 1.23 V would appear to be sufficient, in practice a much wider bandgap is needed, as a significant driving force is required for both H_2 and O_2 evolution for kinetic reasons. Not even the prototypical wide-bandgap material TiO_2 (with a bandgap E_g in the range 3.0–3.2 eV) is capable of full water splitting without any external bias, because its conduction band level is too close to the redox potential for water reduction. In fact, only $SrTiO_3$ ($E_g = 3.2$ eV) and $KTaO_3$ ($E_g = 3.5$ eV) have been shown so far to be able to perform photoelectrolysis of water with no external bias [196–198]. In both cases, however, the photocurrents and therefore the solar-to-hydrogen efficiencies achievable are very low, because of the absorption of these materials being limited to the UV region of the solar spectrum.

Because of this situation, which prevents to effectively perform both half-reactions using a wide-bandgap semiconductor, the recent efforts in the field of photoelectrochemical cells have been oriented toward identifying suitable photoanode materials capable of performing water oxidation as well as suitable photocathode materials to perform the other half-reaction, water reduction. The idea underlying this concept is that if a suitable SCLJ can be established in the photoelectrochemical device, separation of the photogenerated charge carriers can be more effective, ultimately leading to more efficient water reduction and oxidation. As highlighted in Section 1.1.3.2, this favorable band bending situation is generally achieved using p-type semiconductors as photocathodes and n-type semiconductors as photoanodes, respectively. In the first case, the photogenerated electron will be driven by the built-in field toward the electrolyte, where water can be reduced to hydrogen, while the hole will be transferred to the back contact. In the second case, the opposite phenomenon will occur, that is, the photogenerated hole will be driven toward the electrolyte by the built-in field, where water can be oxidized to oxygen, while the photogenerated electron will travel across the material and get collected at the back contact. Excellent reviews describing the fundamentals of photoelectrochemical water splitting as well as the progress achieved so far with various photoanode and photocathode materials have been recently published by several authors, including N. Lewis, K. Sivula, and R. van de Krol [119,199–201].

1.3.2.1 Photoanode Materials

Several n-type semiconductors have been investigated as photoanode materials for oxygen evolution from water in the last few decades. A fundamental condition to be fulfilled by these materials is that the valence band edge should be more positive than the redox potential for oxygen evolution from water. TiO_2 is by far the best characterized photocatalytic material and was long considered to be a potential photoanode material. In the seminal publication by Fujishima and Honda, single-crystal anatase was shown to successfully split water, although with very low efficiency and in the presence of a pH-gradient between two half-cells [25]. After several decades of studies, it is now clear that pristine TiO_2 has a bandgap way too high ($E_g = 3.2$ eV) to perform efficient water splitting; indeed, as its absorption characteristics are roughly confined to the UV region (corresponding to about 4% of the solar spectrum), the solar-to-hydrogen conversion efficiency achievable with this material appears to be theoretically limited to about 2% [199]. Approaches to decrease its bandgap, for instance, by introduction of dopants [71,76] or by disorder engineering [102], only partially translated into improved water-splitting performances so far, and the water oxidation photocurrents obtained under visible-light illumination are usually limited to a few hundreds of $\mu W/cm^2$ [202,203]. This can be associated to the tendency that by decreasing the bandgap of TiO_2 or the relative energetics of the holes formed in dopant-related energy levels, the photogenerated holes generally turn out to have a lower driving force to perform oxidation reactions [186]. Only one work has shown so far significantly higher conversion efficiencies (over 8.3% solar-to-hydrogen) using C-doped TiO_2 [204], although later reports and comments suggested that this value was probably a gross overestimation [72].

Because of their extended solar light absorption, lower bandgap oxide materials may be unable to generate the necessary driving force for full water splitting in real systems, where kinetic overpotentials and ohmic resistances play a significant role. However, this class of materials can lead to much higher theoretical solar-to-hydrogen efficiencies and has therefore been extensively investigated in the last decade as photoanode materials. The three most promising oxides investigated so far are the binary oxides WO_3 ($E_g = 2.7$ eV) and $\alpha\text{-}Fe_2O_3$ ($E_g = 2.0$ eV) and the more recently reported ternary oxide $BiVO_4$ ($E_g = 2.4$ eV), with maximum theoretical efficiencies η_{STH} equal to 6%, 15%, and 9%, respectively [199]. With nanostructured $\alpha\text{-}Fe_2O_3$, very promising photocurrents of over 3.0 mA/cm^2 at 1.23 V versus RHE have been recently achieved, despite the limited charge transfer performances shown by this material [94]. It should be highlighted that $\alpha\text{-}Fe_2O_3$ is a particularly interesting material for practical applications, because of its abundance, lack of toxicity, very low cost, and foremost because of its stability. Similar performances have been obtained using WO_3, with reported photocurrents for oxygen evolution of 2–3 mA/cm^2 at 1.23 V versus RHE, although this material seems to bear less potential for further improvements [205]. Using $BiVO_4$ photoanodes, photocurrents of 3.6 mA/cm^2 at 1.23 V versus RHE have already been demonstrated, in combination with a Co–Pi WOC; this is particularly remarkable considering that this material has been thoroughly investigated only in the last few years [101].

Because of the harsh conditions in which photoelectrochemical water oxidation takes place, metal oxides appear to be the natural choice for a photoanode material because of their superior stability. However, other materials have recently shown to bear some potential: among non-oxide materials, earth-abundant silicon could be a potential photoanode material if properly protected against photocorrosion. An n-Si photoanode protected by atomic layer deposition of TiO_2 was recently shown to generate over 10 mA/cm^2 at the reversible potential for oxygen evolution under AM1.5G illumination in acidic conditions, and slightly lower currents in basic and neutral conditions [206]. The nitride Ta_3N_5 also shows some potential as photoanode material; upon nanostructuring and deposition of active WOCs to improve its activity and stability, water oxidation photocurrents in the order of 1 mA/cm^2 have been achieved [90,91]. Metal oxynitrides are another class of emerging photoanode materials, typically featuring bandgaps smaller than those of the corresponding oxides and thus appropriate for visible energy conversion. TaON is the most investigated member of this class, for which promising photocurrents on the order of 1 mA/cm^2 have been reported, while its natural tendency to undergo oxidative photodecomposition has been significantly reduced [207].

1.3.2.2   Photocathode Materials

Because the conditions at which water reduction takes place are less harsh than those required for water oxidation, stability concerns might appear to be less stringent for photocathode materials. However, a very limited number of metal oxides have a conduction band edge sufficiently negative to efficiently perform water reduction; conversely, various non-oxide semiconductors turn out to have a conduction band sufficiently energetic to reduce water. As a consequence, III–V semiconductors (in particular phosphides) and metal chalcogenides (in particular sulfides and selenides) have been the most widely investigated photocathode materials so far.

Among III–V semiconductors, InP, GaP [146], and GaInP$_2$ have long been known to be highly photoactive p-type materials and have been used in a number of extremely efficient tandem devices (as described in Sections 1.3.3 and 1.3.4). However, these materials are composed of scarce elements (in particular indium) and need to be made single-crystalline and defect-free. As a consequence, material and manufacturing costs appear to be a critical shortcoming of this class of materials, which will most likely rule out their use in any large-scale commercial product. Among p-type chalcogenides, CdTe and CuIn$_{1-x}$Ga$_x$Se$_2$ (CIGS) have been known to be photoactive for many decades; while limited performances were achieved using CdTe, pretty high solar-to-hydrogen conversion efficiencies have been shown using CIGS [208]. These two materials also have the important advantage of established production techniques, because of their use as thin film photovoltaic materials. However, these are not ideal materials from a large-scale commercial point of view: the scarcity and the high cost of indium represent a problem for CIGS, while the environmental and health issues related to highly toxic cadmium and the supply concerns of tellurium are important drawbacks of CdTe. However, the alternate view is that cadmium is a by-product of metals mining and in its free state, storage is actually the real toxic risk; but in compound CdTe form it is highly stable and its use is therefore recommendable. From a commercial point of view, indium-free CuGaSe$_2$ is more promising and very high water reduction photocurrents (around 13 mA/cm^2) have been already shown with this material [209]. Similarly, WS$_2$, WSe$_2$, and the most recently studied Cu$_2$ZnSnS$_4$ (CZTS) are all composed of inexpensive, abundant, and nontoxic elements and are therefore attractive materials. Very impressive solar-to-hydrogen conversion efficiencies around 6%–7% have been obtained with both WS$_2$ [210] and WSe$_2$ [211]. Photocathodes based on Cu$_2$ZnSnS$_4$, which only attracted attention in the last few years, have already shown solar-to-hydrogen conversion efficiencies above 1%, in combination with platinum as a catalyst and n-type CdS as a buffer layer to reduce charge carriers recombination [212]. Stability, however, appears to be an important issue with this promising material, although atomic layer deposition of nanoscopic TiO$_2$ protective layers has been suggested as a promising strategy to overcome this issue [213].

In the last 5 years, CuO$_2$ and p-silicon have emerged as the two most promising photocathode materials, although their long-term instability has not been addressed yet. Inexpensive and environmentally friendly silicon-based materials appeared to show much promise as photocathode materials in the last couple of years. p-doped Si ($E_g = 1.2$ eV) has recently yielded solar-to-hydrogen efficiencies of 5%–6% (and good stability within the first few hours operation) when nanostructured into arrays of high-aspect-ratio nanopillars using Pt [214] and even using noble-metal free molecular Mo_3S_4 [149] as WRC. More recently, similar solar-to-hydrogen efficiencies have been obtained even with low-cost amorphous silicon (a-Si), which also shows much improved stability upon deposition of TiO_2 protective layers [215]. Among the few metal oxides having a conduction band sufficiently negative to perform water reduction, CuO_2 ($E_g = 2.0$ eV) appears to be the best candidate photocathode material, particularly because of its very low cost and nontoxicity, but also due to its remarkable energy conversion efficiency. Very high photoactivity, with water reduction photocurrents up to 7.6 mA/cm² at 0 V versus RHE (though accompanied by very low stability against photocorrosion), has been demonstrated using Pt catalyst and an atomic layer deposited protective coating [216]. More recently, improved stability has been obtained with this material, at the expense of slightly reduced photocurrents, using alternative WRCs [131], including noble-metal free amorphous MoS_x [141].

Some visible-light active ternary oxides have also been recently shown to bear some potential as photocathodes for water splitting; in particular, promising cathodic photocurrents were demonstrated using p-doped strontium titanates, copper niobates, and copper tantalates, in the orders of hundreds of µA/cm² for Rh-doped $SrTiO_3$ [217] and $CuNbO_3$ [218,219] and over 1 mA/cm² for $Cu_5Ta_{11}O_{30}$ [220]. It can be expected that some new multinary oxides having suitable properties to act as photocathode materials will be developed in the near future using high-throughput combinatorial approaches, which have already allowed the identification of a few potential materials [221–223]. On the other hand, novel polymeric metal-free semiconductor materials may also turn out to be of potential interest for photoelectrochemical (or photocatalytic) water splitting in the near future. Very recently, graphitic carbon nitride (g-C_3N_4) and graphite oxide (GO) have been reported to show cathodic photocurrents toward water reduction [224,225]. Although the photocurrents generated so far are quite modest, this class of metal-free polymeric materials can be of high promise for photoelectrochemical water splitting, owing to their extraordinary structural and electronic properties and their predicted low cost.

1.3.3 TANDEM PHOTOELECTROCHEMICAL SYSTEMS

No single absorber material has been shown so far to evolve oxygen and hydrogen from water under solar irradiation with reasonable conversion efficiency and the required prolonged stability. This is not surprising, in view of the natural trade-off between the absorption spectrum of semiconductor materials and the driving force of the photogenerated charge carriers: materials that can absorb a significant fraction of the solar light tend to generate charge carriers having a small driving force for water reduction and oxidation, while materials that can generate electron–hole pairs with a sufficient driving force to split water can generally only absorb a small fraction of the incident solar light. An approach to overcome this trade-off consists in using two (or more) different absorber materials in the same device, to constitute what is generally known as a tandem cell. In this configuration, the photovoltage generated by each absorber material (related to its bandgap) is additive, and therefore a sufficient driving force to split water at a reasonable rate can be achieved even using materials having a relatively small bandgap, which in turn implies that a significant fraction of the solar spectrum can be effectively harvested. The drawback, however, is that more photons are required per each electron–hole pair (and therefore per each molecule of hydrogen) generated. In the prototypical case of a dual (D) absorber configuration, four photons will be required to evolve a molecule of hydrogen; this configuration is therefore designated as D4 tandem cell, in contrast to the single (S) absorber configuration, which requires two photons only to generate a molecule of hydrogen and is designated as S2 approach, accordingly [226]. The two "components" of the tandem cell can be either stacked in a monolithic wireless device or can be physically separated and simply electrically connected through wires. Regardless of the specific design of the device, the general idea is that the two absorbers will be positioned with respect to the incident solar light in such a way that as many photons as possible which are not absorbed by the (first) higher bandgap absorber will be harvested by the (second) lower bandgap absorber.

Two major classes of photoelectrochemical tandem cells can be distinguished: photoanode–photocathode tandem cells and photoelectrode–photovoltaic (or hybrid) tandem cells. In the former type, an n-type material is responsible for the oxygen evolution half-reaction (photoanode), while a p-type semiconductor is responsible for the hydrogen evolution half-reaction (photocathode). Both electrodes for hydrogen and oxygen evolution are therefore based on the formation of an SCLJ. The latter type, conversely, usually involves only one photoelectrode (which is an n-type photoanode in the majority of the systems reported so far) in combination with a photovoltaic cell. In this "hybrid" configuration, the driving force (photovoltage) for water splitting is only partially generated by an SCLJ, while the remaining driving force is due to the photovoltage generated by the photovoltaic cell. Besides these two classes, several other configurations of tandem water-splitting cells can be envisaged. For example, two photoactive materials can be combined in the same photoelectrode to form a monolithic heterojunction photoanode or photocathode. In addition to improved sunlight absorption, this configuration can yield enhanced charge separation owing to the electrical field generated within the junction between the two semiconductors. This junction can be made between a p-type and an n-type semiconductor, such as in the recently reported p-$CaFe_2O_4$/n-TaON heterojunction photoanode, showing a solar-to-hydrogen conversion efficiency of 0.5% [227], but even between two n-type (or two p-type) materials, such as in the recently investigated n-Si/n-Fe_2O_3 heterojunction photoanode [228,229]. Finally, tandem water-splitting cells can also be prepared based solely on photovoltaic cells, that is, without any active SCLJ. This configuration, however, is not an actual photoelectrochemical cell, but is rather a photovoltaic-based photoelectrolysis cell and will be described in Section 1.3.4.

1.3.3.1 Photoanode–Photocathode Tandem Cells

The most straightforward design for a photoelectrochemical tandem cell consists of an n-type semiconductor and a p-type semiconductor electrically connected (either through wires or through an ohmic contact) and immersed in an aqueous electrolyte. This configuration was already suggested in the late 1970s by A. J. Nozik, who called it photoelectrochemical diode [230,231], and is solely based on the formation of SCLJs, which are in principle comparatively more straightforward to form (and generally more inexpensive) than solid-state p–n junctions. In practice, however, solar-to-hydrogen conversion efficiencies reported so far using this concept are generally below 1%. One remarkable exception is a device reported in 1987, which consisted of an n-type GaAs photoanode protected with a thin layer of MnO_2 and electrically connected to a p-type InP photocathode, yielding a solar-to-hydrogen conversion efficiency of over 8% [232]. An obvious drawback of this device, however, is its very high cost of manufacturing. Apart from this outstanding example, the vast majority of the devices reported so far generally involve the use of TiO_2, $SrTiO_3$, Fe_2O_3, or WO_3 as photoanode materials in combination with either GaP or $GaInP_2$ photocathodes and the efficiencies obtained are generally very low [230,231,233–237]. Examples of photoelectrochemical diodes where the p-type material is not a III–V semiconductor are very rare; among these, a device based on a nanostructured p-type Cu-Ti-O photocathode has been reported to yield a photoconversion efficiency of 0.30% in combination with a TiO_2 photoanode [238].

1.3.3.2 Photoelectrode–Photovoltaic Hybrid Tandem Cells

In the last couple of years, hybrid tandem cells involving a photoelectrode and a photovoltaic device have emerged as a promising alternative to photoelectrochemical diodes. Despite the fact that the first successful example of a hybrid tandem cell was demonstrated with a p-type photocathode [35], the vast majority of the most recent and promising hybrid cells consist of an n-type photoanode electrically connected to a photovoltaic cell. This is because, as pointed out in Section 1.3.2, a few reasonably stable and active photoanode materials (such as Fe_2O_3, WO_3, and $BiVO_4$) have already been identified. However, the recent progress related to surface protected CuO_2 and p-Si photocathodes suggests that the development of photocathode–photovoltaic tandem cells could again become of interest in the near future despite the ongoing challenges in terms of long-term stability.

The most efficient hybrid tandem cell reported so far (which was also among the first ever reported) involves a p–n GaAs photovoltaic cell and a p-$GaInP_2$ photocathode integrated into a monolithic device. This device, developed by O. Khaselev and J. A. Turner in 1998, could achieve an impressive photoconversion efficiency of 12.4%, owing to the significant bias generated by the p–n junction and to the complementary absorption of p–n GaAs and p-$GaInP_2$ [35]. This device still represents the state of the art in terms of photoconversion efficiency; however, the limited stability of p-$GaInP_2$ and in particular the very high manufacturing cost obviously prevents its large-scale feasibility. Potentially much cheaper hybrid tandem cells have been recently developed based on other photovoltaic materials such as CIGS a-Si, and DSSCs. For example, a monolithic device consisting of a CIGS/CdS photovoltaic cell deposited onto a TiO_2 photoanode has been shown to yield conversion efficiencies around 1% [239]. While the problem of poor stability of CIGS cells in aqueous conditions was tackled here using Nb–TiO_2 protective layers, the potential for large-scale commercialization of CIGS remains unknown because of the cost of indium.

Comparatively, a-Si solar cells and DSSCs currently appear as the most promising and inexpensive alternative for hybrid tandem devices.

Monolithic devices involving one or more a-Si photovoltaic junctions combined to an oxide photoanode have been described by E. L. Miller in an early work in 2003, where Fe_2O_3, WO_3, and TiO_2 were pointed out as potential photoanodes for constructing multijunction cells [240]. Besides the requirement of current matching between the two components of the tandem device, a major drawback of this particular combination is the requirement for a low-temperature deposition technique for the metal oxide, in order to avoid degradation of the underlying a-Si layers. While no net water splitting could be achieved with a Fe_2O_3 photoanode [241], solar-to-hydrogen conversion efficiencies over 3% and stable operation under outdoor conditions were achieved using WO_3 photoanodes [242,243]. Very recently, a new record efficiency for metal oxide–based hybrid tandem cells was set using the novel photoanode material $BiVO_4$. A monolithic device consisting of a layer of this oxide deposited on top of a double-junction a-Si cell yielded a solar-to-hydrogen conversion efficiency of 4.9% [101]. It is understood that one of the key property of $BiVO_4$ which makes it more favorable than WO_3 to be combined in a hybrid tandem cell is its bandgap. While the ideal bandgap for such a configuration appears to be 1.8–2.0 eV, in order to achieve solar-to-hydrogen conversion efficiencies around 10% a bandgap of at most 2.3–2.4 eV is mandatory, which is quite close to the bandgap of $BiVO_4$, but significantly smaller than the bandgap of WO_3 [101].

An idea of combining one or more DSSCs to a metal oxide photoanode in the same device was suggested by M. Grätzel and J. Augustinky in the mid-1990s and conversion efficiencies up to 4.5% were predicted using WO_3 as photoanode [244]. Ten years later, a monolithic device based on two bipolar WO_3/Pt and dye-sensitized TiO_2/Pt photoelectrodes was demonstrated to yield 1.9% solar-to-hydrogen conversion efficiency [245] connected through a Z-scheme akin to natural photosynthetic energetic principles, while conversion efficiencies up to 2.8% were obtained using two DSSCs connected in series [246]. Later, a hybrid cell composed of a Fe_2O_3 photoanode and two series-connected DSSCs was demonstrated to split water with over 1.3% solar-to-hydrogen efficiency [247]. More recently, a record efficiency of 3.1% was achieved with a device combining a WO_3 photoanode with only one DSSC, while an efficiency of almost 1.2% was demonstrated in the same work using Fe_2O_3 photoanode. The use of only one solar cell, which greatly simplifies the overall device with respect to a construction requiring two cells, was allowed by the particularly high open-circuit voltage of the DSSC utilized, where a cobalt-based electrolyte was employed instead of the conventional iodine/tri-iodide redox shuttle [248].

1.3.4 PHOTOVOLTAIC-BASED WATER SPLITTING

As already pointed out, solid-state junctions are obviously more difficult to manufacture when compared to solid–liquid junctions; therefore, water-splitting cells based on solid-state photovoltaic junctions are in principle expected to be more expensive than photoelectrochemical cells. However, photovoltaic-based water splitting has the important advantage of being based on technologies and manufacturing techniques already well-established for photovoltaic production of electricity. The simplest way to drive water splitting using photovoltaic power is to utilize a commercial PV producing a sufficient voltage to bias a conventional electrolyzer (EC). Currently, this "brute-force" PV–EC approach would be, however, too expensive, as the estimated price of hydrogen produced would range around 8–20 $/kg. Although the cost of PVs is expected to significantly decrease in the near future, it is clear that a substantial cost reduction is required in order to make photovoltaic-based water splitting competitive. It is reasonable to expect that such a cost reduction could be achieved with an integrated stand-alone device, combining a photovoltaic absorber unit for sunlight harvesting and an electrocatalytic unit for water reduction and oxidation. This simple stand-alone device would potentially remove the cost related to glass or other substrates, frames, and wires associated with a conventional "brute force" PV–EC approach. An additional advantage of an integrated approach for solar-driven water splitting comes from the fact that given the low current densities allowed by the solar emission spectrum, much lower overpotentials are required to drive the water-splitting reaction, and therefore high overall efficiencies can be achieved, with respect to a commercial electrolyzer connected to a photovoltaic panel, or more generally to the grid. An important drawback of photovoltaic-driven water splitting, however, is that the voltage generated by a traditional p–n junction solar cell strongly varies according to the light intensity; in practical situations, as recently pointed out by K. Sivula, this could lead to significant losses and/or the need for a complex electronic switching mechanism in order to optimize the efficiency to cope with the varying light intensity caused by haze, clouds, and time of day [249].

Analogously to what happened for the development of hybrid tandem cells, the photovoltaic materials of choice for the first photovoltaic-based water-splitting devices reported were based on III–V semiconductors and on a-Si. In fact, some devices presented in this section feature many similarities to some other systems, which are classified as hybrid tandem cells. To set a strict classification between the various configurations of solar water-splitting cells is obviously to some extent arbitrary. However, the convention followed here is based on the presence/absence of an active SCLJ in the design of the device; in principle, in a photovoltaic-driven water-splitting cell, the photovoltage generated by the photoactive assembly is not related to the formation of any SCLJ.

Not surprisingly, the highest efficiencies so far have been achieved with devices based on III–V semiconductors: in the early 2000s, a monolithic dual-bandgap device based on the multijunction AlGaAs/Si system yielded a solar-to-hydrogen efficiency of 18.3%, which after more than a decade remains the ultimate record efficiency for solar energy conversion [250]. In addition to high-quality crystalline photovoltaic materials and a very complex multilayer structure, high-surface area electrodes were employed to achieve such very high performances; moreover, the electrodes were coated with highly active catalysts Pt-black and RuO_2 to promote hydrogen and oxygen evolution, respectively. An analogous dual-bandgap multilayer device based on the GaInP/GaAs photovoltaic system and Pt as both WRC and WOC showed conversion efficiencies over 16% [251]. These two studies importantly showed that very high conversion efficiencies could be achieved with photovoltaic-driven water splitting; however, the cost of these devices based on III–V semiconductors is obviously far too elevated for large-scale production and commercialization.

Multijunction a-Si solar cells have also been shown to split water with high efficiency under sunlight illumination. Because single-junction a-Si solar cells generally produce a voltage of about 0.6–0.7 V, at least two cells connected in series (or rather a double-junction cell) are needed to provide the thermodynamic driving force to split water. In practice, however, because of the kinetic overpotentials required, three cells are generally required in order to obtain reasonable photocurrents and conversion efficiencies. As already mentioned in Section 1.3.3, silicon is inherently unstable in aqueous electrolytes, and particularly in basic environments. Therefore, if the solar cell has to be immersed in water (as is the case for a monolithic stand-alone device) it must be either encapsulated or else a protective coating has to be deposited. Typically, transparent and conductive indium tin oxide (ITO) is the material of choice; however, its effectiveness as protective layer strongly depends on its annealing temperature, which is limited by the fact that the underlying a-Si cannot withstand high temperatures. A solar-to-hydrogen conversion efficiency of 5% was already demonstrated in 1989 using a triple stack of p–i–n a-Si cells, using RuO_2 and Pt catalyst deposited on the back and top cell, respectively [252]. Later, a conversion efficiency of 7.8% was achieved with a similar construction using Pt [251] and even using earth-abundant materials such as $NiFe_yO_x$ and Co–Mo alloy [253] to catalyze the two half-reactions. This is quite impressive, considering that the conversion efficiency of the a-Si solar cell itself was about 10%. Remarkably, an excellent long-term stability was achieved in the second case, in both 1 M KOH electrolyte and drinking water, although the a-Si photovoltaic cell was not physically immersed in the electrolyte. A comparable conversion efficiency and excellent stability were also demonstrated using a slightly different configuration, where the triple junction a-Si cell coated with ITO was immersed in the electrolyte but encapsulated by an epoxy resin. The stability of this device was quite poor due to rapid corrosion of the ITO layer by the harsh electrolyte (and subsequent exposure of Si to the latter), but was later significantly enhanced by deposition of an additional F-doped tin oxide (FTO) layer [254].

More recently, a significant breakthrough was achieved as efficient photovoltaic-driven water splitting was demonstrated for the first time at neutral pH. This achievement was made possible by the development of a novel and inexpensive cobalt-based (Co–Pi) WOC, which efficiently operates under neutral pH conditions [37]. Naturally, the fact that operation can be made at neutral pH instead of alkaline conditions allows for increased stability as well as flexibility in terms of materials choice; in particular, an earth-abundant NiMoZn alloy can be used in place of Pt as the hydrogen evolution catalyst. In the early 2010s, Steven Y. Reece et al. reported a water-splitting cell based on a commercial triple-junction a-Si cell in combination with inexpensive Co–Pi and NiMoZn catalysts and yielding a solar-to-hydrogen conversion efficiency of 4.7% in a wired configuration [169]. An analogous wireless monolithic device instead yielded only 2.5% conversion efficiency, due to increased electrolyte resistance; the light-to-electricity efficiency of the photovoltaic cell were 7.7% for the wired device and 6.2% for the wireless device. Excellent stability was demonstrated even in seawater, which is unique to the Co–Pi catalyst and probably related to its repair mechanism [255]. Owing to the facts that this stand-alone wireless system is based on earth-abundant materials and that it can work efficiently in natural waters and under ambient conditions, this device was termed the artificial leaf, to underline the similarities it shares with the natural photosynthetic system. The Co–Pi WOC can also be deposited on crystalline silicon solar cells, to yield a very similar device; although crystalline and a-Si cells generally yield a comparable voltage, crystalline silicon has the advantage that it can withstand higher temperatures than a-Si. The important practical implication is that the protective ITO layer deposited on the crystalline Si cell can be annealed to temperatures up to 400°C, which renders it much more effective and significantly enhances the lifetime of the device [170]. The efficiencies achieved so far using Si-based absorbers are promising, yet still moderate and far behind those demonstrated using III–V semiconductors. A theoretical study recently introduced an improved framework to model integrated photovoltaic–based water-splitting cells and surprisingly showed that up to 16% conversion efficiency can be achieved using crystalline silicon, provided that the coupling between the photovoltaic cell and the electrochemical process is optimized, the solution resistance is minimized, and high performance catalysts are utilized [256]. Considering the relatively low cost of silicon-based photovoltaics, this value is astonishingly close to the maximum conversion efficiency achievable using GaAs solar cells, which is slightly above 18% according to the same study. From a commercial and engineering point of view, it will be interesting to compare these efficiency numbers and related payback periods to those of a simple photovoltaic device powering an EC. Ultimately, such considerations come into play when taking technological selections in the commercial world.

Clearly, silicon-based photovoltaics in combination with earth-abundant catalysts bear much promise as low-cost alternatives for photovoltaic-driven water splitting. However, the highest conversion efficiency demonstrated with relatively stable devices are currently less than half of what would be required for its viable commercialization. On the other hand, impressive efficiencies have been demonstrated using III–V semiconductors; devices based on this class of materials, however, are of limited practical use, because of their limited stability and their very high cost. An absorber material with intermediate cost and efficiency between a-Si and III–V semiconductors is CIGS. Furthermore, the bandgap of this material can be easily tuned between 1.0 and 1.7 eV by simply changing the In:Ga ratio [257]; this is a very important advantage for water splitting, as the bandgap can be optimized with respect to the driving force required to perform the electrochemical reaction. Very recently, an impressive 10% solar-to-hydrogen efficiency has been reported using this material in a photovoltaic-based water-splitting cell, in both monolithic and wired configurations [258]. Besides the absorber material used, another interesting concept of this work was the use of interconnected cells in series instead of a multijunction stack to generate the required voltage. Because of the low bandgap of CIGS, the voltage that is generated by a CIGS cell is insufficient to drive the water-splitting reaction and three cells were therefore connected in series. The requirement of current matching between the three series-connected cells, which imposes severe constraints and limitations in multijunction stacks, was straightforward in the design utilized here, as the three cells were not stacked on top of each other but were rather placed side-by-side, therefore naturally experiencing the same illumination. Instability of CIGS in aqueous environment was overcome here by encapsulating the cell with glass and a polymer film, and the cell was claimed to most likely pass a 5000-hour stability test [258]. A limitation of this system is that large platinum foils covered by platinum nanoparticles were used as the two electrodes; however, cheap materials having catalytic water reduction and oxidation activity comparable or even higher than platinum have already been demonstrated (as described in Section 1.2.3). Clearly, the scarcity of indium and the presence of a CdS buffer layer remain two important limitations using CIGS. On the other hand, highly efficient Cd-free buffer layers such as zinc oxysulfide [Zn(O,S)] have already been identified and implemented in commercial solar cells [259,260]. Furthermore, the absorber material CZTS, with which an encouraging solar-to-electricity efficiency of 12% has already been demonstrated, could prove to be an inexpensive alternative to CIGS in the near future [261].

The photovoltaic-driven water-splitting approach has been demonstrated with moderate conversion efficiencies even using DSSCs. As for silicon and CIGS cells, three DSSCs are generally required, although two high-voltage DSSCs based on cobalt redox couple could be sufficient to provide the required voltage. In an early report by Park et al., two different designs—both based on bipolar dye-sensitized TiO_2/Pt panels—were reported: one design consists of a stack of three connected cells, where the light flux is directed at 45° with respect to the plane of the cells, while the other simply consists of three cells connected in series side-by-side. Here again, current matching is straightforward, as all the cells experience the same illumination and therefore generate (in principle) the same photocurrent. While solar energy conversion efficiency with the first design was 2.2%, an efficiency of 3.7% was achieved using the side-by-side configuration [262]. This difference is mostly due to the lower amount of light absorbed using the first configuration; the fact that such efficiencies can be achieved even when light is not shined directly through the front panel of the cell is a unique feature of DSSCs, which makes this option particularly promising under low light conditions.

Finally, a novel miniaturized version of photovoltaic-based water splitting has been recently reported. In this original design, the absorber photovoltaic material is made into nanostructured particles suspended in the electrolyte and then protected with a nanoporous anodic aluminum oxide membrane, in order to avoid photocorrosion and short-circuit, such that only the catalysts for water oxidation and water reduction are exposed to the electrolyte [263].

1.3.5   Dye-Sensitized Water Splitting

A recent approach toward solar water splitting is based on dye sensitization of oxide semiconductors. Contrary to all previously described schemes, in this approach the semiconductor material does not act as absorber but rather as scaffold for a molecular chromophore and as charge transfer mediator. Although this concept of dye sensitization can also be applied to particle photocatalysts, it has mainly been applied so far in the context of photoelectrochemical cells. In this design, the structure of the water-splitting cell is entirely analogous to the well-known DSSCs, where the sensitized oxide represents the photoanode and an inert metallic counter-electrode represents the cathode. The fundamental difference between DSSCs and dye-sensitized water-splitting cells is that while in the former the electrolyte contains a reversible redox couple which is continuously regenerated, in the latter a net chemical transformation is taking place: oxygen is evolved at the photoanode and hydrogen is evolved at the cathode. The first type of cell is said to be regenerative, while the second is classified as a photosynthetic cell [80].

In analogy to DSSCs, and similarly to what is also encountered in photocatalytic systems, this kind of water-splitting cells have to overcome a major challenge, that is, no electrical field is established at the surface of the semiconductor upon contact with the surrounding electrolyte. Once again, this is because the particles constituting the oxide film are too small to sustain this field. As a consequence, charge recombination and detrimental back electron transfer would readily occur and severely limit the performances of this device, unless some specific design principles are undertaken to promote sustained charge separation. In DSSCs, effective kinetic competition against recombination and back electron transfer is successfully achieved by introducing a suitable redox mediator (such as I_3^-/I^-), which can readily reduce the oxidized state of the photoexcited dye. Obviously, this is not possible in the context of water splitting, where the relevant redox couples are strictly H^+/H_2 and O_2/OH^-. Therefore, very fast water oxidation/reduction catalysis and an architecture promoting vectorial charge transfer are compulsory to achieve sustained charge separation and thus sufficient energy conversion efficiencies in dye-sensitized water-splitting cells.

For almost two decades, the group of Thomas E. Mallouk has been the most active in the quest toward these two goals. They initially took inspiration from biological systems to design an artificial multicomponent electron transfer chain, capable of vectorial electron transfer and based on layer-by-layer assembly of solids. The first generation of layered solids, based on metal-organic phosphonates (derivatized with an inorganic dye and methyl-viologen as electron acceptor) lead to poor compartmentalization and mixing of successive layers within the multilayer architecture, thus leading to insufficient vectorial charge transfer [264]. Improved compartmentalization and vectorial electron transfer were however achieved with a second generation of layered solids, based on two-dimensional sheets made by exfoliation of layered metal oxide semiconductors (LMOS), such as titanoniobates [265,266]. By functionalizing the layered solid with Pt nanoparticles, HI splitting was demonstrated with this system [267], while sacrificial hydrogen evolution was recently achieved with an analogous photocatalytic system, where the two-dimensional titanoniobate sheets were replaced by niobate nanosheets, acting as electron mediator [268]. The progress on sacrificial hydrogen evolution using exfoliated LMOS sheets is summarized in an excellent review by the Mallouk group [269]. Recently, sacrificial hydrogen evolution from a dye-sensitized photocatalytic system was achieved even using [NiFeSe]-hydrogenase enzyme in place of platinum and TiO_2 as scaffold and electron mediator. Remarkably, this construct was shown to be particularly robust for a biological system, showing good stability even upon exposure to air condition and solar irradiation [270,271].

Overall water splitting was first achieved using dye-sensitized photoelectrochemical cells in the late 2000s. Remarkably, this was achieved using a simple mesoporous TiO_2 scaffold (similar to the one used to construct DSSCs) instead of the more complex multilayer assembly of LMOS sheets. The TiO_2 film was sensitized with a Ru-based dye, which was in turn covalently attached to hydrated IrO_2 nanoparticles, acting as the WOC. The obtained TiO_2–Ru-sensitizer–IrO_2 system was used as photoanode in combination with a Pt counter-electrode, to drive water splitting with an internal quantum yield of 0.9% in the presence of a bias voltage [50]. Besides the required external bias, two major drawbacks limit the performances of this construct: first, several dye molecules are not utilized as they turn out not to be linked to the TiO_2 scaffold, but rather to multiple catalyst particles; second, the excited state of the dye is susceptible to be quenched by the oxidized form of the catalyst. A possible solution to both these problems has been recently developed, through the introduction of a biomimetic electron transfer relay between the dye and the catalyst. With this approach, internal quantum yields over 2% have been recently demonstrated [272]. Finally, a modification of this system has been developed, where the noble-metal-based catalyst is replaced by bio-inspired Mn-oxo clusters, impregnated into a proton-conducting Nafion membrane, itself supported on sensitized TiO_2. This system has been shown to split water in the absence of an external bias, although photocurrent and stability were very modest, revealing photodecomposition of one of the components of the construct [273].

Despite the significant advances with dye-sensitized water splitting, the conversion efficiency obtained so far with these devices is very far from what is required for a practical application. This is mainly due to the very low internal quantum yield, determined by sustained charge recombination. To improve the quantum yield, a better WOC and an improved charge separation are still needed. Regarding the first issue, despite the recent progress with novel and inexpensive catalyst materials, it seems quite difficult that catalysts significantly more efficient that IrO_2 will be developed. Contrary to photoelectrochemical or photovoltaic-driven water splitting, where the TOF required from a catalyst is simply determined by the current density available from the solar spectrum, in dye-sensitized water splitting, the TOF has to be much higher, in order to compete with back electron transfer. Trying to improve charge separation appears therefore to be a more promising approach; recently, a core–shell architecture, where the TiO_2 particles are covered by another wide-bandgap oxide such as ZrO_2 or Nb_2O_5 (a concept already applied to DSSCs) has shown to be a potential avenue [274]. Furthermore, as pointed out by R. Swierk and T. E. Mallouk, because of the low conduction band of TiO_2, a dual-absorber Z-scheme will be compulsory to achieve water splitting without bias and with reasonable efficiencies [275]. A further serious issue that is yet to be addressed is stability: while for a practical application a TON of 10^9 would be needed, this value is currently eight orders of magnitude lower. This is partly due to the instability of the oxidized dye, which tends to decompose in aqueous solutions. A repair mechanism, analogous to the one taking place in natural photosynthesis, could be envisaged, but appears to be very challenging from a practical point of view; on the other hand, it has been put forward that a simple protective strategy, involving encapsulation of the photoredox assembly with a thin insulating coating, could be the best avenue to be followed [275]. Nevertheless, the problems highlighted here are highly challenging and require an all-round success in the most difficult and mutually competing drawbacks.

1.4   CONCLUSIONS AND OUTLOOK

Over the last four decades, several approaches have been followed to achieve water splitting using sunlight and semiconductor systems. The scientific community has put so much effort to achieve this goal and rightly it earned the status of being termed as the "Holy Grail" of chemistry. This distinction also underlines the impact that a viable solar water-splitting technology would have on the society: hydrogen, a simple and energy-rich fuel, would be available at a competitive cost and from the most abundant renewable source of energy, the Sun. Mankind will also earn the credit of finally achieving artificial photosynthesis in a practically sustainable manner, reminding us the prime position of natural photosynthesis in originating and sustaining life on this planet. The harvested energy in the form of hydrogen could be released by combining it with plentiful oxygen and the by-product of this reaction would be water again. In an engineering parlance, achieving photoinduced water splitting through photoelectrochemical or photosynthetic techniques can be considered as electrolyzing water without actually exporting electrons (electricity) outside of the system as otherwise familiar with the case of photovoltaic-cell-powered powered water electrolyzers. This view should be understood from the basic fact that even if scientists achieve the Holy Grail of chemistry in an efficient manner, engineering it into large-scale operating systems is another major challenge to climb over. Clearly, none of the approaches described has already reached this stage; however, some appear to be closer than the others to this goal. On the other hand, after four decades of studies, some conclusions and research directions common to all various approaches can be drawn.

Photovoltaic-driven water splitting and photoelectrochemical water splitting clearly appear to be the most promising approaches, both because of their conversion efficiency and stability. Interestingly, but perhaps not surprisingly, these two approaches both involve the presence of a rectifying junction at their very "heart." This can be either a SCLJ, such as in photoelectrochemical devices, or a junction between two semiconductors, such as in photovoltaic-driven water splitting, or again a combination of the two, as is the case of hybrid tandem cells. These junctions are of paramount importance in these devices as they establish an electrical field within the semiconductor material, which promotes the separation of the charges photogenerated by the absorbed photons. On the other hand, photocatalytic and dye-sensitized systems, due to the size of the semiconductor particles involved, lack the electrical field required by engineers to build up a large system. Hence these approaches require more complex techniques toward building efficient charge separation. Despite strenuous efforts, this challenge has not been overcome yet and the energy conversion efficiencies obtained with these types of devices are still significantly lower than what has been achieved with photoelectrochemical and photovoltaic-based water-splitting cells.

In the last few years, photovoltaic-based systems and hybrid tandem cells have emerged as the most promising technologies toward water splitting. In both cases, solar-to-hydrogen efficiencies of 5% have already been achieved using inexpensive metal oxides (such as $BiVO_4$) and a-Si. The 10% conversion efficiency threshold, often seen as the minimum required for a commercially viable technology, has already been demonstrated using more expensive and scarce materials, such as CIGS. Furthermore, outstanding efficiencies above 18% have been obtained using expensive III–V semiconductors; despite the limited practical relevance of these materials, this represented an important landmark, demonstrating the potential of these approaches. The long-term stability of these systems is still to be demonstrated, although oxide semiconductors such as Fe_2O_3 appear to be sufficiently robust for practical applications. The well-known instability of silicon-based materials in water under illumination has been significantly mitigated using conformal protective coating, while encapsulation techniques using polymers, although not necessarily straightforward, could provide the required long-term stability; this approach has already been demonstrated in a few instances and could be applied to a-Si, crystalline Si, CIGS cells, and DSSCs. These technologies also show high promise in view of criteria such as toxicity and scalability: the former should not be an important issue with materials such as Fe_2O_3, $BiVO_4$, and silicon, while relatively scalable techniques (such as spray-pyrolysis for Fe_2O_3 and $BiVO_4$ or chemical vapor deposition for silicon) are typically employed to synthesize these materials. Clearly, these direct water-splitting technologies would ultimately have to positively compare, in terms of system cost and overall conversion efficiency, with hydrogen produced from electrolyzers powered (either directly or indirectly) by photovoltaic solar cells. Interestingly, given the moderate current densities which can be achieved using direct solar water splitting (inherently limited by the solar flux), which in turn translate into moderate overpotential for hydrogen and oxygen evolution, higher efficiencies could indeed be achieved in principle using these technologies. Reassuringly, this has also been demonstrated to be the case in practice, over 20 years ago [251], clearly showing the enormous potential of integrated solar water-splitting cells, in particular for decentralized power generation.

In comparison to hybrid tandem cells and photovoltaic-based devices, the development of photocathode–photoanode tandem devices is clearly lagging behind, which is essentially due to the lack of efficient and stable photocathode materials. However, the recent advances in the protection of unstable photocathode materials such as CuO_2 and p-Si indicate that the interest for photocathode–photoanode devices may rise again in the near future. Very often, protective coatings to effectively stabilize these materials are made by atomic layer deposition; it is worth pointing out that this technique, which is not suitable for high-throughput production in its conventional form, has been recently adapted for high throughput processing and implemented in the mass production of buffer layers for CIGS cells.

Photocatalytic systems would certainly be a simple and low-cost water-splitting technology; unfortunately, however, this approach appears to be far from a practical reality, as solar-to-hydrogen efficiencies are currently over one order of magnitude lower to what would be needed for practical applications. In addition, this approach has generally yielded poor stability so far; a remarkable exception is GaN:ZnO nanoparticles, which besides being the most efficient single-step photocatalytic system has shown rather promising stability. Dye-sensitized water splitting is a very elegant approach to achieve the Holy Grail, which can be realized using both photocatalytic and photoelectrochemical systems; however, and despite the significant progress achieved in this field, the difficulties associated with this approach have so far yielded very low conversion efficiency and very limited stability. The subtle kinetic competition between the (desired) forward and the (detrimental) back electron transfer, which has been successfully mastered in DSSCs, offer very little potential for succesfully using this system for the much more complicated water-splitting reactions.

Regardless of the approach followed to achieve the Holy Grail, impressive advances have been achieved in the last few years in the development of visible-light harvesting materials, nanostructured architectures, and novel catalysts. A series of highly active and earth-abundant catalysts for both water oxidation and water reduction has been developed in the last 5 years. The development of multinary oxides, doped materials, solid solutions, plasmonics, disorder-engineering, and other bandgap engineering techniques has allowed expanding the spectrum of available visible-light harvesting materials. High-throughput screening methods and computational techniques have played and will play an increasingly important role in this endeavor. Novel deposition techniques have permitted a high degree of control on the morphology and the composition of these materials at the nanoscale, as well as the construction of complex assemblies for sustained charge carrier separation and reduced recombination.

Furthermore, theoretical frameworks have been developed, indicating the maximum achievable efficiencies for the various schemes as well as the various factors that can limit the efficiency of a system and, most importantly, identifying the critical loss mechanisms to be tackled in order to optimize the current systems. For instance, early theoretical studies have been of paramount importance in underlining the overwhelming advantages of dual-absorber systems [226]; after almost 30 years from these studies, it is now quite clear that significant promise toward practical applications is indeed limited to dual-absorber schemes—whether photocatalytic Z-schemes or tandem photoelectrochemical cells. More recently, theoretical frameworks have allowed to point out the unexpected potential of earth-abundant photovoltaic-based water-splitting techniques and to focus on the work yet to be done to realize that potential. In particular, it appears that the importance of optimizing the design of the water splitting cell in order to minimize the losses associated with solution resistance has been widely underestimated so far. A further surprising result of this analysis is that a photovoltaic-based water-splitting device based on crystalline silicon and cobalt-based catalyst, once the aforementioned optimization has been implemented, could lead to a solar-to-hydrogen efficiency of 16% [256]. It is not clear as of now how close to this value we can reach with real systems. Certainly, if ever this predictions were realized, we will be able to conclude that the Holy Grail exists and is indeed attainable.

REFERENCES

1. A. J. Bard and M. A. Fox, "Artificial Photosynthesis: Solar Splitting of Water to Hydrogen and Oxygen," *Acc. Chem. Res.*, vol. 28, no. 3, pp. 141–145, 1995.
2. R. E. Smalley, "Top Ten Problems of Humanity for Next 50 Years," *Energy & NanoTechnology Conference,* Rice University, Houston, TX, 2003.
3. Core Writing Team, R. K. Pachauri, and A. Reisinger, *Climate Change 2007: Synthesis Report. Contribution of Working Groups I, II and III to the Fourth Assessment Report of the Intergovernmental Panel on Climate Change*, IPCC, Geneva, 2007.
4. J. A. Turner, "Sustainable Hydrogen Production," *Science*, vol. 305, no. 5686, pp. 972–974, 2004.
5. A. E. Becquerel, "Mémoire sur les effets électriques produits sous l'influence des rayons solaires," *Comptes Rendus*, vol. 9, pp. 561–567, 1839.
6. R. W. Gurney and N. F. Mott, "The Theory of the Photolysis of Silver Bromide and the Photographic Latent Image," *Proc. R. Soc. A Math. Phys. Eng. Sci.*, vol. 164, no. 917, pp. 151–167, 1938.
7. G. Ciamician, "The Photochemistry of the Future," *Science*, vol. 36, no. 926, pp. 385–94, 1912.
8. H. Dau and I. Zaharieva, "Principles, Efficiency, and Blueprint Character of Solar-Energy Conversion in Photosynthetic Water Oxidation," *Acc. Chem. Res.*, vol. 42, no. 12, pp. 1861–1870, 2009.

9. Y.-C. Cheng and G. R. Fleming, "Dynamics of Light Harvesting in Photosynthesis," *Annu. Rev. Phys. Chem.*, vol. 60, pp. 241–262, 2009.

10. D. R. Turner, "The Anode Behavior of Germanium in Aqueous Solutions," *J. Electrochem. Soc.*, vol. 103, no. 4, p. 252, 1956.

11. D. M. Chapin, C. S. Fuller, and G. L. Pearson, "A New Silicon p-n Junction Photocell for Converting Solar Radiation into Electrical Power," *J. Appl. Phys.*, vol. 25, no. 5, p. 676, 1954.

12. S. C. Markham, "Photocatalytic Properties of Oxides," *J. Chem. Educ.*, vol. 32, no. 10, p. 540, 1955.

13. P. J. Boddy, "Oxygen Evolution on Semiconducting TiO_2," *J. Electrochem. Soc.*, vol. 115, no. 2, p. 199, 1968.

14. J. F. Dewald, "The Charge Distribution at the Zinc Oxide-Electrolyte Interface," *J. Phys. Chem. Solids*, vol. 14, pp. 155–161, 1960.

15. S. R. Morrison, "Chemical Role of Holes and Electrons in ZnO Photocatalysis," *J. Chem. Phys.*, vol. 47, no. 4, p. 1543, 1967.

16. W. P. Gomes, T. Freund, and S. R. Morrison, "Chemical Reactions Involving Holes at the Zinc Oxide Single Crystal Anode," *J. Electrochem. Soc.*, vol. 115, no. 8, p. 818, 1968.

17. R. Williams, "Becquerel Photovoltaic Effect in Binary Compounds," *J. Chem. Phys.*, vol. 32, no. 5, p. 1505, 1960.

18. R. Memming and G. Schwandt, "Electrochemical Properties of Gallium Phosphide in Aqueous Solutions," *Electrochim. Acta*, vol. 13, no. 6, pp. 1299–1310, 1968.

19. H. Gerischer, "Über den Ablauf von Redoxreaktionen an Metallen und an Halbleitern," *Zeitschrift für Phys. Chemie*, vol. 26, no. 3_4, pp. 223–247, 1960.

20. H. Gerischer, "Über den Ablauf von Redoxreaktionen an Metallen und an Halbleitern," *Zeitschrift für Phys. Chemie*, vol. 27, no. 1_2, pp. 48–79, 1961.

21. H. Gerischer and W. Mindt, "The Mechanisms of the Decomposition of Semiconductors by Electrochemical Oxidation and Reduction," *Electrochim. Acta*, vol. 13, no. 6, pp. 1329–1341, 1968.

22. H. Gerischer, "Electrochemical Photo and Solar Cells Principles and Some Experiments," *J. Electroanal. Chem. Interfacial Electrochem.*, vol. 58, no. 1, pp. 263–274, 1975.

23. H. Gerischer and H. Tributsch, "Electrochemical Studies on the Spectral Sensitization of Zinc Oxide Single Crystals," *Berichte der Bunsengesellschaft für Phys. Chemie*, vol. 72, p. 437, 1968.

24. H. Gerischer, "Solar Photoelectrolysis with Semiconductor Electrodes," in *Topics in Applied Physics (Vol. 31)—Solar Energy Conversion*, B. O. Seraphin, Ed. Berlin: Springer-Verlag, 1979.

25. A. Fujishima and K. Honda, "Electrochemical Photolysis of Water at a Semiconductor Electrode," *Nature*, vol. 238, no. 5358, pp. 37–38, 1972.

26. H. Kallmann and M. Pope, "Decomposition of Water by Light," *Nature*, vol. 188, no. 4754, pp. 935–936, 1960.

27. P. A. Kohl and A. J. Bard, "Semiconductor Electrodes. 13. Characterization and Behavior of n-Type Zinc Oxide, Cadmium Sulfide, and Gallium Phosphide Electrodes in Acetonitrile Solutions," *J. Am. Chem. Soc.*, vol. 99, no. 23, pp. 7531–7539, 1977.

28. P. A. Kohl, "Semiconductor Electrodes," *J. Electrochem. Soc.*, vol. 126, no. 1, p. 59, 1979.

29. P. A. Kohl, "Semiconductor Electrodes," *J. Electrochem. Soc.*, vol. 126, no. 4, p. 598, 1979.

30. R. E. Malpas, K. Itaya, and A. J. Bard, "Semiconductor Electrodes. 20. Photogeneration of Solvated Electrons on p-Type Gallium Arsenide Electrodes in Liquid Ammonia," *J. Am. Chem. Soc.*, vol. 101, no. 10, pp. 2535–2537, 1979.

31. R. N. Dominey, N. S. Lewis, and M. S. Wrighton, "Fermi Level Pinning of p-Type Semiconducting Indium Phosphide Contacting Liquid Electrolyte Solutions: Rationale for Efficient Photoelectrochemical Energy Conversion," *J. Am. Chem. Soc.*, vol. 103, no. 5, pp. 1261–1263, 1981.

32. C. M. Gronet, "n-Type GaAs Photoanodes in Acetonitrile: Design of a 10.0% Efficient Photoelectrode," *Appl. Phys. Lett.*, vol. 43, no. 1, p. 115, 1983.

33. M. J. Heben, A. Kumar, C. Zheng, and N. S. Lewis, "Efficient Photovoltaic Devices for InP Semiconductor/Liquid Junctions," *Nature*, vol. 340, no. 6235, pp. 621–623, 1989.

34. D. S. Ginley, "Interfacial Chemistry at p-GaP Photoelectrodes," *J. Electrochem. Soc.*, vol. 129, no. 9, p. 2141, 1982.

35. O. Khaselev and J. A. Turner, "A Monolithic Photovoltaic-Photoelectrochemical Device for Hydrogen Production via Water Splitting," *Science*, vol. 280, no. 5362, pp. 425–427, 1998.

36. B. C. O'Regan and M. Grätzel, "A Low-Cost, High-Efficiency Solar Cell Based on Dye-Sensitized Colloidal TiO_2 Films," *Nature*, vol. 353, no. 6346, pp. 737–740, 1991.

37. M. W. Kanan and D. G. Nocera, "In Situ Formation of an Oxygen-Evolving Catalyst in Neutral Water Containing Phosphate and Co^{2+}," *Science*, vol. 321, no. 5892, pp. 1072–1075, 2008.

38. A. J. Nozik and R. Memming, "Physical Chemistry of Semiconductor–Liquid Interfaces," *J. Phys. Chem.*, vol. 100, no. 31, pp. 13061–13078, 1996.

39. A. J. Nozik, "Photoelectrochemistry: Applications to Solar Energy Conversion," *Annu. Rev. Phys. Chem.*, pp. 189–222, 1978.

40. K. Rajeshwar, P. Singh, and J. DuBow, "Energy Conversion in Photoelectrochemical Systems—A Review," *Electrochim. Acta*, vol. 23, no. 11, pp. 1117–1144, 1978.

41. R. A. Marcus, "On the Theory of Oxidation-Reduction Reactions Involving Electron Transfer. I," *J. Chem. Phys.*, vol. 24, no. 5, p. 966, 1956.

42. K. Rajeshwar, "Hydrogen Generation at Irradiated Oxide Semiconductor–Solution Interfaces," *J. Appl. Electrochem.*, vol. 37, no. 7, pp. 765–787, 2007.

43. O. K. Varghese and C. A. Grimes, "Appropriate Strategies for Determining the Photoconversion Efficiency of Water Photoelectrolysis Cells: A Review with Examples Using Titania Nanotube Array Photoanodes," *Sol. Energy Mater. Sol. Cells*, vol. 92, no. 4, pp. 374–384, 2008.

44. Z. Chen, T. F. Jaramillo, T. G. Deutsch, A. Kleiman-Shwarsctein, A. J. Forman, N. Gaillard, R. Garland et al., "Accelerating Materials Development for Photoelectrochemical Hydrogen Production: Standards for Methods, Definitions, and Reporting Protocols," *J. Mater. Res.*, vol. 25, no. 1, pp. 3–16, 2011.

45. M. A. Butler and D. S. Ginley, "Principles of Photoelectrochemical, Solar Energy Conversion," *J. Mater. Sci.*, vol. 15, no. 1, pp. 1–19, 1980.

46. A. Yella, H.-W. Lee, H. N. Tsao, C. Yi, A. K. Chandiran, M. K. Nazeeruddin, E. W.-G. Diau, C.-Y. Yeh, S. M. Zakeeruddin, and M. Grätzel, "Porphyrin-Sensitized Solar Cells with Cobalt (II/III)-Based Redox Electrolyte Exceed 12 Percent Efficiency," *Science*, vol. 334, no. 6056, pp. 629–634, 2011.

47. H. Vogel, "Ueber die Lichtempfindlichkeit des Bromsilbers für die sogenannten chemisch unwirksamen Farben," *Berichte der Dtsch. Chem. Gesellschaft*, vol. 6, no. 2, pp. 1302–1306, 1873.

48. J. Moser, "Notiz über Verstärkung photoelektrischer Ströme durch optische Sensibilisirung," *Monatshefte für Chemie*, vol. 8, no. 1, p. 373, 1887.

49. D. Gust, T. A. Moore, and A. L. Moore, "Realizing Artificial Photosynthesis," *Faraday Discuss.*, vol. 155, p. 9, 2012.

50. W. J. Youngblood, S.-H. A. Lee, Y. Kobayashi, E. A. Hernandez-Pagan, P. G. Hoertz, T. A Moore, A. L. Moore, D. Gust, and T. E. Mallouk, "Photoassisted Overall Water Splitting in a Visible Light-Absorbing Dye-Sensitized Photoelectrochemical Cell," *J. Am. Chem. Soc.*, vol. 131, no. 3, pp. 926–927, 2009.

51. F. Lakadamyali, M. Kato, and E. Reisner, "Colloidal Metal Oxide Particles Loaded with Synthetic Catalysts for Solar H_2 Production," *Faraday Discuss.*, vol. 155, p. 191, 2012.

52. Y. Il Kim, S. W. Keller, J. S. Krueger, E. H. Yonemoto, G. B. Saupe, and T. E. Mallouk, "Photochemical Charge Transfer and Hydrogen Evolution Mediated by Oxide Semiconductor Particles in Zeolite-Based Molecular Assemblies," *J. Phys. Chem. B*, vol. 101, no. 14, pp. 2491–2500, 1997.

53. K. Maeda, M. Eguchi, W. J. Youngblood, and T. E. Mallouk, "Niobium Oxide Nanoscrolls as Building Blocks for Dye-Sensitized Hydrogen Production from Water under Visible Light Irradiation," *Chem. Mater.*, vol. 20, no. 21, pp. 6770–6778, 2008.

54. P. Wang, B. Huang, Y. Dai, and M.-H. Whangbo, "Plasmonic Photocatalysts: Harvesting Visible Light with Noble Metal Nanoparticles," *Nanoscale*, vol. 14, no. 28, pp. 9813–9825, 2012.

55. S. C. Warren and E. Thimsen, "Plasmonic Solar Water Splitting," *Energy Environ. Sci.*, vol. 5, no. 1, p. 5133, 2012.

56. S. Linic, P. Christopher, and D. B. Ingram, "Plasmonic-Metal Nanostructures for Efficient Conversion of Solar to Chemical Energy," *Nat. Mater.*, vol. 10, no. 12, pp. 911–921, 2011.

57. Y. Tian and T. Tatsuma, "Plasmon-Induced Photoelectrochemistry at Metal Nanoparticles Supported on Nanoporous TiO_2," *Chem. Commun. (Camb).*, vol. 21, no. 16, pp. 1810–1811, 2004.

58. J. S. Jang, K. Y. Yoon, X. Xiao, F.-R. F. Fan, and A. J. Bard, "Development of a Potential Fe_2O_3-Based Photocatalyst Thin Film for Water Oxidation by Scanning Electrochemical Microscopy: Effects of $Ag–Fe_2O_3$ Nanocomposite and Sn Doping," *Chem. Mater.*, vol. 21, no. 20, pp. 4803–4810, 2009.

59. E. Thimsen, F. Le Formal, M. Grätzel, and S. C. Warren, "Influence of Plasmonic Au Nanoparticles on the Photoactivity of Fe_2O_3 Electrodes for Water Splitting," *Nano Lett.*, vol. 11, no. 1, pp. 35–43, 2011.

60. I. Thomann, B. A. Pinaud, Z. Chen, B. M. Clemens, T. F. Jaramillo, and M. L. Brongersma, "Plasmon Enhanced Solar-to-Fuel Energy Conversion," *Nano Lett.*, vol. 11, no. 8, pp. 3440–3446, 2011.

61. D. B. Ingram and S. Linic, "Water Splitting on Composite Plasmonic-Metal/Semiconductor Photoelectrodes: Evidence for Selective Plasmon-Induced Formation of Charge Carriers Near the Semiconductor Surface," *J. Am. Chem. Soc.*, vol. 133, no. 14, pp. 5202–5205, 2011.

62. D. B. Ingram, P. Christopher, J. L. Bauer, and S. Linic, "Predictive Model for the Design of Plasmonic Metal/Semiconductor Composite Photocatalysts," *ACS Catal.*, vol. 1, no. 10, pp. 1441–1447, 2011.

63. Z. Liu, W. Hou, P. Pavaskar, M. Aykol, and S. B. Cronin, "Plasmon Resonant Enhancement of Photocatalytic Water Splitting under Visible Illumination," *Nano Lett.*, vol. 11, no. 3, pp. 1111–1116, 2011.

64. C. G. Silva, R. Juárez, T. Marino, R. Molinari, and H. García, "Influence of Excitation Wavelength (UV or Visible Light) on the Photocatalytic Activity of Titania Containing Gold Nanoparticles for the Generation of Hydrogen or Oxygen from Water," *J. Am. Chem. Soc.*, vol. 133, no. 3, pp. 595–602, 2011.

65. A. Primo, T. Marino, A. Corma, R. Molinari, and H. García, "Efficient Visible-Light Photocatalytic Water Splitting by Minute Amounts of Gold Supported on Nanoparticulate CeO_2 Obtained by a Biopolymer Templating Method," *J. Am. Chem. Soc.*, vol. 133, no. 18, pp. 6930–6933, 2011.

66. T. Torimoto, H. Horibe, T. Kameyama, K. Okazaki, S. Ikeda, M. Matsumura, A. Ishikawa, and H. Ishihara, "Plasmon-Enhanced Photocatalytic Activity of Cadmium Sulfide Nanoparticle Immobilized on Silica-Coated Gold Particles," *J. Phys. Chem. Lett.*, vol. 2, no. 16, pp. 2057–2062, 2011.

67. R. Solarska, A. Królikowska, and J. Augustyński, "Silver Nanoparticle Induced Photocurrent Enhancement at WO_3 Photoanodes," *Angew. Chem. Int. Ed. Engl.*, vol. 49, no. 43, pp. 7980–7983, 2010.
68. S. Rehman, R. Ullah, A. M. Butt, and N. D. Gohar, "Strategies of Making TiO_2 and ZnO Visible Light Active," *J. Hazard. Mater.*, vol. 170, no. 2/3, pp. 560–569, 2009.
69. S. G. Kumar and L. G. Devi, "Review on Modified TiO_2 Photocatalysis under UV/ Visible Light: Selected Results and Related Mechanisms on Interfacial Charge Carrier Transfer Dynamics," *J. Phys. Chem. A*, vol. 115, no. 46, pp. 13211–13241, 2011.
70. X. Z. Fujishima and D. Tryk, "TiO_2 Photocatalysis and Related Surface Phenomena," *Surf. Sci. Rep.*, vol. 63, no. 12, pp. 515–582, 2008.
71. R. Asahi, T. Morikawa, T. Ohwaki, K. Aoki, and Y. Taga, "Visible-Light Photocatalysis in Nitrogen-Doped Titanium Oxides," *Science*, vol. 293, pp. 269–271, 2001.
72. C. Hägglund, M. Grätzel, and B. Kasemo, "Comment on 'Efficient Photochemical Water Splitting by a Chemically Modified n-TiO_2' (II)," *Science*, vol. 301, no. 5640, p. 1673; discussion 1673, 2003.
73. E. M. Neville, M. J. Mattle, D. Loughrey, B. Rajesh, M. Rahman, J. M. D. MacElroy, J. A. Sullivan, and K. R. Thampi, "Carbon-Doped TiO_2 and Carbon, Tungsten-Codoped TiO_2 through Sol–Gel Processes in the Presence of Melamine Borate: Reflections through Photocatalysis," *J. Phys. Chem. C*, vol. 116, no. 31, pp. 16511–16521, 2012.
74. Y. Park, W. Kim, H. Park, T. Tachikawa, T. Majima, and W. Choi, "Carbon-Doped TiO_2 Photocatalyst Synthesized without Using an External Carbon Precursor and the Visible Light Activity," *Appl. Catal. B Environ.*, vol. 91, no. 1/2, pp. 355–361, 2009.
75. M. Ni, M. K. H. Leung, D. Y. C. Leung, and K. Sumathy, "A Review and Recent Developments in Photocatalytic Water-Splitting Using TiO_2 for Hydrogen Production," *Renew. Sustain. Energy Rev.*, vol. 11, no. 3, pp. 401–425, 2007.
76. G. Liu, L.-C. Yin, J. Wang, P. Niu, C. Zhen, Y. Xie, and H.-M. Cheng, "A Red Anatase TiO_2 Photocatalyst for Solar Energy Conversion," *Energy Environ. Sci.*, vol. 5, no. 11, p. 9603, 2012.
77. W.-J. Yin, H. Tang, S.-H. Wei, M. M. Al-Jassim, J. A. Turner, and Y. Yan, "Band Structure Engineering of Semiconductors for Enhanced Photoelectrochemical Water Splitting: The Case of TiO_2," *Phys. Rev. B*, vol. 82, no. 4, p. 045106, 2010.
78. J. Tang, A. J. Cowan, J. R. Durrant, and D. R. Klug, "Mechanism of O_2 Production from Water Splitting: Nature of Charge Carriers in Nitrogen Doped Nanocrystalline TiO_2 Films and Factors Limiting O_2 Production," *J. Phys. Chem. C*, vol. 115, no. 7, pp. 3143–3150, 2011.
79. X. Zong, C. Sun, H. Yu, Z. G. Chen, Z. Xing, D. Ye, G. Q. M. Lu, X. Li, and L. Wang, "Activation of Photocatalytic Water Oxidation on N-Doped ZnO Bundle-Like Nanoparticles under Visible Light," *J. Phys. Chem. C*, vol. 117, no. 10, pp. 4937–4942, 2013.
80. M. Grätzel, "Photoelectrochemical Cells," *Nature*, vol. 414, no. 6861, pp. 338–344, 2001.
81. C. Liu, N. P. Dasgupta, and P. Yang, "Semiconductor Nanowires for Artificial Photosynthesis," *Chem. Mater.*, p. 131002125040005, 2013.
82. Z. Zhang and P. Wang, "Optimization of Photoelectrochemical Water Splitting Performance on Hierarchical TiO_2 Nanotube Arrays," *Energy Environ. Sci.*, vol. 5, no. 4, p. 6506, 2012.
83. X. Feng, K. Shankar, O. K. Varghese, M. Paulose, T. J. Latempa, and C. A. Grimes, "Vertically Aligned Single Crystal TiO_2 Nanowire Arrays Grown Directly on Transparent Conducting Oxide Coated Glass: Synthesis Details and Applications," *Nano Lett.*, vol. 8, no. 11, pp. 3781–3786, 2008.
84. G. Wang, H. Wang, Y. Ling, Y. Tang, X. Yang, R. C. Fitzmorris, C. Wang, J. Z. Zhang, and Y. Li, "Hydrogen-Treated TiO_2 Nanowire Arrays for Photoelectrochemical Water Splitting," *Nano Lett.*, vol. 11, no. 7, pp. 3026–3033, 2011.

85. S. W. Boettcher, J. M. Spurgeon, M. C. Putnam, E. L. Warren, D. B. Turner-Evans, M. D. Kelzenberg, J. R. Maiolo, H. A. Atwater, and N. S. Lewis, "Energy-Conversion Properties of Vapor-Liquid-Solid-Grown Silicon Wire-Array Photocathodes," *Science*, vol. 327, no. 5962, pp. 185–187, 2010.

86. Y. J. Hwang, A. Boukai, and P. Yang, "High Density n-Si/n-TiO$_2$ Core/Shell Nanowire Arrays with Enhanced Photoactivity," *Nano Lett.*, vol. 9, no. 1, pp. 410–415, 2009.

87. C. Liu, J. Tang, H. M. Chen, B. Liu, and P. Yang, "A Fully Integrated Nanosystem of Semiconductor Nanowires for Direct Solar Water Splitting," *Nano Lett.*, vol. 13, no. 6, pp. 2989–2992, 2013.

88. K. Sun, Y. Jing, C. Li, X. Zhang, R. Aguinaldo, A. Kargar, K. Madsen et al., "3D Branched Nanowire Heterojunction Photoelectrodes for High-Efficiency Solar Water Splitting and H$_2$ Generation," *Nanoscale*, vol. 4, no. 5, pp. 1515–1521, 2012.

89. G. K. Mor, H. E. Prakasam, O. K. Varghese, K. Shankar, and C. A. Grimes, "Vertically Oriented Ti-Fe-O Nanotube Array Films: Toward a Useful Material Architecture for Solar Spectrum Water Photoelectrolysis," *Nano Lett.*, vol. 7, no. 8, pp. 2356–2364, 2007.

90. Y. Cong, H. S. Park, S. Wang, H. X. Dang, F. F. Fan, C. B. Mullins, and A. J. Bard, "Synthesis of Ta3N5 Nanotube Arrays Modified with Electrocatalysts for Photoelectrochemical Water Oxidation," *J. Phys. Chem. C*, vol. 116, no. 27, pp. 14541–14550, 2012.

91. H. X. Dang, N. T. Hahn, H. S. Park, A. J. Bard, and C. B. Mullins, "Nanostructured Ta3N5 Films as Visible-Light Active Photoanodes for Water Oxidation," *J. Phys. Chem. C*, vol. 116, no. 36, pp. 19225–19232, 2012.

92. J. Y. Kim, G. Magesh, D. H. Youn, J.-W. Jang, J. Kubota, K. Domen, and J. S. Lee, "Single-Crystalline, Wormlike Hematite Photoanodes for Efficient Solar Water Splitting," *Sci. Rep.*, vol. 3, p. 2681, 2013.

93. A. Kay, I. Cesar, and M. Grätzel, "New Benchmark for Water Photooxidation by Nanostructured Alpha-Fe$_2$O$_3$ Films," *J. Am. Chem. Soc.*, vol. 128, no. 49, pp. 15714–15721, 2006.

94. S. D. Tilley, M. Cornuz, K. Sivula, and M. Grätzel, "Light-Induced Water Splitting with Hematite: Improved Nanostructure and Iridium Oxide Catalysis," *Angew. Chem. Int. Ed. Engl.*, vol. 49, pp. 6405–6408, 2010.

95. D. Wang, T. Hisatomi, T. Takata, C. Pan, M. Katayama, J. Kubota, and K. Domen, "Core/Shell Photocatalyst with Spatially Separated Co-Catalysts for Efficient Reduction and Oxidation of Water," *Angew. Chem. Int. Ed. Engl.*, vol. 52, no. 43, pp. 11252–11256, 2013.

96. D. M. Kaschak, S. A. Johnson, C. C. Waraksa, J. Pogue, and T. E. Mallouk, "Artificial Photosynthesis in Lamellar Assemblies of Metal Poly(pyridyl) Complexes and Metalloporphyrins," *Coord. Chem. Rev.*, vol. 185–186, pp. 403–416, 1999.

97. P. G. Hoertz and T. E. Mallouk, "Light-to-Chemical Energy Conversion in Lamellar Solids and Thin Films," *Inorg. Chem.*, vol. 44, no. 20, pp. 6828–6840, 2005.

98. K. Sivula, F. Le Formal, and M. Grätzel, "WO$_3$–Fe$_2$O$_3$ Photoanodes for Water Splitting: A Host Scaffold, Guest Absorber Approach," *Chem. Mater.*, vol. 21, no. 13, pp. 2862–2867, 2009.

99. T. Hisatomi, J. Brillet, M. Cornuz, F. Le Formal, N. Tétreault, K. Sivula, and M. Grätzel, "A Ga$_2$O$_3$ Underlayer as an Isomorphic Template for Ultrathin Hematite Films toward Efficient Photoelectrochemical Water Splitting," *Faraday Discuss.*, vol. 155, p. 223, 2012.

100. M. Stefik, M. Cornuz, N. Mathews, T. Hisatomi, S. Mhaisalkar, and M. Grätzel, "Transparent, Conducting Nb:SnO$_2$ for Host-Guest Photoelectrochemistry," *Nano Lett.*, vol. 12, no. 10, pp. 5431–5435, 2012.

101. F. F. Abdi, L. Han, A. H. M. Smets, M. Zeman, B. Dam, and R. van de Krol, "Efficient Solar Water Splitting by Enhanced Charge Separation in a Bismuth Vanadate-Silicon Tandem Photoelectrode," *Nat. Commun.*, vol. 4, p. 2195, 2013.

102. X. Chen, L. Liu, P. Y. Yu, and S. S. Mao, "Increasing Solar Absorption for Photocatalysis with Black Hydrogenated Titanium Dioxide Nanocrystals," *Science*, vol. 331, no. 6018, pp. 746–750, 2011.

103. K. Maeda, K. Teramura, D. Lu, T. Takata, N. Saito, Y. Inoue, and K. Domen, "Photocatalyst Releasing Hydrogen from Water," *Nature*, vol. 440, no. 7082, p. 295, 2006.

104. C. Aprile, A. Corma, and H. Garcia, "Enhancement of the Photocatalytic Activity of TiO_2 through Spatial Structuring and Particle Size Control: From Subnanometric to Submillimetric Length Scale," *Phys. Chem. Chem. Phys.*, vol. 10, no. 6, pp. 769–783, 2008.

105. H. Dotan, O. Kfir, E. Sharlin, O. Blank, M. Gross, I. Dumchin, G. Ankonina, and A. Rothschild, "Resonant Light Trapping in Ultrathin Films for Water Splitting," *Nat. Mater.*, vol. 12, no. 2, pp. 158–164, 2013.

106. G. I. N. Waterhouse, A. K. Wahab, M. Al-Oufi, V. Jovic, D. H. Anjum, D. Sun-Waterhouse, J. Llorca, and H. Idriss, "Hydrogen Production by Tuning the Photonic Band Gap with the Electronic Band Gap of TiO_2," *Sci. Rep.*, vol. 3, p. 2849, 2013.

107. L. J. Heidt and A. F. McMillan, "Conversion of Sunlight into Chemical Energy Available in Storage for Man's Use," *Science*, vol. 117, no. 3030, pp. 75–76, 1953.

108. L. J. Heidt and A. F. McMillan, "Influence of Perchloric Acid and Cerous Perchlorate upon the Photochemical Oxidation of Cerous to Ceric Perchlorate in Dilute Aqueous Perchloric Acid," *J. Am. Chem. Soc.*, vol. 76, no. 8, pp. 2135–2139, Apr. 1954.

109. J. Kiwi and M. Grätzel, "Oxygen Evolution from Water via Redox Catalysis," *Angew. Chemie Int. Ed. English*, vol. 17, no. 11, pp. 860–861, 1978.

110. W. Vonach and N. Getoff, "Photocatalytic Splitting of Liquid Water by n-TiO_2 Suspension," *Zeitschrift für Naturforsch. A*, vol. 36a, p. 876, 1981.

111. G. N. Schrauzer and T. D. Guth, "Hydrogen Evolving Systems. 1. The Formation of Molecular Hydrogen from Aqueous Suspensions of Iron(II) Hydroxide and Reactions with Reducible Substrates, Including Molecular Nitrogen," *J. Am. Chem. Soc.*, vol. 98, no. 12, pp. 3508–3513, 1976.

112. T. Kawai and T. Sakata, "Hydrogen Evolution from Water Using Solid Carbon and Light Energy," *Nature*, vol. 282, 1979.

113. T. Kawai and T. Sakata, "Photocatalytic Decomposition of Gaseous Water over TiO_2 and TiO_2–RuO_2 Surfaces," *Chem. Phys. Lett.*, vol. 72, no. 1, pp. 87–89, 1980.

114. S. Sato, "Photocatalytic Water Decomposition and Water-Gas Shift Reactions over NaOH-Coated, Platinized TiO_2," *J. Catal.*, vol. 69, no. 1, pp. 128–139, 1981.

115. K. R. Thampi, M. S. Rao, W. Schwarz, M. Grätzel, and J. Kiwi, "Preparation of $SrTiO_3$ by Sol–Gel Techniques for the Photoinduced Production of H_2 and Surface Peroxides from Water," *J. Chem. Soc. Faraday Trans. 1 Phys. Chem. Condens. Phases*, vol. 84, no. 5, p. 1703, 1988.

116. A. V. Bulatov and M. L. Khidekel', "Decomposition of Water Exposed to UV Light in the Presence of Platinized Titanium Dioxide," *Bull. Acad. Sci. USSR Div. Chem. Sci.*, vol. 25, no. 8, p. 1794, 1976.

117. J. M. Lehn, J. P. Sauvage, R. Zlessel, and L. Hilaire, "Water Photolysis by UV Irradiation of Rhodium Loaded Strontium Titanate Catalysts. Relation between Catalytic Activity and Nature of the Deposit from Combined Photolysis and ESCA Studies," *Isr. J. Chem.*, vol. 22, no. 2, pp. 168–172, 1982.

118. E. Borgarello, J. Kiwi, E. Pelizzetti, M. Visca, and M. Grätzel, "Photochemical Cleavage of Water by Photocatalysis," *Nature*, vol. 289, no. 5794, pp. 158–160, 1981.

119. M. G. Walter, E. L. Warren, J. R. McKone, S. W. Boettcher, Q. Mi, E. A. Santori, and N. S. Lewis, "Solar Water Splitting Cells," *Chem. Rev.*, vol. 110, no. 11, pp. 6446–6473, 2010.

120. B. E. Conway and J. O. M. Bockris, "The Adsorption of Hydrogen and the Mechanism of the Electrolytic Hydrogen Evolution Reaction," *Naturwissenschaften*, vol. 43, no. 19, p. 446, 1956.

121. P. Sabatier, "Hydrogénations et déshydrogénations par catalyse," *Berichte der Dtsch. Chem. Gesellschaft*, vol. 44, no. 3, pp. 1984–2001, 1911.

122. A. Balandin, "Modern State of the Multiplet Theory of Heterogeneous Catalysis," *Adv. Catal.*, vol. 19, pp. 1–210, 1969.

123. S. Trasatti, "Work Function, Electronegativity, and Electrochemical Behaviour of Metals," *J. Electroanal. Chem. Interfacial Electrochem.*, vol. 39, no. 1, pp. 163–184, 1972.

124. H. Gerischer, "Mechanismus der Elektrolytischen Wasserstoffabscheidung und Adsorptionsenergie von Atomarem Wasserstoff," *Bull. des Sociétés Chim. Belges*, vol. 67, no. 7/8, pp. 506–527, 2010.

125. R. Parsons, "The Rate of Electrolytic Hydrogen Evolution and the Heat of Adsorption of Hydrogen," *Trans. Faraday Soc.*, vol. 54, p. 1053, 1958.

126. J. K. Nørskov, T. Bligaard, A. Logadottir, J. R. Kitchin, J. G. Chen, S. Pandelov, and U. Stimming, "Trends in the Exchange Current for Hydrogen Evolution," *J. Electrochem. Soc.*, vol. 152, no. 3, p. J23, 2005.

127. J. Greeley, T. F. Jaramillo, J. Bonde, I. B. Chorkendorff, and J. K. Nørskov, "Computational High-Throughput Screening of Electrocatalytic Materials for Hydrogen Evolution," *Nat. Mater.*, vol. 5, no. 11, pp. 909–913, 2006.

128. N. P. Dasgupta, C. Liu, S. Andrews, F. B. Prinz, and P. Yang, "Atomic Layer Deposition of Platinum Catalysts on Nanowire Surfaces for Photoelectrochemical Water Reduction," *J. Am. Chem. Soc.*, vol. 135, no. 35, pp. 12932–12935, 2013.

129. E. Navarro-Flores, Z. Chong, and S. Omanovic, "Characterization of Ni, NiMo, NiW and NiFe Electroactive Coatings as Electrocatalysts for Hydrogen Evolution in an Acidic Medium," *J. Mol. Catal. A Chem.*, vol. 226, no. 2, pp. 179–197, 2005.

130. H. Over, "Surface Chemistry of Ruthenium Dioxide in Heterogeneous Catalysis and Electrocatalysis: From Fundamental to Applied Research," *Chem. Rev.*, vol. 112, no. 6, pp. 3356–3426, 2012.

131. S. D. Tilley, M. Schreier, J. Azevedo, M. Stefik, and M. Grätzel, "Ruthenium Oxide Hydrogen Evolution Catalysis on Composite Cuprous Oxide Water-Splitting Photocathodes," *Adv. Funct. Mater.*, vol. 24, no. 3, pp. 303–311.

132. S. Fukuzumi and Y. Yamada, "Catalytic Activity of Metal-Based Nanoparticles for Photocatalytic Water Oxidation and Reduction," *J. Mater. Chem.*, vol. 22, no. 46, p. 24284, 2012.

133. N. Zhang, S. Liu, and Y.-J. Xu, "Recent Progress on Metal Core@Semiconductor Shell Nanocomposites as a Promising Type of Photocatalyst," *Nanoscale*, vol. 4, no. 7, pp. 2227–2238, 2012.

134. W. Jaegermann and H. Tributsch, "Interfacial Properties of Semiconducting Transition Metal Chalcogenides," *Prog. Surf. Sci.*, vol. 29, no. 1/2, pp. 1–167, 1988.

135. B. Hinnemann, P. G. Moses, J. Bonde, K. P. Jørgensen, J. H. Nielsen, S. Horch, I. Chorkendorff, and J. K. Nørskov, "Biomimetic Hydrogen Evolution: MoS_2 Nanoparticles as Catalyst for Hydrogen Evolution," *J. Am. Chem. Soc.*, vol. 127, no. 15, pp. 5308–5309, 2005.

136. T. F. Jaramillo, K. P. Jørgensen, J. Bonde, J. H. Nielsen, S. Horch, and I. Chorkendorff, "Identification of Active Edge Sites for Electrochemical H_2 Evolution from MoS_2 Nanocatalysts," *Science*, vol. 317, no. 5834, pp. 100–102, 2007.

137. X. Zong, H. Yan, G. Wu, G. Ma, F. Wen, L. Wang, and C. Li, "Enhancement of Photocatalytic H_2 Evolution on CdS by Loading MoS_2 as Cocatalyst under Visible Light Irradiation," *J. Am. Chem. Soc.*, vol. 130, no. 23, pp. 7176–7177, 2008.

138. X. Zong, G. Wu, H. Yan, G. Ma, J. Shi, F. Wen, L. Wang, and C. Li, "Photocatalytic H_2 Evolution on MoS_2/CdS Catalysts under Visible Light Irradiation," *J. Phys. Chem. C*, vol. 114, no. 4, pp. 1963–1968, 2010.

139. D. Merki, S. Fierro, H. Vrubel, and X. Hu, "Amorphous Molybdenum Sulfide Films as Catalysts for Electrochemical Hydrogen Production in Water," *Chem. Sci.*, vol. 2, no. 7, p. 1262, 2011.

140. D. Merki, H. Vrubel, L. Rovelli, S. Fierro, and X. Hu, "Fe, Co, and Ni Ions Promote the Catalytic Activity of Amorphous Molybdenum Sulfide Films for Hydrogen Evolution," *Chem. Sci.*, vol. 3, no. 8, p. 2515, 2012.

141. C. G. Morales-Guio, S. D. Tilley, H. Vrubel, M. Grätzel, and X. Hu, "Hydrogen Evolution from a Copper(I) Oxide Photocathode Coated with an Amorphous Molybdenum Sulphide Catalyst," *Nat. Commun.*, vol. 5, no. I, p. 3059, 2014.

142. H. Vrubel and X. Hu, "Molybdenum Boride and Carbide Catalyze Hydrogen Evolution in Both Acidic and Basic Solutions," *Angew. Chem. Int. Ed. Engl.*, vol. 51, no. 51, pp. 12703–12706, 2012.

143. P. Du and R. Eisenberg, "Catalysts Made of Earth-Abundant Elements (Co, Ni, Fe) for Water Splitting: Recent Progress and Future Challenges," *Energy Environ. Sci.*, vol. 5, no. 3, p. 6012, 2012.

144. Z. Han, F. Qiu, R. Eisenberg, P. L. Holland, and T. D. Krauss, "Robust Photogeneration of H_2 in Water Using Semiconductor Nanocrystals and a Nickel Catalyst," *Science*, vol. 338, no. 6112, pp. 1321–1324, 2012.

145. V. Artero, M. Chavarot-Kerlidou, and M. Fontecave, "Splitting Water with Cobalt," *Angew. Chem. Int. Ed. Engl.*, vol. 50, no. 32, pp. 7238–7266, 2011.

146. A. Krawicz, J. Yang, E. Anzenberg, J. Yano, I. D. Sharp, and G. F. Moore, "Photofunctional Construct That Interfaces Molecular Cobalt-Based Catalysts for H_2 Production to a Visible-Light-Absorbing Semiconductor," *J. Am. Chem. Soc.*, vol. 135, no. 32, pp. 11861–11868, 2013.

147. H. I. Karunadasa, E. Montalvo, Y. Sun, M. Majda, J. R. Long, and C. J. Chang, "A Molecular MoS_2 Edge Site Mimic for Catalytic Hydrogen Generation," *Science*, vol. 335, no. 6069, pp. 698–702, 2012.

148. H. I. Karunadasa, C. J. Chang, and J. R. Long, "A Molecular Molybdenum-Oxo Catalyst for Generating Hydrogen from Water," *Nature*, vol. 464, no. 7293, pp. 1329–1333, 2010.

149. Y. Hou, B. L. Abrams, P. C. K. Vesborg, M. E. Björketun, K. Herbst, L. Bech, A. M. Setti et al., "Bioinspired Molecular Co-Catalysts Bonded to a Silicon Photocathode for Solar Hydrogen Evolution," *Nat. Mater.*, vol. 10, no. 6, pp. 434–438, 2011.

150. B. Conway and M. Salomon, "Electrochemical Reaction Orders: Applications to the Hydrogen- and Oxygen-Evolution Reactions," *Electrochim. Acta*, vol. 9, no. 12, pp. 1599–1615, 1964.

151. S. Trasatti, "Transition Metal Oxides: Versatile Materials for Electrocatalysis," in *The Electrochemistry of Novel Materials*, J. Lipkowski and P. N. Ross, Eds. New York: VCH Publishers, 1994, pp. 207–295.

152. S. Trasatti, "Electrocatalysis by Oxides—Attempt at a Unifying Approach," *J. Electroanal. Chem. Interfacial Electrochem.*, vol. 111, no. 1, pp. 125–131, 1980.

153. J. Suntivich, K. J. May, H. A. Gasteiger, J. B. Goodenough, and Y. Shao-Horn, "A Perovskite Oxide Optimized for Oxygen Evolution Catalysis from Molecular Orbital Principles," *Science*, vol. 334, no. 6061, pp. 1383–1385, 2011.

154. H. B. Gray, "Powering the Planet with Solar Fuel," *Nat. Chem.*, vol. 1, no. 2, p. 112, 2009.

155. A. Harriman, J. M. Thomas, and G. R. Milward, "Catalytic and Structural Properties of Iridium-Iridium Dioxide Colloids," *New J. Chem.*, vol. 11, no. 11/12, pp. 757–762, 1987.

156. N. D. Morris and T. E. Mallouk, "A High-Throughput Optical Screening Method for the Optimization of Colloidal Water Oxidation Catalysts," *J. Am. Chem. Soc.*, vol. 124, no. 37, pp. 11114–11121, 2002.

157. A. Harriman, I. J. Pickering, J. M. Thomas, and P. A. Christensen, "Metal Oxides as Heterogeneous Catalysts for Oxygen Evolution under Photochemical Conditions," *J. Chem. Soc. Faraday Trans. 1 Phys. Chem. Condens. Phases*, vol. 84, no. 8, p. 2795, 1988.

158. J. O. Bockris, "The Electrocatalysis of Oxygen Evolution on Perovskites," *J. Electrochem. Soc.*, vol. 131, no. 2, p. 290, 1984.

159. J. O. Bockris and T. Otagawa, "Mechanism of Oxygen Evolution on Perovskites," *J. Phys. Chem.*, vol. 87, no. 15, pp. 2960–2971, 1983.

160. K. Maeda, A. Xiong, T. Yoshinaga, T. Ikeda, N. Sakamoto, T. Hisatomi, M. Takashima et al., "Photocatalytic Overall Water Splitting Promoted by Two Different Cocatalysts for Hydrogen and Oxygen Evolution under Visible Light," *Angew. Chem. Int. Ed. Engl.*, vol. 49, no. 24, pp. 4096–4099, 2010.

161. Y. Matsumoto and E. Sato, "Electrocatalytic Properties of Transition Metal Oxides for Oxygen Evolution Reaction," *Mater. Chem. Phys.*, vol. 14, no. 5, pp. 397–426, 1986.

162. S. Trasatti, "Electrocatalysis in the Anodic Evolution of Oxygen and Chlorine," *Electrochim. Acta*, vol. 29, no. 11, pp. 1503–1512, 1984.

163. D. A. Lutterman, Y. Surendranath, and D. G. Nocera, "A Self-Healing Oxygen-Evolving Catalyst," *J. Am. Chem. Soc.*, vol. 131, no. 11, pp. 3838–3839, 2009.

164. Y. Surendranath, M. Dinca, and D. G. Nocera, "Electrolyte-Dependent Electrosynthesis and Activity of Cobalt-Based Water Oxidation Catalysts," *J. Am. Chem. Soc.*, vol. 131, no. 7, pp. 2615–2620, 2009.

165. E. M. P. Steinmiller and K.-S. Choi, "Photochemical Deposition of Cobalt-Based Oxygen Evolving Catalyst on a Semiconductor Photoanode for Solar Oxygen Production," *Proc. Natl. Acad. Sci. USA*, vol. 106, no. 49, pp. 20633–20636, 2009.

166. J. A. Seabold and K.-S. Choi, "Effect of a Cobalt-Based Oxygen Evolution Catalyst on the Stability and the Selectivity of Photo-Oxidation Reactions of a WO_3 Photoanode," *Chem. Mater.*, vol. 23, no. 5, pp. 1105–1112, 2011.

167. D. K. Zhong, M. Cornuz, K. Sivula, M. Grätzel, and D. R. Gamelin, "Photo-Assisted Electrodeposition of Cobalt–Phosphate (Co–Pi) Catalyst on Hematite Photoanodes for Solar Water Oxidation," *Energy Environ. Sci.*, vol. 4, no. 5, p. 1759, 2011.

168. K. J. McDonald and K.-S. Choi, "Photodeposition of Co-Based Oxygen Evolution Catalysts on α-Fe_2O_3 Photoanodes," *Chem. Mater.*, vol. 23, no. 7, pp. 1686–1693, 2011.

169. S. Y. Reece, J. A. Hamel, K. Sung, T. D. Jarvi, A. J. Esswein, J. J. H. Pijpers, and D. G. Nocera, "Wireless Solar Water Splitting Using Silicon-Based Semiconductors and Earth-Abundant Catalysts," *Science*, vol. 334, no. 6056, pp. 645–648, 2011.

170. J. J. H. Pijpers, M. T. Winkler, Y. Surendranath, T. Buonassisi, and D. G. Nocera, "Light-Induced Water Oxidation at Silicon Electrodes Functionalized with a Cobalt Oxygen-Evolving Catalyst," *Proc. Natl. Acad. Sci. USA*, vol. 108, no. 25, pp. 10056–10061, 2011.

171. E. R. Young, R. Costi, S. Paydavosi, D. G. Nocera, and V. Bulović, "Photo-Assisted Water Oxidation with Cobalt-Based Catalyst Formed from Thin-Film Cobalt Metal on Silicon Photoanodes," *Energy Environ. Sci.*, vol. 4, no. 6, p. 2058, 2011.

172. M. Dincă, Y. Surendranath, and D. G. Nocera, "Nickel-Borate Oxygen-Evolving Catalyst That Functions under Benign Conditions," *Proc. Natl. Acad. Sci. USA*, vol. 107, no. 23, pp. 10337–10341, 2010.

173. R. D. L. Smith, M. S. Prévot, R. D. Fagan, Z. Zhang, P. A. Sedach, M. K. J. Siu, S. Trudel, and C. P. Berlinguette, "Photochemical Route for Accessing Amorphous Metal Oxide Materials for Water Oxidation Catalysis," *Science*, vol. 340, no. 6128, pp. 60–63, 2013.

174. Y. V Geletii, B. Botar, P. Kögerler, D. A. Hillesheim, D. G. Musaev, and C. L. Hill, "An All-Inorganic, Stable, and Highly Active Tetraruthenium Homogeneous Catalyst for Water Oxidation," *Angew. Chem. Int. Ed. Engl.*, vol. 47, no. 21, pp. 3896–3899, 2008.

175. T. Kawai and T. Sakata, "Conversion of Carbohydrate into Hydrogen Fuel by a Photocatalytic Process," *Nature*, vol. 286, no. 5772, pp. 474–476, 1980.

176. P. Pichat, J.-M. Herrmann, J. Disdier, H. Courbon, and M.-N. Mozzanega, "Photocatalytic Hydrogen Production from Aliphatic Alcohols over a Bifunctional Platinum on Titanium Dioxide Catalyst," *Nouv. J. Chim.*, vol. 5, pp. 627–636, 1981.

177. D. H. M. W. Thewissen, K. Timmer, M. Eeuwhorst-Reinten, A. H. A. Tinnemans, and A. Mackor, "Photo(Electro)Chemical Oxidation of Water by the Persulfate Ion over Aqueous Suspensions of Strontium Titanate $SrTiO_3$ Containing Lanthanum Chromite $LaCrO_3$," *Isr. J. Chem.*, vol. 22, no. 2, pp. 173–176, 1982.

178. A. Kudo and Y. Miseki, "Heterogeneous Photocatalyst Materials for Water Splitting," *Chem. Soc. Rev.*, vol. 38, no. 1, pp. 253–278, 2009.

179. X. Chen, S. Shen, L. Guo, and S. S. Mao, "Semiconductor-Based Photocatalytic Hydrogen Generation," *Chem. Rev.*, vol. 110, no. 11, pp. 6503–6570, 2010.

180. J. Zhu and M. Zäch, "Nanostructured Materials for Photocatalytic Hydrogen Production," *Curr. Opin. Colloid Interface Sci.*, vol. 14, no. 4, pp. 260–269, 2009.

181. K. Maeda and K. Domen, "Photocatalytic Water Splitting: Recent Progress and Future Challenges," *J. Phys. Chem. Lett.*, vol. 1, no. 18, pp. 2655–2661, 2010.

182. H. Kato and A. Kudo, "Photocatalytic Water Splitting into H_2 and O_2 over Various Tantalate Photocatalysts," *Catal. Today*, vol. 78, no. 1–4, pp. 561–569, 2003.

183. Z. Zou and H. Arakawa, "Direct Water Splitting into H_2 and O_2 under Visible Light Irradiation with a New Series of Mixed Oxide Semiconductor Photocatalysts," *J. Photochem. Photobiol. A Chem.*, vol. 158, no. 2/3, pp. 145–162, 2003.

184. K. Maeda and K. Domen, "New Non-Oxide Photocatalysts Designed for Overall Water Splitting under Visible Light," *J. Phys. Chem. C*, vol. 111, no. 22, pp. 7851–7861, 2007.

185. Z. Zou, J. Ye, K. Sayama, and H. Arakawa, "Direct Splitting of Water under Visible Light Irradiation with an Oxide Semiconductor Photocatalyst," *Nature*, vol. 414, no. 6864, pp. 625–627, 2001.

186. N. Serpone and A. V. Emeline, "Semiconductor Photocatalysis—Past, Present, and Future Outlook," *J. Phys. Chem. Lett.*, vol. 3, no. 5, pp. 673–677, 2012.

187. P. Ritterskamp, A. Kuklya, M.-A. Wüstkamp, K. Kerpen, C. Weidenthaler, and M. Demuth, "A Titanium Disilicide Derived Semiconducting Catalyst for Water Splitting under Solar Radiation-Reversible Storage of Oxygen and Hydrogen," *Angew. Chem. Int. Ed. Engl.*, vol. 46, no. 41, pp. 7770–7774, 2007.

188. J. Sato, N. Saito, Y. Yamada, K. Maeda, T. Takata, J. N. Kondo, M. Hara, H. Kobayashi, K. Domen, and Y. Inoue, "RuO_2-Loaded Beta-Ge_3N_4 as a Non-Oxide Photocatalyst for Overall Water Splitting," *J. Am. Chem. Soc.*, vol. 127, no. 12, pp. 4150–4151, 2005.

189. Y. Lee, H. Terashima, Y. Shimodaira, K. Teramura, M. Hara, H. Kobayashi, K. Domen, and M. Yashima, "Zinc Germanium Oxynitride as a Photocatalyst for Overall Water Splitting under Visible Light," *J. Phys. Chem. C*, vol. 111, no. 2, pp. 1042–1048, 2007.

190. K. Maeda, T. Takata, M. Hara, N. Saito, Y. Inoue, H. Kobayashi, and K. Domen, "GaN:ZnO Solid Solution as a Photocatalyst for Visible-Light-Driven Overall Water Splitting," *J. Am. Chem. Soc.*, vol. 127, no. 23, pp. 8286–8287, 2005.

191. K. Maeda, K. Teramura, and K. Domen, "Effect of Post-Calcination on Photocatalytic Activity of (Ga1-xZnx)(N1-xOx) Solid Solution for Overall Water Splitting under Visible Light," *J. Catal.*, vol. 254, no. 2, pp. 198–204, 2008.

192. T. Ohno, L. Bai, T. Hisatomi, K. Maeda, and K. Domen, "Photocatalytic Water Splitting Using Modified GaN:ZnO Solid Solution under Visible Light: Long-Time Operation and Regeneration of Activity," *J. Am. Chem. Soc.*, vol. 134, no. 19, pp. 8254–8259, 2012.

193. K. Maeda, "Z-Scheme Water Splitting Using Two Different Semiconductor Photocatalysts," *ACS Catal.*, vol. 3, no. 7, pp. 1486–1503, 2013.

194. K. Maeda, M. Higashi, D. Lu, R. Abe, and K. Domen, "Efficient Nonsacrificial Water Splitting through Two-Step Photoexcitation by Visible Light Using a Modified Oxynitride as a Hydrogen Evolution Photocatalyst," *J. Am. Chem. Soc.*, vol. 132, no. 16, pp. 5858–5868, 2010.

195. Y. Sasaki, H. Nemoto, K. Saito, and A. Kudo, "Solar Water Splitting Using Powdered Photocatalysts Driven by Z-Schematic Interparticle Electron Transfer without an Electron Mediator," *J. Phys. Chem. C*, vol. 113, no. 40, pp. 17536–17542, 2009.

196. J. G. Mavroides, J. A. Kafalas, and D. F. Kolesar, "Photoelectrolysis of Water in Cells with SrTiO$_3$ Anodes," *Appl. Phys. Lett.*, vol. 28, no. 5, p. 241, 1976.

197. M. S. Wrighton, A. B. Ellis, P. T. Wolczanski, D. L. Morse, H. B. Abrahamson, and D. S. Ginley, "Strontium Titanate Photoelectrodes. Efficient Photoassisted Electrolysis of Water at Zero Applied Potential," *J. Am. Chem. Soc.*, vol. 98, no. 10, pp. 2774–2779, 1976.

198. A. B. Ellis, S. W. Kaiser, and M. S. Wrighton, "Semiconducting Potassium Tantalate Electrodes. Photoassistance Agents for the Efficient Electrolysis of Water," *J. Phys. Chem.*, vol. 80, no. 12, pp. 1325–1328, 1976.

199. M. S. Prévot and K. Sivula, "Photoelectrochemical Tandem Cells for Solar Water Splitting," *J. Phys. Chem. C*, vol. 117, no. 35, pp. 17879–17893, 2013.

200. K. Sivula, "Metal Oxide Photoelectrodes for Solar Fuel Production, Surface Traps, and Catalysis," *J. Phys. Chem. Lett.*, vol. 4, no. 10, pp. 1624–1633, 2013.

201. R. van de Krol, Y. Liang, and J. Schoonman, "Solar Hydrogen Production with Nanostructured Metal Oxides," *J. Mater. Chem.*, vol. 18, no. 20, p. 2311, 2008.

202. Q. Peng, B. Kalanyan, P. G. Hoertz, A. Miller, D. H. Kim, K. Hanson, L. Alibabaei et al., "Solution-Processed, Antimony-Doped Tin Oxide Colloid Films Enable High-Performance TiO$_2$ Photoanodes for Water Splitting," *Nano Lett.*, vol. 13, no. 4, pp. 1481–1488, 2013.

203. S. Hoang, S. Guo, N. T. Hahn, A. J. Bard, and C. B. Mullins, "Visible Light Driven Photoelectrochemical Water Oxidation on Nitrogen-Modified TiO$_2$ Nanowires," *Nano Lett.*, vol. 12, no. 1, pp. 26–32, 2012.

204. S. U. M. Khan, M. Al-Shahry, and W. B. Ingler, "Efficient Photochemical Water Splitting by a Chemically Modified n-TiO$_2$," *Science*, vol. 297, no. 5590, pp. 2243–2245, 2002.

205. R. Solarska, B. D. Alexander, A. Braun, R. Jurczakowski, G. Fortunato, M. Stiefel, T. Graule, and J. Augustynski, "Tailoring the Morphology of WO$_3$ Films with Substitutional Cation Doping: Effect on the Photoelectrochemical Properties," *Electrochim. Acta*, vol. 55, no. 26, pp. 7780–7787, 2010.

206. Y. W. Chen, J. D. Prange, S. Dühnen, Y. Park, M. Gunji, C. E. D. Chidsey, and P. C. McIntyre, "Atomic Layer-Deposited Tunnel Oxide Stabilizes Silicon Photoanodes for Water Oxidation," *Nat. Mater.*, vol. 10, no. 7, pp. 539–544, 2011.

207. M. Higashi, K. Domen, and R. Abe, "Highly Stable Water Splitting on Oxynitride TaON Photoanode System under Visible Light Irradiation," *J. Am. Chem. Soc.*, vol. 134, no. 16, pp. 6968–6971, 2012.

208. D. Yokoyama, T. Minegishi, K. Maeda, M. Katayama, J. Kubota, A. Yamada, M. Konagai, and K. Domen, "Photoelectrochemical Water Splitting Using a Cu(In,Ga)Se$_2$ Thin Film," *Electrochem. Commun.*, vol. 12, no. 6, pp. 851–853, 2010.

209. B. Marsen, B. Cole, and E. L. Miller, "Photoelectrolysis of Water Using Thin Copper Gallium Diselenide Electrodes," *Sol. Energy Mater. Sol. Cells*, vol. 92, no. 9, pp. 1054–1058, 2008.

210. J. A. Baglio, G. S. Calabrese, D. J. Harrison, E. Kamienicki, A. J. Ricco, M. S. Wrighton, and G. D. Zoski, "Electrochemical Characterization of p-Type Semiconducting Tungsten Disulfide Photocathodes: Efficient Photoreduction Processes at Semiconductor/Liquid Electrolyte Interfaces," *J. Am. Chem. Soc.*, vol. 105, no. 8, pp. 2246–2256, 1983.

211. J. R. McKone, A. P. Pieterick, H. B. Gray, and N. S. Lewis, "Hydrogen Evolution from Pt/Ru-Coated p-Type WSe$_2$ Photocathodes," *J. Am. Chem. Soc.*, vol. 135, no. 1, pp. 223–231, 2013.

212. D. Yokoyama, T. Minegishi, K. Jimbo, T. Hisatomi, G. Ma, M. Katayama, J. Kubota, H. Katagiri, and K. Domen, "H$_2$ Evolution from Water on Modified Cu$_2$ZnSnS$_4$ Photoelectrode under Solar Light," *Appl. Phys. Express*, vol. 3, no. 10, p. 101202, 2010.

213. L. Rovelli, S. D. Tilley, and K. Sivula, "Optimization and Stabilization of Electrodeposited Cu$_2$ZnSnS$_4$ Photocathodes for Solar Water Reduction," *ACS Appl. Mater. Interfaces*, vol. 5, pp. 8018–8024, 2013.

214. S. W. Boettcher, E. L. Warren, M. C. Putnam, E. A. Santori, D. Turner-Evans, M. D. Kelzenberg, M. G. Walter et al., "Photoelectrochemical Hydrogen Evolution Using Si Microwire Arrays," *J. Am. Chem. Soc.*, vol. 133, no. 5, pp. 1216–1219, 2011.

215. Y. Lin, C. Battaglia, M. Boccard, M. Hettick, Z. Yu, C. Ballif, J. W. Ager, and A. Javey, "Amorphous Si Thin Film Based Photocathodes with High Photovoltage for Efficient Hydrogen Production," *Nano Lett.*, vol. 13, no. 11, pp. 5615–5618, 2013.

216. A. Paracchino, V. Laporte, K. Sivula, M. Grätzel, and E. Thimsen, "Highly Active Oxide Photocathode for Photoelectrochemical Water Reduction," *Nat. Mater.*, vol. 10, no. 6, pp. 456–461, 2011.

217. K. Iwashina and A. Kudo, "Rh-Doped SrTiO$_3$ Photocatalyst Electrode Showing Cathodic Photocurrent for Water Splitting under Visible-Light Irradiation," *J. Am. Chem. Soc.*, vol. 133, no. 34, pp. 13272–13275, 2011.

218. U. A. Joshi, A. M. Palasyuk, and P. A. Maggard, "Photoelectrochemical Investigation and Electronic Structure of a p-Type CuNbO$_3$ Photocathode," *J. Phys. Chem. C*, vol. 115, no. 27, pp. 13534–13539, 2011.

219. U. A. Joshi and P. A. Maggard, "CuNb$_3$O$_8$: A p-Type Semiconducting Metal Oxide Photoelectrode," *J. Phys. Chem. Lett.*, vol. 3, no. 11, pp. 1577–1581, 2012.

220. L. Fuoco, U. A. Joshi, and P. A. Maggard, "Preparation and Photoelectrochemical Properties of p-Type Cu$_5$Ta$_{11}$O$_{30}$ and Cu$_3$Ta$_7$O$_{19}$ Semiconducting Polycrystalline Films," *J. Phys. Chem. C*, vol. 116, no. 19, pp. 10490–10497, 2012.

221. M. Woodhouse and B. A. Parkinson, "Combinatorial Discovery and Optimization of a Complex Oxide with Water Photoelectrolysis Activity," *Chem. Mater.*, vol. 20, no. 7, pp. 2495–2502, 2008.

222. J. E. Katz, T. R. Gingrich, E. A. Santori, and N. S. Lewis, "Combinatorial Synthesis and High-Throughput Photopotential and Photocurrent Screening of Mixed-Metal Oxides for Photoelectrochemical Water Splitting," *Energy Environ. Sci.*, vol. 2, no. 1, p. 103, 2009.

223. M. Woodhouse and B. A. Parkinson, "Combinatorial Approaches for the Identification and Optimization of Oxide Semiconductors for Efficient Solar Photoelectrolysis," *Chem. Soc. Rev.*, vol. 38, no. 1, pp. 197–210, 2009.

224. Y. Zhang, Z. Schnepp, J. Cao, S. Ouyang, Y. Li, J. Ye, and S. Liu, "Biopolymer-Activated Graphitic Carbon Nitride toward a Sustainable Photocathode Material," *Sci. Rep.*, vol. 3, p. 2163, 2013.

225. T.-F. Yeh, S.-J. Chen, C.-S. Yeh, and H. Teng, "Tuning the Electronic Structure of Graphite Oxide through Ammonia Treatment for Photocatalytic Generation of H$_2$ and O$_2$ from Water Splitting," *J. Phys. Chem. C*, vol. 117, no. 13, pp. 6516–6524, 2013.

226. J. R. Bolton, S. J. Strickler, and J. S. Connolly, "Limiting and Realizable Efficiencies of Solar Photolysis of Water," *Nature*, vol. 316, no. 6028, pp. 495–500, 1985.

227. E. S. Kim, N. Nishimura, G. Magesh, J. Y. Kim, J. Jang, H. Jun, J. Kubota, K. Domen, and J. S. Lee, "Fabrication of CaFe$_2$O$_4$/TaON Heterojunction Photoanode for Photoelectrochemical Water Oxidation," *J. Am. Chem. Soc.*, vol. 135, no. 14, pp. 5375–5383, 2013.

228. M. T. Mayer, C. Du, and D. Wang, "Hematite/Si Nanowire Dual-Absorber System for Photoelectrochemical Water Splitting at Low Applied Potentials," *J. Am. Chem. Soc.*, vol. 134, no. 30, pp. 12406–12409, 2012.

229. R. van de Krol and Y. Liang, "An n-Si/n-Fe$_2$O$_3$ Heterojunction Tandem Photoanode for Solar Water Splitting," *Chimia (Aarau)*, vol. 67, no. 3, pp. 168–171, 2013.

230. A. J. Nozik, "p-n Photoelectrolysis Cells," *Appl. Phys. Lett.*, vol. 29, no. 3, p. 150, 1976.

231. A. J. Nozik, "Photochemical Diodes," *Appl. Phys. Lett.*, vol. 30, no. 11, p. 567, 1977.

232. R. C. Kainthla, "Significant Efficiency Increase in Self-Driven Photoelectrochemical Cell for Water Photoelectrolysis," *J. Electrochem. Soc.*, vol. 134, no. 4, p. 841, 1987.

233. H. Wang, T. G. Deutsch, and J. A. Turner, "Direct Water Splitting under Visible Light with Nanostructured Hematite and WO$_3$ Photoanodes and a GaInP$_2$ Photocathode," *J. Electrochem. Soc.*, vol. 155, no. 5, p. F91, 2008.

234. H. Wang, T. G. Deutsch, and J. A. Turner, "Direct Water Splitting under Visible Light with a Nanostructured Photoanode and GaInP$_2$ Photocathode," *ECS Transactions*, 2008, vol. 6, no. 17, pp. 37–44.

235. H. Wang and J. A. Turner, "Characterization of Hematite Thin Films for Photoelectrochemical Water Splitting in a Dual Photoelectrode Device," *J. Electrochem. Soc.*, vol. 157, no. 11, p. F173, 2010.

236. H. Mettee, J. W. Otvos, and M. Calvin, "Solar Induced Water Splitting with p/n Heterotype Photochemical Diodes: n-Fe$_2$O$_3$/p-GaP," *Sol. Energy Mater.*, vol. 4, no. 4, pp. 443–453, 1981.

237. K. Ohashi, J. McCann, and J. O. Bockris, "Stable Photoelectrochemical Cells for the Splitting of Water," *Nature*, vol. 266, no. 5603, pp. 610–611, 1977.

238. G. K. Mor, O. K. Varghese, R. H. T. Wilke, S. Sharma, K. Shankar, T. J. Latempa, K.-S. Choi, and C. A. Grimes, " P-Type Cu-Ti-O Nanotube Arrays and Their Use in Self-Biased Heterojunction Photoelectrochemical Diodes for Hydrogen Generation," *Nano Lett.*, vol. 8, no. 7, pp. 1906–1911, 2008.

239. B. Neumann, P. Bogdanoff, and H. Tributsch, "TiO$_2$-Protected Photoelectrochemical Tandem Cu(In,Ga)Se$_2$ Thin Film Membrane for Light-Induced Water Splitting and Hydrogen Evolution," *J. Phys. Chem. C*, vol. 113, no. 49, pp. 20980–20989, 2009.

240. E. L. Miller, R. E. Rocheleau, and X. Deng, "Design Considerations for a Hybrid Amorphous Silicon/Photoelectrochemical Multijunction Cell for Hydrogen Production," *Int. J. Hydrogen Energy*, vol. 28, no. 6, pp. 615–623, 2003.

241. E. Miller, "A Hybrid Multijunction Photoelectrode for Hydrogen Production Fabricated with Amorphous Silicon/Germanium and Iron Oxide Thin Films," *Int. J. Hydrogen Energy*, vol. 29, no. 9, pp. 907–914, 2004.

242. A. Stavrides, A. Kunrath, J. Hu, R. Treglio, A. Feldman, B. Marsen, B. Cole, E. Miller, and A. Madan, "Use of Amorphous Silicon Tandem Junction Solar Cells for Hydrogen Production in a Photoelectrochemical Cell," *Proceedings of SPIE 6340, Solar Hydrogen and Nanotechnology*, p. 63400K, http://dx.doi.org/10.1117/12.678870, 2006.

243. E. L. Miller, D. Paluselli, B. Marsen, and R. E. Rocheleau, "Development of Reactively Sputtered Metal Oxide Films for Hydrogen-Producing Hybrid Multijunction Photoelectrodes," *Sol. Energy Mater. Sol. Cells*, vol. 88, no. 2, pp. 131–144, 2005.

244. J. Augustynski, G. Calzaferri, J. C. Courvoisier, and M. Grätzel, "Photoelectrochemical Hydrogen production: State of the Art with Special Reference to IEA's Hydrogen Programme," in *Hydrogen Energy Progress XI: Proceedings of the 11th World Hydrogen Energy Conference*, T. N. Veziroğlu, Ed. Stuttgart, Germany, 1996, p. 2379.

245. J. H. Park and A. J. Bard, "Photoelectrochemical Tandem Cell with Bipolar Dye-Sensitized Electrodes for Vectorial Electron Transfer for Water Splitting," *Electrochem. Solid-State Lett.*, vol. 9, no. 2, pp. E5–E8, 2006.

246. H. Arakawa, C. Shiraishi, M. Tatemoto, H. Kishida, D. Usui, A. Suma, A. Takamisawa, and T. Yamaguchi, "Solar Hydrogen Production by Tandem Cell System Composed of Metal Oxide Semiconductor Film Photoelectrode and Dye-Sensitized Solar Cell," in *Proceedings of SPIE 6650, Solar Hydrogen and Nanotechnology II*, p. 665003, http://dx.doi.org/10.1117/12.773366, 2007.

247. J. Brillet, M. Cornuz, F. Le Formal, J.-H. Yum, M. Grätzel, and K. Sivula, "Examining Architectures of Photoanode–Photovoltaic Tandem Cells for Solar Water Splitting," *J. Mater. Res.*, vol. 25, no. 1, pp. 17–24, 2010.

248. J. Brillet, J.-H. Yum, M. Cornuz, T. Hisatomi, R. Solarska, J. Augustynski, M. Graetzel, and K. Sivula, "Highly Efficient Water Splitting by a Dual-Absorber Tandem Cell," *Nat. Photonics*, vol. 6, no. 12, pp. 824–828, 2012.

249. K. Sivula, "Solar-to-Chemical Energy Conversion with Photoelectrochemical Tandem Cells," *Chimia (Aarau)*, vol. 67, no. 3, pp. 155–161, 2013.

250. S. Licht, B. Wang, S. Mukerji, T. Soga, M. Umeno, and H. Tributsch, "Efficient Solar Water Splitting, Exemplified by RuO_2-Catalyzed AlGaAs/Si Photoelectrolysis," *J. Phys. Chem. B*, vol. 104, no. 38, pp. 8920–8924, 2000.

251. O. Khaselev, A. Bansal, and J. A. Turner, "High-Efficiency Integrated Multijunction Photovoltaic/Electrolysis Systems for Hydrogen Production," *Int. J. Hydrogen Energy*, vol. 26, no. 2, pp. 127–132, 2001.

252. G. H. Lin, M. Kapur, R. C. Kainthla, and J. O. Bockris, "One Step Method to Produce Hydrogen by a Triple Stack Amorphous Silicon Solar Cell," *Appl. Phys. Lett.*, vol. 55, no. 4, p. 386, 1989.

253. R. E. Rocheleau, E. L. Miller, and A. Misra, "High-Efficiency Photoelectrochemical Hydrogen Production Using Multijunction Amorphous Silicon Photoelectrodes," *Energy & Fuels*, vol. 12, no. 1, pp. 3–10, 1998.

254. N. Kelly and T. Gibson, "Design and Characterization of a Robust Photoelectrochemical Device to Generate Hydrogen Using Solar Water Splitting," *Int. J. Hydrogen Energy*, vol. 31, no. 12, pp. 1658–1673, 2006.

255. D. G. Nocera, "The Artificial Leaf," *Acc. Chem. Res.*, vol. 45, no. 5, pp. 767–776, 2012.

256. M. T. Winkler, C. R. Cox, D. G. Nocera, and T. Buonassisi, "Modeling Integrated Photovoltaic-Electrochemical Devices Using Steady-State Equivalent Circuits," *Proc. Natl. Acad. Sci. USA*, vol. 110, no. 12, pp. E1076–E1082, 2013.

257. S.-H. Wei, S. B. Zhang, and A. Zunger, "Effects of Ga Addition to $CuInSe_2$ on Its Electronic, Structural, and Defect Properties," *Appl. Phys. Lett.*, vol. 72, no. 24, p. 3199, 1998.

258. T. J. Jacobsson, V. Fjällström, M. Sahlberg, M. Edoff, and T. Edvinsson, "A Monolithic Device for Solar Water Splitting Based on Series Interconnected Thin Film Absorbers Reaching over 10% Solar-to-Hydrogen Efficiency," *Energy Environ. Sci.*, vol. 6, no. 12, pp. 3676–3683, 2013.

259. N. Naghavi, D. Abou-Ras, N. Allsop, N. Barreau, S. Bücheler, A. Ennaoui, C.-H. Fischer et al., "Buffer Layers and Transparent Conducting Oxides for Chalcopyrite Cu(In,Ga) $(S,Se)_2$ Based Thin Film Photovoltaics: Present Status and Current Developments," *Prog. Photovoltaics Res. Appl.*, vol. 18, no. 6, pp. 411–433, 2010.

260. K. Ramanathan, J. Mann, S. Glynn, S. Christensen, J. Pankow, J. Li, J. Scharf, L. Mansfield, M. Contreras, and R. Noufi, "A Comparative Study of Zn(O,S) Buffer Layers and CIGS Solar Cells Fabricated by CBD, ALD, and Sputtering," in *Proceedings of the 2012 38th IEEE Photovoltaic Specialists Conference*, 2012, pp. 001677–001681, IEEE: Austin, TX, http://dx.doi.org/10.1109/PVSC.2012.6317918.

261. M. A. Green, K. Emery, Y. Hishikawa, W. Warta, and E. D. Dunlop, "Solar Cell Efficiency Tables (Version 43)," *Prog. Photovoltaics Res. Appl.*, vol. 22, no. 1, pp. 1–9, 2014.

262. J. H. Park and A. J. Bard, "Unassisted Water Splitting from Bipolar Pt/Dye-Sensitized TiO_2 Photoelectrode Arrays," *Electrochem. Solid-State Lett.*, vol. 8, no. 12, p. G371, 2005.

263. S. M. Jawahar Hussaini, N. Singh, J. Lee, G. Stucky, M. Moskovits, and E. McFarland, "An Autonomous Solar-to-Chemical Energy Conversion System," *Meet. Abstr.*, vol. MA2013–02, no. 44, p. 2565, 2013.

264. G. Decher, "Fuzzy Nanoassemblies: Toward Layered Polymeric Multicomposites," *Science*, vol. 277, no. 5330, pp. 1232–1237, 1997.

265. S. W. Keller, S. A. Johnson, E. S. Brigham, E. H. Yonemoto, and T. E. Mallouk, "Photoinduced Charge Separation in Multilayer Thin Films Grown by Sequential Adsorption of Polyelectrolytes," *J. Am. Chem. Soc.*, vol. 117, no. 51, pp. 12879–12880, 1995.

266. D. M. Kaschak, J. T. Lean, C. C. Waraksa, G. B. Saupe, H. Usami, and T. E. Mallouk, "Photoinduced Energy and Electron Transfer Reactions in Lamellar Polyanion/Polycation Thin Films: Toward an Inorganic 'Leaf,'" *J. Am. Chem. Soc.*, vol. 121, no. 14, pp. 3435–3445, 1999.

267. Y. Il Kim, S. Salim, M. J. Huq, and T. E. Mallouk, "Visible-Light Photolysis of Hydrogen Iodide Using Sensitized Layered Semiconductor Particles," *J. Am. Chem. Soc.*, vol. 113, no. 25, pp. 9561–9563, 1991.

268. K. Maeda, M. Eguchi, S. A. Lee, W. J. Youngblood, H. Hata, and T. E. Mallouk, "Photocatalytic Hydrogen Evolution from Hexaniobate Nanoscrolls and Calcium Niobate Nanosheets Sensitized by Ruthenium(II) Bipyridyl Complexes," *J. Phys. Chem. C*, vol. 113, no. 18, pp. 7962–7969, 2009.

269. W. J. Youngblood, S.-H. A. Lee, K. Maeda, and T. E. Mallouk, "Visible Light Water Splitting Using Dye-Sensitized Oxide Semiconductors," *Acc. Chem. Res.*, vol. 42, no. 12, pp. 1966–1973, 2009.

270. E. Reisner, D. J. Powell, C. Cavazza, J. C. Fontecilla-Camps, and F. A. Armstrong, "Visible Light-Driven H_2 Production by Hydrogenases Attached to Dye-Sensitized TiO_2 Nanoparticles," *J. Am. Chem. Soc.*, vol. 131, no. 51, pp. 18457–18466, 2009.

271. E. Reisner, J. C. Fontecilla-Camps, and F. A. Armstrong, "Catalytic Electrochemistry of a [NiFeSe]-Hydrogenase on TiO_2 and Demonstration of Its Suitability for Visible-Light Driven H_2 Production," *Chem. Commun.*, no. 5, pp. 550–552, http://dx.doi.org/10.1039/B817371K, 2009.

272. Y. Zhao, J. R. Swierk, J. D. Megiatto, B. Sherman, W. J. Youngblood, D. Qin, D. M. Lentz et al., "Improving the Efficiency of Water Splitting in Dye-Sensitized Solar Cells by Using a Biomimetic Electron Transfer Mediator," *Proc. Natl. Acad. Sci. USA*, vol. 109, no. 39, pp. 15612–15616, 2012.

273. R. Brimblecombe, A. Koo, G. C. Dismukes, G. F. Swiegers, and L. Spiccia, "Solar Driven Water Oxidation by a Bioinspired Manganese Molecular Catalyst," *J. Am. Chem. Soc.*, vol. 132, no. 9, pp. 2892–2894, 2010.

274. S.-H. A. Lee, Y. Zhao, E. A. Hernandez-Pagan, L. Blasdel, W. J. Youngblood, and T. E. Mallouk, "Electron Transfer Kinetics in Water Splitting Dye-Sensitized Solar Cells Based on Core–Shell Oxide Electrodes," *Faraday Discuss.*, vol. 155, p. 165, 2012.

275. J. R. Swierk and T. E. Mallouk, "Design and Development of Photoanodes for Water-Splitting Dye-Sensitized Photoelectrochemical Cells," *Chem. Soc. Rev.*, vol. 42, no. 6, pp. 2357–2387, 2013.

2 Photocatalytic Hydrogen Evolution

Yusuke Yamada and Shunichi Fukuzumi

CONTENTS

2.1 Introduction 63
2.2 Electron-Transfer Behavior in Photocatalytic Hydrogen Evolution 65
    2.2.1 Intramolecular Electron Transfer in an Electron
        Donor–Acceptor Dyad...66
    2.2.2 Intermolecular Electron Transfer from an Electron Donor
        to a Photoexcited Photosensitizer 70
    2.2.3 Electron Transfer to a Hydrogen-Evolution Catalyst
        and Hydrogen Evolution on the Catalyst 75
2.3 Effect of Size and Shape of Metal Nanoparticles on Photocatalytic
    Hydrogen Evolution 77
    2.3.1 Size and Shape Effects of Pt Nanoparticles 77
    2.3.2 Size and Shape Effects of Ru Nanoparticles 80
    2.3.3 Size and Crystal Structure Effects of Ni Nanoparticles...................84
2.4 Conclusion 89
References..89

This chapter deals with photocatalytic hydrogen-evolution systems comprising a photosensitizer, an electron donor, and a hydrogen evolution catalyst. First, photoinduced electron-transfer behavior is described in an electron donor–acceptor linked dyad used as a photosensitizer. Then, electron transfer to the photosensitizer in the electron-transfer state from a sacrificial electron donor and electron injection from the photosensitizer in the electron-transfer state to metal nanoparticles is summarized. Finally, the hydrogen-evolution catalysis of the metal nanoparticles depending on sizes, shapes, supports, and crystal structures of the particles is reviewed.

2.1 INTRODUCTION

In biological systems, carbon dioxide and water are converted into carbohydrates by utilizing the solar energy in photosynthesis. In this process, water is used as an electron donor to evolve oxygen, which is catalyzed by oxygen-evolving complex (OEC), and the electrons extracted from water are used for reduction of nicotinamide adenine dinucleotide phosphate ($NADP^+$) to form NADPH, which is used in the Calvin

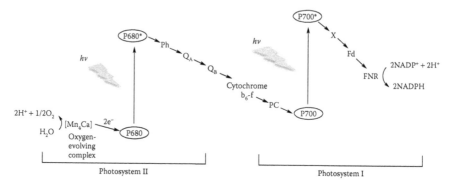

FIGURE 2.1 A schematic diagram of photosynthesis. Two photoexcitation processes are necessary. Fd, ferredoxin; FNR, ferredoxin-NADP⁺ reductase; PC, plastocyanin; Ph, pheophytin; Q_A and Q_B, plastoquinones A and B; X, iron-sulfur protein.

cycle for the synthesis of carbohydrates from carbon dioxide. A schematic diagram of photosynthesis is depicted in Figure 2.1. This process consists of two parts called Photosystem I (PS I) and Photosystem II (PS II). Each part contains a chromophore to carry out water oxidation or NADP⁺ reduction. The photosynthesis is one of the most sophisticated reaction systems; however, a critical drawback is the energy loss during the multiple electron-transfer (ET) steps to achieve charge separation. Thus, two photoexcitation processes are necessary to compensate the energy loss.

Artificial photosynthesis mimics the natural process of photosynthesis, in which light energy is utilized for the production of high-energy chemicals such as hydrogen (H_2). The artificial models of photosynthesis were first reported in the late 1970s by three independent research groups [1–3].

Typically, these models comprise three units: (1) a photosensitizer, (2) an electron relay, and (3) a H_2 evolution catalyst [4]. Figure 2.2 shows the reaction scheme generally accepted for a typical photocatalytic H_2 evolution system using $[Ru(bpy)_3]^{2+}$, methyl viologen (MV^{2+}), and colloidal Pt as a photosensitizer, an electron relay, and a H_2 evolution catalyst, respectively, as well as ethylene diamine tetraacetic acid (EDTA) disodium salt acting as a sacrificial electron donor. The photocatalytic reaction starts by photoirradiation of $[Ru(bpy)_3]^{2+}$ to form a singlet excited state

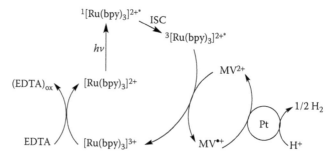

FIGURE 2.2 A typical overall cycle of photocatalytic H_2 evolution using EDTA, $[Ru(bpy)_3]^{2+}$, methyl viologen (MV^{2+}), and colloidal Pt as a sacrificial electron donor, a photosensitizer, an electron relay, and a H_2 evolution catalyst, respectively. ISC, intersystem crossing.

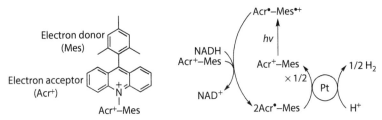

FIGURE 2.3 Chemical structure of 9-mesityl-10-methylacridinium ion (Acr$^+$–Mes) and overall cycle of photocatalytic H$_2$ evolution without electron relay. The photocatalytic system uses NADH, Acr$^+$–Mes, and Pt nanoparticles as a sacrificial electron donor, a photosensitizer, and a H$_2$ evolution catalyst, respectively.

^{1}[Ru(bpy)$_3$]$^{2+*}$. The singlet excited state undergoes intersystem crossing (ISC) providing a triplet excited state ^{3}[Ru(bpy)$_3$]$^{2+*}$ with a relatively long lifetime ($\tau = 600$ ns) [5]. The ^{3}[Ru(bpy)$_3$]$^{2+*}$ species is oxidatively quenched by MV^{2+} used for producing a charge-separation state to form the strong oxidant [Ru(bpy)$_3$]$^{3+}$, which possesses the one-electron reduction potential of 1.29 V versus SCE in CH$_3$CN [5]. Then, the [Ru(bpy)$_3$]$^{3+}$ species is reduced back to [Ru(bpy)$_3$]$^{2+}$ by EDTA. On the other hand, the one-electron reduced MV^{2+} species (MV$^{·+}$) injects the electron to colloidal Pt. On the surface of colloidal Pt, protons in the reaction solution are reduced to evolve H$_2$. This reaction process involves only four ET steps. Thus, energy loss accompanied with multiple ET steps in natural photosynthesis can be dramatically reduced by the appropriate choice of each component.

Recently, a new type of photocatalytic H$_2$ evolution systems without an electron relay has emerged [6–9]. This system consists of only three components: a sacrificial electron donor, a photosensitizer, and a H$_2$ evolution catalyst. For example, an electron donor–acceptor linked dyad such as 9-mesityl-10-methylacridinium ion (Acr$^+$–Mes) is used together with nicotinamide adenine dinucleotide (NADH) and Pt nanoparticles as a sacrificial electron donor and a H$_2$ evolution catalyst, respectively, to evolve H$_2$ following the reaction scheme shown in Figure 2.3 [7]. In this reaction system, Acr$^+$–Mes forms a long-lived ET state under photoirradiation [6,7]. The ET state of Acr$^+$–Mes can directly inject an electron to a H$_2$ evolution catalyst and can be reduced by a sacrificial electron donor owing to the lifetime of the ET state being longer than 10 μs [6,7].

In this chapter, ET behavior in the H$_2$ evolution systems using an electron donor–acceptor linked dyad is described in both intramolecular ET of the electron donor–acceptor linked dyad and the ET from the electron donor–acceptor linked dyad in photoinduced ET state to metal particles. Then, effects of size, shape, support, and crystal structure of metal nanoparticles (MNPs) on the catalytic activity for H$_2$ evolution are discussed.

2.2 ELECTRON-TRANSFER BEHAVIOR IN PHOTOCATALYTIC HYDROGEN EVOLUTION

Photocatalytic H$_2$ evolution is possible with not only the artificial photosynthesis systems but also inorganic semiconductor catalysts such as TiO$_2$. The artificial photosynthesis systems comprising multiple units possess an advantage in that they allow for the improvement by the combination of the optimized units. Both intra- and intermolecular ET behavior of a photosensitizer can be scrutinized by time-resolved transient absorption spectroscopy.

2.2.1 INTRAMOLECULAR ELECTRON TRANSFER IN AN ELECTRON DONOR–ACCEPTOR DYAD

Fast ET from an electron donor to an electron acceptor and their slow back ET are crucial to achieve a long-lived ET state, which allows intermolecular events including chemical reactions. An ET rate from an electron donor to an acceptor can be predicted by the Marcus theory of ET [10,11]. According to the theory, an ET rate is expected to decrease as the ET driving force ($-\Delta G_{ET}^0$), which is expressed as difference between the oxidation potential of an electron donor and the reduction potential of an electron acceptor, increases in the strongly exergonic region where $-\Delta G_{ET}^0 < \lambda$ (λ being the reorganization energy required to reorganize the electron donor, acceptor, and their solvation spheres accompanied with ET). The lifetime of the ET state becomes longer with the larger driving force of the back ET in such a strongly exergonic regime known as the Marcus inverted region. An electron donor–acceptor linked molecule with a small λ value should possess a long lifetime of the ET state as far as it has a high-lying triplet excited state.

9-Mesityl-10-methylacridinium ion (Acr+–Mes) was designed to possess a long-lived ET state based on this theory. The λ value of acridinium for the electron self-exchange between the acridinium ion and the corresponding one-electron reduced neutral radical is very small (0.3 eV) among the redox active organic compounds [12]. At the 9-position of the acridinium ion, the electron-donor moiety (mesityl group, Mes) is directly bonded to yield Acr+–Mes. Solvent reorganization energy of Acr+–Mes accompanied with ET should be small, because the overall charge of Acr+–Mes remains +1 even after ET (Acr•–Mes•+).

The X-ray crystal structure of Acr+–Mes is shown in Figure 2.4a [13]. The nearly perpendicular dihedral angle between aromatic ring planes suggests little interactions between highest occupied molecular orbital (HOMO) and lowest unoccupied molecular orbital (LUMO). The HOMO and LUMO of Acr+–Mes calculated by density functional theory (DFT) methods are localized on mesitylene and acridinium moieties, respectively, as shown in Figure 2.4b (HOMO) and c (LUMO).

Formation of the ET state (Acr•–Mes•+) via photoinduced ET from the mesitylene moiety to the singlet excited state of the acridinium ion moiety (^{1}Acr+*–Mes) was observed by photoirradiation of a deaerated acetonitrile (MeCN) solution of Acr+–Mes by nanosecond laser excitation at 430 nm. Only one apparent peak around 500 nm suggested that the absorption band of the radical cation of the mesitylene moiety ($\lambda_{max} = 480$ nm) [14] overlaps with that of the acridinyl radical moiety ($\lambda_{max} = 520$ nm) (Figure 2.5a) [12]. The overlap was confirmed by the photoirradiation of an MeCN solution containing Acr+–Mes in the presence of anthracene. The reduction of Mes•+ moiety by anthracene to Mes moiety resulted in the decrease in the absorbance at 480 nm (Figure 2.5a, closed circle) and arising the absorbance at 720 nm assigned to anthracene radical cation [15]. The remained absorption band at 520 nm can be assigned to Acr• moiety. Figure 2.5b shows the time courses of the

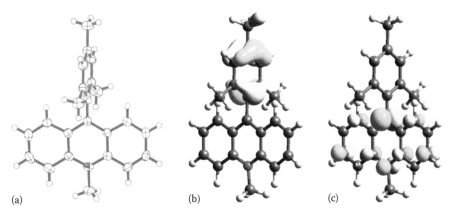

FIGURE 2.4 (a) ORTEP drawing of 9-mesityl-10-methylacridinium ion (Acr⁺–Mes). (b) HOMO and (c) LUMO calculated by a DFT method with Gaussian 98 (B3LYP/6-31G* basis set). (Reprinted with permission from Fukuzumi, S. et al., *J. Am. Chem. Soc.*, 126, 1600, 2004. Copyright 2004 American Chemical Society.)

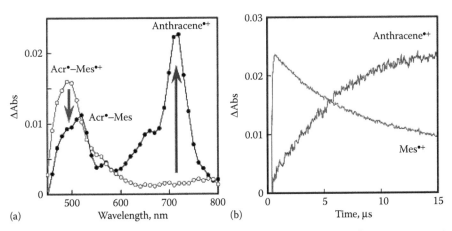

FIGURE 2.5 (a) Transient absorption spectra of Acr⁺–Mes (5.0×10^{-5} M) in deaerated MeCN at 298 K taken 15 μs after laser excitation at 430 nm in the absence (○) and presence (●) of anthracene (5.0×10^{-5} M). (b) Time profiles of the absorbance decay at 480 nm and the rise at 720 nm. (Reprinted with permission from Fukuzumi, S. et al., *J. Am. Chem. Soc.*, 126, 1600, 2004. Copyright 2004 American Chemical Society.)

absorbance owing to Mes•⁺ and anthracene radical cation, in which the absorbance at 720 nm increased with a simultaneous decrease in the absorption band due to Mes•⁺ moiety [13].

Typically, back ET in electron donor–acceptor linked molecules is an intramolecular event; however, back ET of the ET state of Acr⁺–Mes (Acr•–Mes•⁺) is an intermolecular event, not an intramolecular event. Figure 2.6a shows the decay time profile at 500 nm assignable to Acr• moiety after laser excitation (355 nm) in benzonitrile (PhCN). The decay of the ET state obeyed second-order kinetics in PhCN as

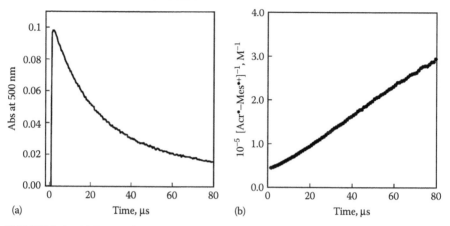

FIGURE 2.6   (a) Decay time profile at 500 nm of Acr⁺–Mes (5.0 × 10⁻⁵ M) in deaerated PhCN at 298 K after laser excitation at 355 nm. (b) Second-order plot of the decay of transient absorption spectrum of Acr⁺–Mes. (Reprinted with permission from Fukuzumi, S. et al., *J. Am. Chem. Soc.*, 126, 1600, 2004. Copyright 2004 American Chemical Society.)

shown in Figure 2.6b. As predicted from the Marcus theory of ET, intramolecular back ET of Acr˙–Mes˙⁺ becomes very slow; thus, intermolecular back ET proceeds predominantly. Photoirradiation of molecularly isolated Acr⁺–Mes supported on silica-alumina surfaces by cation exchange, in which no intermolecular interaction event is expected, resulted in long lifetime of more than 1 s, which is longer than the lifetime of the charge-separated state in natural photosynthesis [16].

2-Phenyl-4-(1-naphthyl)quinolinium ion (QuPh⁺–NA) is another electron donor–acceptor linked dyad based on the same design concept for Acr⁺–Mes [17]. 2-Phenyl quinolinium and naphthalene moieties act as an electron acceptor and donor, respectively. The X-ray crystallographic analysis of QuPh⁺–NA revealed that the dihedral angle between the NA and quinolinium (QuPh⁺) moieties was nearly perpendicular (87°) as shown in Figure 2.7a [17]. DFT calculations based on the crystal structure of QuPh⁺–NA at the B3LYP/6-31G level suggested little orbital interaction between the NA and QuPh⁺ moieties, because the HOMO and LUMO of QuPh⁺–NA were localized on the NA and QuPh⁺ moieties, respectively (Figure 2.7b and c) [17].

Formation of the photoexcited ET state (QuPh˙–NA˙⁺) was confirmed by femtosecond laser excitation at 390 nm of a deaerated MeCN solution of QuPh⁺–NA (1.0 × 10⁻⁴ M), which has an absorption band at 420 nm due to QuPh˙ and also at 690 nm due to NA˙⁺ [18,19]. The transient absorption spectrum of QuPh˙–NA˙⁺ taken at 10 ps after laser excitation is shown in Figure 2.8a [17]. Time courses of the absorbance at 420 and 690 nm due to QuPh˙–NA˙⁺ are shown in Figure 2.8b [17]. No decay of absorption at 690 nm due to QuPh˙–NA˙⁺ was observed within 1500 ps [17].

The long-lived ET state of QuPh⁺–NA was observed by nanosecond laser excitation (355 nm) of a deaerated MeCN solution of QuPh⁺–NA. Two characteristic absorption bands at 420 and 690 nm due to QuPh˙–NA˙⁺, which are also observed by femtosecond laser excitation (Figure 2.8) and a new broad band at 1050 nm, can be assigned to the naphthalene π-dimer radical cation [18–20] as shown in Figure 2.9a.

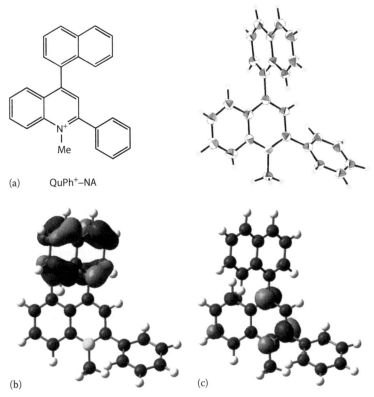

(a)   QuPh$^+$–NA

(b)   (c)

FIGURE 2.7   (a) Chemical structure and an ORTEP drawing of QuPh$^+$–NA. (b) HOMO and (c) LUMO of QuPh$^+$–NA calculated by a DFT method with Gaussian 98 (B3LYP/6-31G basis set). (From Kotani, H. et al., *Faraday Discuss.*, 2012, 155: p. 89. Copyright 2012 The Royal Society of Chemistry, reproduced with permission.)

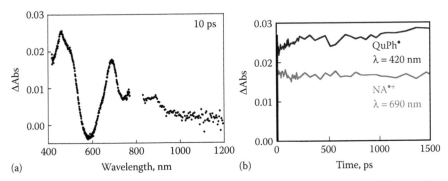

(a)   (b)

FIGURE 2.8   (a) Transient absorption spectrum observed by the femtosecond laser excitation ($\lambda = 390$ nm) of a deaerated MeCN solution containing QuPh$^+$–NA (1.0×10^{-3} M) taken at 10 ps after laser excitation at 295 K. (b) Decay time profiles at 420 nm due to QuPh$^\bullet$ and 690 nm due to NA$^{\bullet+}$. (From Kotani, H. et al., *Faraday Discuss.*, 2012, 155: p. 89. Copyright 2012 The Royal Society of Chemistry, reproduced with permission.)

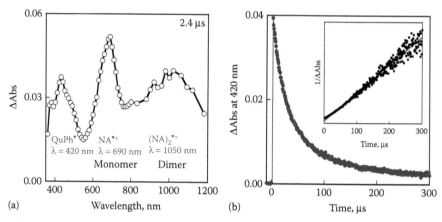

FIGURE 2.9 (a) Transient absorption spectra of QuPh⁺–NA (1.0×10^{-4} M) in deaerated MeCN at 298 K taken at 2.4 μs after laser excitation at 355 nm. (b) Decay time profile of absorbance at 420 nm due to QuPh⋅. Inset: Second-order plot. (From Kotani, H. et al., *Faraday Discuss.*, 2012, 155: p. 89. Copyright 2012 The Royal Society of Chemistry, reproduced with permission.)

The long-lived ET state (QuPh⋅–NA⋅⁺) forms the π-complex with QuPh⁺–NA to afford the π-dimer radical cation [(QuPh⋅–NA⋅⁺)(QuPh⁺–NA)], in which the NA⋅⁺ and QuPh⋅ moieties interact by π-bonding with the NA and QuPh⁺ moieties, respectively [17].

The intramolecular back ET in the π-dimer radical cation is also too slow to allow the intermolecular back ET between two π-dimer radical cations. The intermolecular back ET is evidenced by the second-order decay of absorbance at 420 nm due to the QuPh⋅ moiety as shown in Figure 2.9b (inset) [17]. The rate constant of intermolecular back ET was 3.3×10^9 M^{-1} s^{-1} which was determined from the slope of the second-order plot and molar absorption coefficient of QuPh⋅ at 420 nm ($\varepsilon = 1500$ M^{-1} cm^{-1}) [17]. The quantum yield of the π-dimer radical cation of the long-lived ET state of QuPh⁺–NA was also determined by the comparative method to be 83% [17].

There has been discussion on the photoinduced state of these electron donor–acceptor linked dyads, which is either the ET state or the triplet excited state [21,22]. However, it has been demonstrated that the ET state is indeed produced, followed by formation of the π-dimer radical cation and its ET state, which was also detected by the near infrared absorption due to the π–π* transition of the π-dimer radical cation [20].

2.2.2 INTERMOLECULAR ELECTRON TRANSFER FROM AN ELECTRON DONOR TO A PHOTOEXCITED PHOTOSENSITIZER

Intermolecular ET from an electron donor to a photosensitizer was scrutinized by transient absorption spectroscopy. The photoexcited ET state of Acr⁺–Mes (Acr⋅–Mes⋅⁺) is able to oxidize NADH ($E^{\circ}_{ox} = 0.76$ V vs. SCE) [23,24], which is employed as a sacrificial electron donor for the photocatalytic H$_2$ evolution, by the Mes⋅⁺ moiety ($E^{\circ}_{red} = 1.88$ V vs. SCE) [13,25]. Nanosecond laser excitation at 430 nm of a deaerated mixed solution of H_2O and MeCN [1:1 (v/v)] containing Acr+–Mes and NADH clearly showed the transient absorption band to be at 520 nm owing to formation of Acr⋅ as shown in Figure 2.10a [7]. The absorbance at 520 nm gradually increased up to 10 μs after the photoexcitation suggests the further formation of Acr⋅–Mes. On the other hand, the transient absorption due to NADH⋅+ was not observed, but instead the initial bleaching band at 420 nm increased, accompanied by increase in the absorption band at 520 nm due to Acr⋅–Mes as shown in Figure 2.10b. This indicates that the additional formation of Acr⋅–Mes is owing to ET from NAD⋅, which can be formed by rapid deprotonation of NADH⋅+ [7] to Acr+–Mes as shown in Scheme 2.1.

The slow increase of the transient absorption band at 520 nm depends on the concentration of Acr+–Mes as shown in Figure 2.11a. The pseudo-first-order rate constant (k_{obs}) for Acr⋅–Mes formation increases in proportion to the concentration of Acr+–Mes and the second-order rate constant (k_{red}) of ET from NAD⋅ to Acr+–Mes was determined from the slope of the linear plot in Figure 2.11b to be 3.7×10^9 M^{-1} s^{-1} [7]. The large k_{red} value close to the diffusion-limited value [26] is reasonable because of the large driving force (0.44 eV) of the ET from NAD⋅ ($E°_{ox} = -1.1$ V vs. SCE) [24] to the Acr+ moiety in Acr+–Mes ($E°_{red} = -0.66$ V vs. SCE) [13,27].

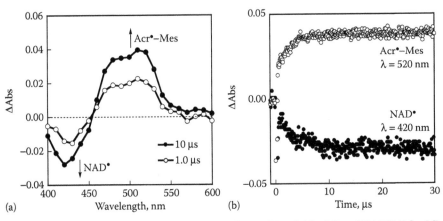

FIGURE 2.10 (a) Transient absorption spectra of Acr+–Mes (0.10 mM) and NADH (1.0 mM) in a deaerated mixed solution of H_2O and MeCN [1:1 (v/v)] mixed solution (2.0 mL) at 298 K taken at 1.0 μs (●) and 10 μs (○) after nanosecond laser excitation at 430 nm. (b) Time profiles of formation of Acr⋅–Mes at 520 nm and decay of NAD⋅ at 420 nm. (From Kotani, H. et al., *Phys. Chem. Chem. Phys.*, 2007, 9: p. 1487. Copyright 2007 The Royal Society of Chemistry, reproduced with permission.)

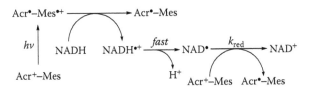

SCHEME 2.1 Photoinduced ET from NADH to Acr+–Mes.

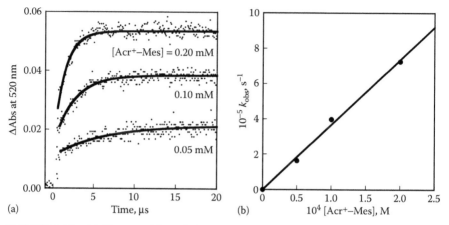

(a)   (b)

FIGURE 2.11 (a) Time profiles of Acr˙–Mes at 520 nm at various concentrations of Acr⁺–Mes. (b) Plot of the pseudo-first-order rate constant (k_{obs}) for ET from NAD˙ to Acr⁺–Mes versus [Acr⁺–Mes]. (From Kotani, H. et al., *Phys. Chem. Chem. Phys.*, 2007, 9: p. 1487. Copyright 2007 The Royal Society of Chemistry, reproduced with permission.)

Similarly, photodynamics between a photoexcited QuPh⁺–NA (QuPh˙–NA) and oxalate were investigated by nanosecond laser flash photolysis. Oxalate can also act as two-electron donor similar to NADH in the photocatalytic H_2 evolution. Laser excitation at 355 nm of a deaerated mixed solution of a phosphate buffer (pH 6.0) and MeCN [1:1 (v/v)] containing QuPh⁺–NA (0.056 mM) results in the formation of ET state (QuPh˙–NA˙⁺), which was confirmed by the transient absorption bands at 420 nm (QuPh˙ moiety), 690 nm (NA˙⁺ moiety), and 1000 nm [(QuPh˙–NA˙⁺)(QuPh⁺–NA)] at 4.0 μs after laser excitation as shown in Figure 2.12a [17,28]. These transient absorption bands decayed monotonously at

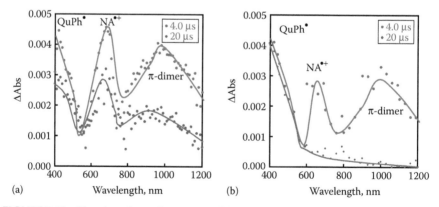

(a)   (b)

FIGURE 2.12 Transient absorption spectra of QuPh⁺–NA (0.056 mM) in a deaerated mixed solution of a phosphate buffer (pH 6.0) and MeCN [1:1 (v/v)] at 298 K taken at 4.0 μs (red) and 20 μs (blue) after nanosecond laser excitation at 355 nm in the (a) absence and (b) presence of oxalate (6.0 mM). (From Yamada, Y. et al., *Phys. Chem. Chem. Phys.*, 2012, 14: p. 10564. Copyright 2012 The Royal Society of Chemistry, reproduced with permission.)

a prolonged time after photoexcitation in the absence of oxalate (Figure 2.12a, blue). On the other hand, in the presence of oxalate, the absorption band at 420 nm remained even at 20 μs after photoexcitation, whereas the absorption bands at 690 and 1000 nm assigned to the NA$^{•+}$ and π-dimer radical cation, [(QuPh$^•$–NA$^{•+}$) (QuPh$^+$–NA)], respectively, disappeared in this time period [28]. This observation suggested that ET from oxalate to the NA$^{•+}$ moiety of the π-dimer radical cation occurs as expected by the large driving force (1.07 V) of the ET from oxalate (E_{ox} = 0.80 V vs. SCE) [28] to NA$^{•+}$ moiety of QuPh$^•$–NA$^{•+}$ (E_{red} = 1.87 V vs. SCE) [17].

The ET from oxalate to the NA$^{•+}$ moiety of QuPh$^•$–NA$^{•+}$ was monitored by the decay of absorption band at 690 nm with various concentrations of oxalate as shown in Figure 2.13a [28]. The rate obeyed pseudo-first-order kinetics and the pseudo-first-order rate constant (k_{obs}) increased linearly in proportion to the concentrations of oxalate as shown in Figure 2.13b. From the slope of a linear plot in Figure 2.13b, the second-order rate constant (k_{ox}) of ET from oxalate to the NA$^{•+}$ moiety was determined to be 9.1 × 10^6 M^{-1} s^{-1} [28].

In contrast to the decay of the absorption band at 690 nm due to the NA$^{•+}$ moiety, the absorption band at 420 nm due to the QuPh$^•$ moiety remained at 0.3 ms or longer time after laser excitation (Figure 2.14a) [28]. The yields of remained QuPh$^•$–NA at 0.8 ms after laser excitation were determined based on the maximum absorbance achieved in the absence of oxalate and plotted against concentrations of oxalate as shown in Figure 2.14b [28]. The yield of remained QuPh$^•$–NA at 0.8 ms after laser excitation linearly increased in proportion to concentration of sodium oxalate up to 6.0 mM, which is the highest concentration to a mixed solution of a phosphate buffer (pH 6.0) and MeCN [1:1(v/v)] [28]. When the concentration of oxalate was increased to 35 mM by using tetra-n-butylammonium salt, the absorbance at 420 nm assigned

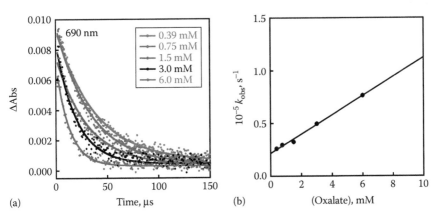

(a)

(b)

FIGURE 2.13 (a) Decay time profile of absorption at 690 nm due to QuPh$^•$–NA$^{•+}$ with various concentrations of oxalate (0.39 mM, red; 0.75 mM, blue; 1.5 mM, green; 3.0 mM, black; 6.0 mM, purple) in the presence of QuPh$^+$–NA (0.056 mM). QuPh$^•$–NA$^{•+}$ was produced by the laser excitation (λ = 355 nm) of a deaerated mixed solution of a phosphate buffer (pH 6.0) and MeCN [1:1 (v/v)]. (b) Plot of the pseudo-first-order rate constant (k_{obs}) for ET from oxalate to QuPh$^•$–NA$^{•+}$ versus [oxalate]. (From Yamada, Y. et al., *Phys. Chem. Chem. Phys.*, 2012, 14: p. 10564. Copyright 2012 The Royal Society of Chemistry, reproduced with permission.)

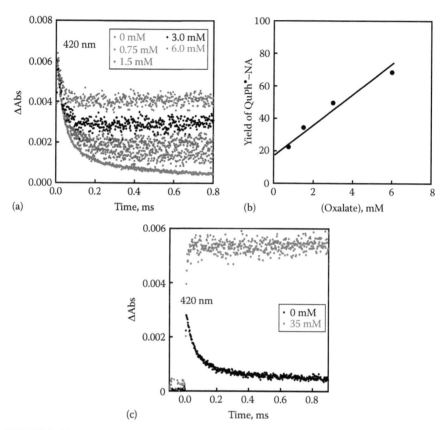

(a)

(b)

(c)

FIGURE 2.14   (a) Time profiles of decay of absorption at 420 nm due to QuPh•–NA detected by nanosecond laser excitation at 355 nm of a deaerated mixed solution of a phosphate buffer (pH 6.0) and MeCN [1:1 (v/v)] containing QuPh+–NA (0.056 mM) at 298 K with different concentrations of oxalate (0 mM, red; 0.75 mM, blue; 1.5 mM, green; 3.0 mM, black; 6.0 mM, purple). (b) Yield of QuPh•–NA depending on the concentration of oxalate. (c) Time profiles of formation and decay of absorption at 420 nm due to QuPh•–NA detected by nanosecond laser excitation at 355 nm of a deaerated mixed solution of a phosphate buffer (pH 6.5) and MeCN [1:1 (v/v)] containing QuPh+–NA (0.056 mM) at 298 K in the absence (black) and presence (red) of [COO(n-Bu$_4$N)]$_2$ (35 mM). (From Yamada, Y. et al., *Phys. Chem. Chem. Phys.*, 2012, 14: p. 10564. Copyright 2012 The Royal Society of Chemistry, reproduced with permission.)

to QuPh•–NA was double the maximum absorbance achieved in the absence of oxa-late as shown in Figure 2.14c [28]. There was no further increase in the absorbance at 420 nm by increasing the concentration of oxalate [28]. Thus, two equivalents of QuPh• formed in the presence of high concentrations of oxalate [28].

Scheme 2.2 depicts the overall reaction pathway of formation of QuPh•–NA [28]. Photoexcited QuPh+–NA (QuPh•–NA•+) forms the π-dimer radical cation with QuPh+–NA, [(QuPh•–NA•+)(QuPh+–NA)] [28]. Then, ET from oxalate to the NA•+ moiety of the π-dimer radical cation occurs to produce oxalate radical anion. The oxalate radical anion spontaneously decomposes to CO_2 and $CO_2^{•-}$ with a first-order rate constant

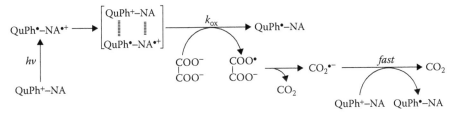

SCHEME 2.2   Photoinduced ET from oxalate to QuPh$^+$–NA. (From Yamada, Y. et al., *Phys. Chem. Chem. Phys.*, 2012, 14: p. 10564. Copyright 2012 The Royal Society of Chemistry, reproduced with permission.)

of 2×10^6 s^{-1} [29–31]. Because CO$_2^{\bullet-}$ is a strong one-electron reductant as evidenced by the high one-electron oxidation potential ($E^\circ_{ox} = -2.2$ V vs. SCE) [32,33], ET from CO$_2^{\bullet-}$ to QuPh$^+$–NA occurs to produce QuPh$^\bullet$–NA by the thermal reaction [28]. Thus, oxalate acts as a two-electron donor in the photocatalytic H$_2$ evolution system.

2.2.3   ELECTRON TRANSFER TO A HYDROGEN-EVOLUTION CATALYST AND HYDROGEN EVOLUTION ON THE CATALYST

After reduction of photoinduced ET state of an electron donor–acceptor linked dyad by a sacrificial electron donor, the reduced species should inject an electron to MNPs employed as a hydrogen evolution catalyst. This ET process can also be monitored by transient absorption spectroscopy in the time range longer than seconds. The reduced form of Acr$^+$–Mes (Acr$^\bullet$–Mes) was generated by photoirradiation of a deaerated mixed solution of a phthalate buffer (pH 4.5) and MeCN containing Acr$^+$–Mes (0.16 mM) and NADH (1.0 mM) for several minutes as evidenced by the emergence of characteristic absorption band at 520 nm (Figure 2.15a, bold line).

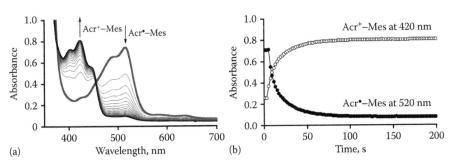

FIGURE 2.15   (a) UV-Vis absorption spectral change observed in ET from Acr$^\bullet$–Mes (0.16 mM) to PtNPs upon addition of PtNPs (2.0 µg) to a deaerated phthalate buffer (pH 4.5, 50 mM) and MeCN [1:1 (v/v)] mixed solution (2.0 mL) of Acr$^\bullet$–Mes. (b) Time profiles of the decay of absorbance at 520 nm due to Acr$^\bullet$–Mes and the rise of absorbance at 420 nm due to Acr$^+$–Mes. (From Kotani, H. et al.: Size- and shape-dependent activity of metal nanoparticles as hydrogen-evolution catalysts: Mechanistic insights into photocatalytic hydrogen evolution. *Chem.–Eur. J.* 2011. 17. 2777. Copyright Wiley-VCH Verlag GmbH & Co. KGaA. Reproduced with permission.)

The addition of an aliquot of a solution containing platinum nanoparticles (PtNPs) (2.0 μg) to the solution containing Acr•–Mes with magnetic stirring resulted in the decrease in the absorption band at 520 nm due to Acr•–Mes accompanied by appearance of the absorption band at 420 nm ($\varepsilon = 4700$ M^{-1} cm^{-1}) due to Acr$^+$–Mes, indicating the efficient ET from Acr•–Mes to PtNPs [6]. The time profile of the decay of absorbance at 520 nm due to Acr•–Mes agreed with that of the rise of absorbance at 420 nm due to Acr$^+$–Mes as shown in Figure 2.15b [6]. The ET rates obeyed pseudo-first-order kinetics, and the pseudo-first-order rate constant (k_{obs}) increased in proportion to the amount of PtNPs added [6].

After injection of the electrons to PtNPs, reduction of protons leading to H$_2$ evolution proceeds on the surface of PtNPs. The rate of H$_2$ evolution should be compared with that of the electron injection to PtNPs to determine which step is the rate-determining step of the photocatalytic H$_2$ evolution. Figure 2.16a shows the decay time profile of absorbance at 420 nm owing to ET from Acr• moiety to PtNPs and the time profile of the amount of evolved H$_2$ quantified by gas chromatography. Their comparison clearly indicated that the H$_2$ evolution rate is comparable to the rate of ET from Acr•–Mes to PtNPs. Thus, ET from Acr•–Mes to PtNPs is the rate-determining step of the H$_2$ evolution reaction. However, QuPh•–NA was employed as a photosensitizer; ET from QuPh•–NA to PtNPs is much faster than H$_2$ evolution because QuPh• is stronger reductant than Acr• (Figure 2.16b) [9]. In this reaction system, the rate-determining step is the protons reduction or H$_2$ evolution step.

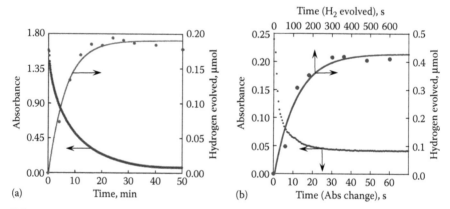

FIGURE 2.16 (a) Time profiles of H$_2$ evolution and absorbance decay at 520 nm due to Acr•–Mes in ET from Acr•–Mes to spherical PtNPs with the diameter of 4.5 nm (0.1 μg) in a mixed solution of an acetate buffer (pH 5.0) and MeCN [1:1 (v/v)]. (From Kotani, H. et al.: Size- and shape-dependent activity of metal nanoparticles as hydrogen-evolution catalysts: Mechanistic insights into photocatalytic hydrogen evolution. *Chem.–Eur. J.* 2011. 17. 2777. Copyright Wiley-VCH Verlag GmbH & Co. KGaA. Reproduced with permission.) (b) Decay time profile of absorption due to QuPh•–NA in ET from QuPh•–NA to PtNPs (1.5 mg L^{-1}) and time profile of hydrogen evolution in a mixed solution of a phthalate buffer (pH 4.5) and MeCN [1:1 (v/v)]. (Reprinted with permission from Yamada, Y. et al., *J. Am. Chem. Soc.*, 133, 16136, 2011. Copyright 2011 American Chemical Society.)

2.3 EFFECT OF SIZE AND SHAPE OF METAL NANOPARTICLES ON PHOTOCATALYTIC HYDROGEN EVOLUTION

2.3.1 SIZE AND SHAPE EFFECTS OF PT NANOPARTICLES

Even if the rate-determining step of photocatalytic H_2 evolution is ET from a photosensitizer to a H_2 evolution catalyst or proton reduction on the catalyst surfaces leading to H_2 evolution, the redox properties or surface conditions of MNPs which are used as H_2 evolution catalysts play a crucial role in improving the efficiency of H_2 evolution. The redox properties and surface conditions of the nanoparticles would be dependent on various features such as particles size, shape, and crystal structure. The recent developments in the inorganic synthesis for size- and/or shape-controlled nanoparticles allow us to evaluate the effects of each factor on the catalysis for H_2 evolution.

Among MNPs, PtNPs have been most widely used as the H_2 evolution catalyst, because of the low overpotential for proton reduction in electrochemical measurements [34–37]. However, the use of Pt metal should be reduced for practical applications because of its high cost and limited supply. The amount of Pt used can be reduced by the enhancement of catalytic activity of PtNPs by controlling the shape and size [6]. The size-controlled spherical or cubic PtNPs were prepared under appropriate preparation conditions [6]. TEM images of size- and shape-controlled PtNPs shown in Figure 2.17 indicate that the mean diameters of spherical and cubic PtNPs are 3.6 ± 0.6 nm and 8.6 ± 0.8 nm (Figure 2.17c and d), respectively [6].

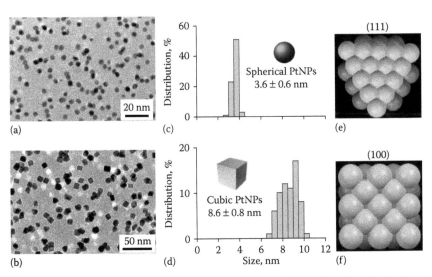

FIGURE 2.17 TEM images of (a) spherical PtNPs and (b) cubic PtNPs. Size distributions of (c) spherical PtNPs and (d) cubic PtNPs. Schematic views of (e) Pt(111) plane and (f) Pt(100) plane. (From Kotani, H. et al.: Size- and shape-dependent activity of metal nanoparticles as hydrogen-evolution catalysts: Mechanistic insights into photocatalytic hydrogen evolution. *Chem.–Eur. J.* 2011. 17. 2777. Copyright Wiley-VCH Verlag GmbH & Co. KGaA. Reproduced with permission.)

TABLE 2.1
Shape and Diameter (2r) of PtNPs Used for the Photocatalytic H_2 Evolution with NADH and Acr+–Mes

Entry	Shape	Diameter (2r) [nm]	$10^{-4} k_{et}$ [s^{-1} mol(Pt)$^{-1}$ L]
a	Sphere	2.1 ± 0.3	2.5
b	Sphere	3.6 ± 0.6	5.2
c	Sphere	4.5 ± 0.7	6.0
d	Sphere	5.1 ± 0.5	4.0
e	Sphere	8.0 ± 1.1	3.5
f	Cube	6.3 ± 0.6	7.0
g	Cube	8.6 ± 0.8	4.1

The size and shape of PtNPs were systematically changed as listed in Table 2.1 by choosing appropriate preparation conditions [6].

The overall catalytic cycle of the photocatalytic H_2 evolution using NADH, Acr+–Mes, and PtNPs as an electron donor, a photosensitizer, and a H_2 evolution catalyst, respectively, is depicted in Figure 2.3. The catalytic activity of PtNPs with various shapes and sizes has been evaluated by determining the rate constant (k_{et}) of ET from Acr•–Mes to PtNPs, which is the rate-determining step of this reaction system (vide supra). The catalytic activity of PtNPs with different shapes and sizes was compared in terms of the rate constant normalized by the number of surface Pt atoms. The number of surface Pt atoms (N_s, mol g^{-1}) was estimated in the following equation [38]:

$$N_s = \frac{1}{N_A} \frac{S d_s}{V d_v} = \frac{2.9 \times 10^{-3} \ (\text{nm mol g}^{-1})}{r \ (\text{nm})} \tag{2.1}$$

where:

N_A is the Avogadro's number

r, S, and V are the radius (nm), the surface area (nm^2), and the volume (nm^3), respectively

d_s and d_v were determined from the density of Pt metals ($d_s = 12.5$ atoms nm^{-2} and $d_v = 2.15 \times 10^{-20}$ g nm^{-3})

The k_{et} values of ET from Acr•–Mes to PtNPs with various shapes and sizes were determined from the slopes of the linear plots of the observed pseudo-first-order rate constant (k_{obs}) versus the molar concentration of Pt atoms [mol(Pt) L^{-1}], which comprise the surface of PtNPs. Among the PtNPs listed in Table 2.1, cubic PtNPs with an average diameter of 6.3 ± 0.6 nm exhibited the maximum k_{et} value in the photocatalytic H_2 evolution [6]. As regards spherical PtNPs, the k_{et} value becomes maximum with the average diameter of 4.5 ± 0.6 nm as shown in Figure 2.18b [6]. Such shape- and size-dependent catalyses of metal NPs have been reported in various catalytic reactions [39–44]. In general, cubic PtNPs are more active than PtNPs

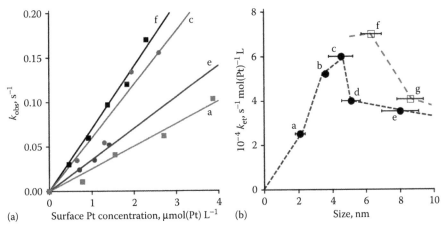

FIGURE 2.18 (a) Plots of pseudo-first-order rate constants (k_{obs}) for ET from Acr•–Mes to PtNPs with various sizes and shapes versus Pt concentrations on the surfaces of PtNPs. (b) Plots of k_{et} versus size of spherical (●) and cubic (□) PtNPs. The lines in (a) and points in (b) denoted by a–g correspond to entries in Table 2.1. (From Kotani, H. et al.: Size- and shape-dependent activity of metal nanoparticles as hydrogen-evolution catalysts: Mechanistic insights into photocatalytic hydrogen evolution. *Chem.–Eur. J.* 2011. 17. 2777. Copyright Wiley-VCH Verlag GmbH & Co. KGaA. Reproduced with permission.)

with other shapes, because the surfaces of cubic PtNPs consist of (100) planes where Pt atoms are loosely aligned as shown in Figure 2.17f. The (111) surface shown in Figure 2.17e is the most thermodynamically stable surface, in which the surface energy is minimum. The loose alignment of Pt atoms on (100) surface increases the surface energy and favors the interaction with a substrate [6].

The k_{et} value of ET from Acr•–Mes to PtNPs increases in proportion to concentration of proton, [H+], as shown in Figure 2.19 (black circles). The ET rate increased linearly with increasing [H+] (Figure 2.19), suggesting that ET from Acr•–Mes to PtNPs is coupled with proton transfer and the proton-coupled ET (PCET) results in formation of a Pt–H bond on the PtNP surface as shown in Scheme 2.3 [6]. H_2 evolution followed by the reduction may result from the reductive elimination of H_2 from two Pt–H species [6].

When the photocatalytic H_2 evolution was performed in D_2O containing CH_3COOD/CH_3COONa buffer, a substantial inverse kinetic isotope effect [KIE = k_{et}(H)/k_{et}(D) = 0.47] was observed in ET from Acr•–Mes to PtNPs (gray circles in Figure 2.19) [17]. In general, an inverse KIE results from a larger zero-point energy difference in the transition state relative to the ground state [45,46]. An inverse KIE is often observed in the surface reactions such as reaction of atomic chlorine with C_2H_4 [46], the hydrogenation of nitrobenzene to aniline using Pd/C catalyst [47], enzymatic reactions involving copper–zinc superoxide dismutase [47], hydrogenation of dienes by iridium phosphine-based complexes [48], and H_2 adsorption on metal surfaces [49,50]. The observation of an inverse KIE in such surface reactions originates from the changes to zero-point energy caused by adsorption on the surface [51,52]. Thus, the substantial inverse KIE in Figure 2.19 may result from the higher zero-point energy of the Pt–H bond formation than the Pt–D bond formation on PtNPs [6].

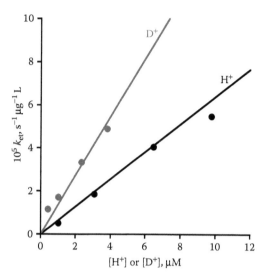

FIGURE 2.19 Dependence of k_{et} on [H+] or [D+] observed in ET from Acr•–Mes to spherical PtNPs with the diameter of 4.5 nm in H_2O/MeCN [1:1 (v/v)] containing CH_3COOH/ CH_3COONa buffer (50 mM) or in D_2O/MeCN [1:1 (v/v)] containing CH_3COOD/CH_3COONa buffer (50 mM) at 298 K. (From Kotani, H. et al.: Size- and shape-dependent activity of metal nanoparticles as hydrogen-evolution catalysts: Mechanistic insights into photocatalytic hydrogen evolution. *Chem.–Eur. J.* 2011. 17. 2777. Copyright Wiley-VCH Verlag GmbH & Co. KGaA. Reproduced with permission.)

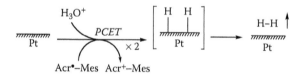

SCHEME 2.3 Proposed mechanism of H_2 evolution on PtNPs. (From Kotani, H. et al.: Size- and shape-dependent activity of metal nanoparticles as hydrogen-evolution catalysts: Mechanistic insights into photocatalytic hydrogen evolution. *Chem.–Eur. J.* 2011. 17. 2777. Copyright Wiley-VCH Verlag GmbH & Co. KGaA. Reproduced with permission.)

2.3.2 SIZE AND SHAPE EFFECTS OF RU NANOPARTICLES

Although the optimization of size and shape of PtNPs resulted in the enhancement of the catalysis in the photocatalytic H_2 evolution, precious Pt is desired to be replaced by metals used in limited applications. Ru is a candidate alternative to Pt, because its high activity for hydrogenation of various olefins [53] implies that H_2 readily interacts with Ru surfaces. RuO_2 nanoparticles have been examined as a H_2 evolution catalyst, resulting in rather modest activity as compared with PtNPs [54–58]. Ru nanoparticles (RuNPs) have been reported to act as an active H_2 evolution catalyst comparable with PtNPs in the photocatalytic H_2 evolution using QuPh+–NA and NADH as a photosensitizer and an electron donor, respectively [9]. The size and support effects of RuNPs on catalytic activity and durability for the photocatalytic H_2 evolution were clarified.

The size of RuNPs used for the photocatalytic H_2 evolution was controlled in the size of 2.0 ± 0.3 nm to 8.0 ± 1.0 nm by choosing appropriate preparation conditions as shown in Figure 2.20 [9]. In order to compare the catalytic activity of the RuNPs with various sizes, the photocatalytic H_2 evolution was performed by photoirradiation ($\lambda > 340$ nm) of a deaerated mixed solution (2.0 mL) of a buffer (pH 4.5) and MeCN [1:1 (v/v)] containing NADH (1.0 mM), QuPh$^+$–NA (0.22 mM), and RuNPs [9]. From the reaction solutions containing RuNPs with the size of 3.3–8.0 nm, nearly stoichiometric amount of H_2 evolution referred to NADH was

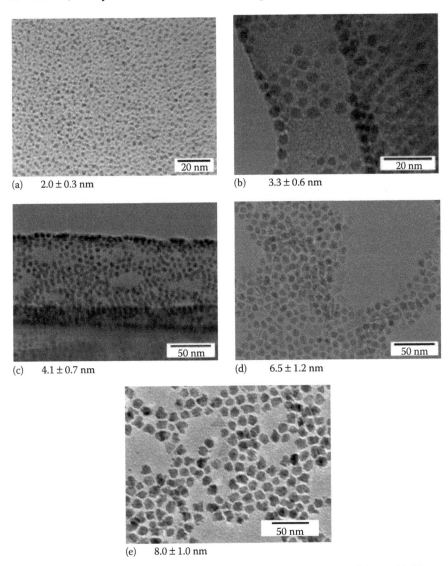

(a) 2.0 ± 0.3 nm   (b) 3.3 ± 0.6 nm

(c) 4.1 ± 0.7 nm   (d) 6.5 ± 1.2 nm

(e) 8.0 ± 1.0 nm

FIGURE 2.20 TEM images of RuNPs with the size of (a) 2.0 nm, (b) 3.3 nm, (c) 4.1 nm, (d) 6.5 nm, and (e) 8.0 nm. (Reprinted with permission from Yamada, Y. et al., *J. Am. Chem. Soc.*, 133, 16136, 2011. Copyright 2011 American Chemical Society.)

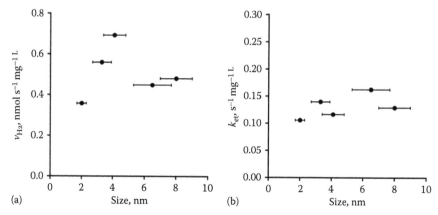

FIGURE 2.21   (a) Plots of H_2 evolution rates depending on the size of RuNPs. H_2 evolution was performed under photoirradiation ($\lambda > 340$ nm) of a mixed solution of a deaerated phthalate buffer (pH 4.5) and MeCN [1:1 (v/v)] containing NADH (1.0 mM), QuPh$^+$–NA (2.2 mM), and RuNPs (12.5 mg L^{-1}) with different sizes. (b) Size dependence of the ET rates (k_{et}) from QuPh$^+$–NA to RuNPs. (Reprinted with permission from Yamada, Y. et al., *J. Am. Chem. Soc.*, 133, 16136, 2011. Copyright 2011 American Chemical Society.)

observed at room temperature [9]. Figure 2.21a shows the plot of H_2 evolution rates (V_{H2}) normalized by the weight concentrations of RuNPs against the size of RuNPs [9]. The maximum H_2 evolution rate was observed for RuNPs with the size of 4.1 ± 0.7 nm [9].

For the reaction system of electrocatalytic H_2 evolution with Ru electrodes, the Volmer–Tafel mechanism has been proposed [59]. The mechanism consists of two steps, that is, "proton reduction" (H$^+$ + e$^-$ → H*; H* is the H adsorbed on the surfaces of the ruthenium electrode) and "hydrogen–atom association" (2H* → H$_2$), where the former and the latter steps are called Volmer and Tafel steps, respectively [59]. In an acidic solution, the Tafel step and following desorption of H_2 are proposed to be the rate-determining step at room temperature [59].

In the photocatalytic H_2 evolution with QuPh$^+$–NA and RuNPs, ET from photoinduced QuPh$^•$–NA to RuNPs is much faster than H_2 evolution as shown in Figure 2.16b [9]. Thus, the rate-determining step of the photocatalytic H_2 evolution follows the ET step, that is, proton reduction (Volmer) or hydrogen–atom association (Tafel) step [9]. Smaller RuNPs have advantage for the proton reduction on the surfaces of RuNPs, because a negative charge of an injected electron is shared by the whole body of RuNPs [9]. Smaller particles have high specific surface area, thus, the surfaces of smaller RuNPs are more negatively charged than those of larger RuNPs. The more negatively charged surfaces of smaller RuNPs readily interact with positively charged protons. On the other hand, larger size may be beneficial for the hydrogen–atom association step, because larger particles can receive more electrons and hydrogen atoms on a single catalytic particle than a smaller particle [9]. Scheme 2.4 summarizes the size effects of RuNPs on the reaction rates for proton reduction and hydrogen–atom association steps. Perhaps the size of 4.1 ± 0.7 nm is balanced well for RuNPs to evolve H_2.

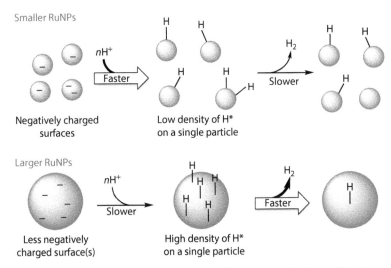

SCHEME 2.4 Effect of RuNPs size on the rates for proton reduction and hydrogen–atom association steps. (Reprinted with permission from Yamada, Y. et al., *J. Am. Chem. Soc.*, 133, 16136, 2011. Copyright 2011 American Chemical Society.)

The main drawback of those MNPs capped with an organic agent for catalytic applications is lack of stability, partly because of agglomerates formation by dissociation of the organic agent during the photocatalytic H$_2$ evolution. The stability of RuNPs capped with polyvinylpyrrolidone has been reported to be less than that of PtNPs capped with polyvinylpyrrolidone in the photocatalytic H$_2$ evolution [60]. The agglomeration of RuNPs can be effectively suppressed by supporting RuNPs on metal oxides [60]. Photocatalytic H$_2$ evolution was examined by using 3 wt% RuNPs supported on metal oxides as H$_2$ evolution catalysts together with NADH and QuPh$^+$–NA as an electron donor and a photosensitizer. As indicated in Figure 2.22, nonstoichiometric amount of H$_2$ evolution referred to NADH was observed from the reaction solution in 30 min using RuNPs supported on MgO (black diamonds), TiO$_2$ (purple reversed triangle), and CeO$_2$ (green triangles) employed as the H$_2$ evolution catalysts, indicating that these metal oxides deactivated RuNPs [60]. On the other hand, the stoichiometric amount of H$_2$ (2.0 μmol) referred to NADH was evolved by photoirradiation of the reaction solution employing Ru/SiO$_2$ and Ru/Al$_2$O$_3$–SiO$_2$ as H$_2$ evolution catalysts [60]. The H$_2$ evolution rate from the reaction system employing Ru/SiO$_2$ is more than double that employing Ru/Al$_2$O$_3$–SiO$_2$; thus, SiO$_2$ is the best support for RuNPs.

The catalytic activity of RuNPs can be improved by employing SiO$_2$ with various morphologies [60]. TEM images of Ru supported on SiO$_2$ with different morphologies are shown in Figure 2.23a–c: The morphology of SiO$_2$ supports is undefined shape (*u*-SiO$_2$), mesoporous structure with hexagonally packed array (*m*-SiO$_2$), and nonporous spherical shape (*s*-SiO$_2$) [60]. Photocatalytic H$_2$ evolution was performed by photoirradiation of the reaction solution employing these Ru/SiO$_2$ catalysts as H$_2$ evolution catalysts in the photocatalytic system. Among these Ru/SiO$_2$ catalysts, the fastest H$_2$ evolution was observed for Ru/*u*-SiO$_2$ (Figure 2.23d, circles) with the H$_2$ evolution rate of 9.1 μmol h^{-1} [60]. Slower H$_2$ evolution was

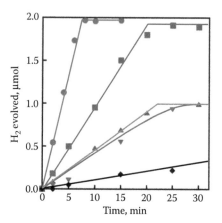

FIGURE 2.22 Time courses of H_2 evolution performed by photoirradiation ($\lambda > 340$ nm) of a deaerated mixed solution (2.0 mL) of a phthalate buffer (pH 4.5) and MeCN [1:1 (v/v)] containing QuPh⁺–NA, NADH, and RuNPs supported on metal oxides (3.0 wt% Ru; 100 mg L⁻¹; Ru/SiO₂, red; Ru/Al₂O₃-SiO₂, blue; Ru/CeO₂, green; Ru/TiO₂, purple; and Ru/MgO, black). (Reprinted with permission from Yamada, Y. et al., *J. Phys. Chem. C*, 117, 13143, 2013. Copyright 2013 American Chemical Society.)

observed for Ru/*s*-SiO₂ (Figure 2.23d, squares) and Ru/*m*-SiO₂ (Figure 2.23d, triangles) with the rates of 6.3 and 4.7 μmol h⁻¹, respectively [60]. The slower H_2 evolution rates for Ru/*m*-SiO₂ and Ru/*s*-SiO₂ originated from the microenvironment of RuNPs surrounded by SiO₂, in which RuNPs are less exposed to the solution compared with Ru/*u*-SiO₂ [60].

The robustness of Ru/*u*-SiO₂ was examined in the repetitive photocatalytic H_2 evolution and compared with that of PtNPs [60]. When the photocatalytic reaction was repeated with Ru/*u*-SiO₂ and PtNPs as H_2 evolution catalysts, H_2 evolution was observed with Ru/*u*-SiO₂ at fourth cycle and PtNPs at fifth cycle as shown in Figure 2.24 [60]. The total amount of evolved H_2 normalized by Ru weight was 1.7 mol g_{-Ru}^{-1}, which is close to the amount with PtNPs (2.0 mol g_{-Pt}^{-1}) [60]. Thus, the durability of RuNPs was improved by optimizing the morphology of an SiO₂ support [60].

2.3.3 SIZE AND CRYSTAL STRUCTURE EFFECTS OF NI NANOPARTICLES

RuNPs act as an efficient H_2 evolution catalyst in the photocatalytic H_2 evolution; however, replacement of Ru with nonprecious and earth-abundant metals is desirable. In natural system, hydrogenases, which are a class of enzyme managing interconversion of H_2 and protons in natural systems, contain Fe and/or Ni in their enzymatic active sites [6,61,62]. In this context, Fe nanoparticles (FeNPs) and Ni nanoparticles (NiNPs) have been examined as H_2 evolution catalysts in the photocatalytic H_2 production [63]. Only little H_2 evolution was observed for FeNPs [6]; however, NiNPs showed certain catalytic activity for the photocatalytic H_2 evolution [8]. This result seems reasonable because a Ru–Ni dinuclear complex has been reported as a functional and structural model of hydrogenases [64].

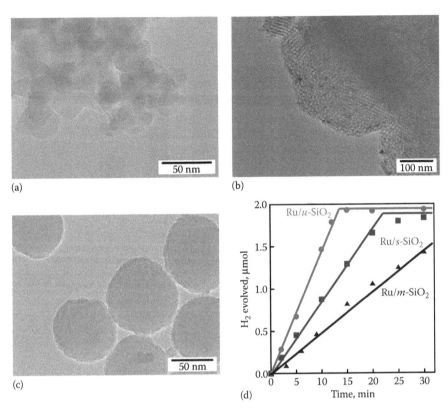

FIGURE 2.23 TEM images of Ru/SiO$_2$ in the shape of (a) undefined, (b) hexagonally packed mesoporous, and (c) nonporous sphere. (d) Time courses of H$_2$ evolution performed by photoirradiation ($\lambda > 340$ nm) of a deaerated mixed solution of a phthalate buffer (pH 4.5) and MeCN [1:1 (v/v)] containing QuPh$^+$–NA, NADH, and RuNPs supported on SiO$_2$ prepared by the CVD method (Ru/SiO$_2$ in the shape of undefined shape, circles; mesoporous SiO$_2$, triangles; and spherical SiO$_2$, squares). (Reprinted with permission from Yamada, Y. et al., *J. Phys. Chem. C*, 117, 13143, 2013. Copyright 2013 American Chemical Society.)

Figure 2.25 compares the time courses of photocatalytic H$_2$ evolution under photoirradiation ($\lambda > 340$ nm) of a mixed solution of phthalate buffer (pH 4.5) and MeCN [1/1 (v/v)] containing NADH, QuPh$^+$–NA, and MNPs (M = Ni, Pt, or Ru) as a sacrificial electron donor, a photocatalyst, and H$_2$ evolution catalysts, respectively [8]. NiNPs provided the evolution of stoichiometric amount of H$_2$ (2.0 μmol) referred to NADH in 8 min with the H$_2$ evolution rate of 11 μmol h^{-1}, suggesting NiNPs act as efficient H$_2$ evolution catalyst. However, the H$_2$ evolution rate of the reaction system using NiNPs is ~40% of those using PtNPs and RuNPs as H$_2$ evolution catalysts [8].

Catalysis of NiNPs employed as a H$_2$ evolution catalyst in the photocatalytic H$_2$ evolution depends on their sizes [8]. The sizes of NiNPs determined by TEM observations and DLS were 6.6 ± 1.6 nm (Figure 2.26a), 11 ± 2 nm (Figure 2.26b), 36 ± 12 (Figure 2.26c), and 210 ± 80 nm (Figure 2.26d) [8]. The crystal structures of these NiNPs were confirmed by powder X-ray diffraction patterns to be hexagonal

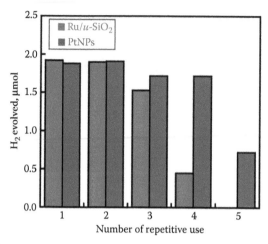

FIGURE 2.24 Amount of H_2 evolved in the repetitive photocatalytic H_2 evolution using Ru/u-SiO$_2$ (red, left) and PtNPs (blue, right). A mixed solution containing NADH was added to the reaction solution after each run. (Reprinted with permission from Yamada, Y. et al., *J. Phys. Chem. C*, 117, 13143, 2013. Copyright 2013 American Chemical Society.)

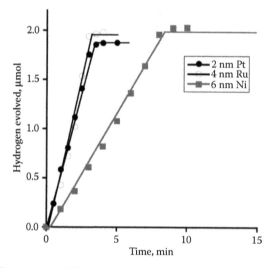

FIGURE 2.25 Time courses of H_2 evolution under photoirradiation ($\lambda > 340$ nm) of a deaerated mixed solution (2 mL) of a phthalate buffer (pH 4.5) and MeCN [1:1 (v/v)] containing QuPh$^+$–NA (0.44 mM), NADH (1.0 mM), and various catalysts [12.5 mg L^{-1}, NiNPs (6 nm, ■), RuNPs (4 nm, ○), and PtNPs (2 nm, ●)] at 298 K. (From Yamada, Y. et al., *Energy Environ. Sci.*, 2012, 5: p. 6111. Copyright 2012 The Royal Society of Chemistry, reproduced with permission.)

close-packed (*hcp*) structure [8]. The photocatalytic H_2 evolution was performed by photoirradiation of a mixed solution containing QuPh$^+$–NA, NADH, *hcp*-NiNPs with different sizes (6.6–210 nm) [8]. As shown in Figure 2.27, the H_2 evolution rate became faster with smaller *hcp*-NiNPs (Figure 2.27a). Other than *hcp* structure, NiNPs is known to have face-centered cubic (*fcc*) structure at room temperature [8].

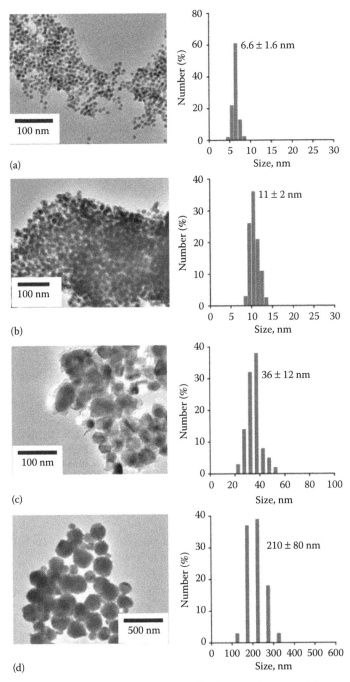

FIGURE 2.26 TEM images and particles size distributions determined by dynamic laser scattering of *hcp*-NiNPs with sizes of (a) 6.6 nm, (b) 11 nm, (c) 36 nm, and (d) 210 nm. (From Yamada, Y. et al., *Energy Environ. Sci.*, 2012, 5: p. 6111. Copyright 2012 The Royal Society of Chemistry, reproduced with permission.)

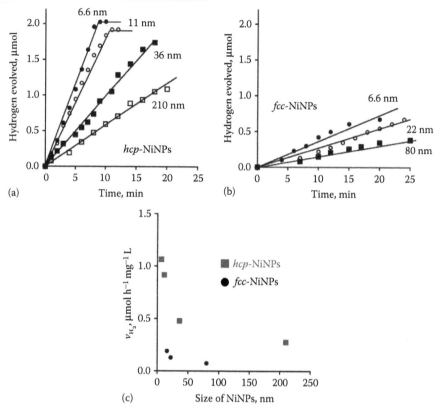

FIGURE 2.27 (a, b) Time courses of H_2 evolution under photoirradiation ($\lambda > 340$ nm) of mixed solutions of deaerated phthalate buffer (pH 4.5) and MeCN [1:1 (v/v)] containing NADH (1.0 mM), QuPh$^+$–NA (0.44 mM), and (a) hcp-NiNPs or (b) fcc-NiNPs with different sizes (12.5 mg L^{-1}) at 298 K. (c) Plots of H_2 evolution rates normalized by weight concentration of NiNPs versus the size of NiNPs. (From Yamada, Y. et al., *Energy Environ. Sci.*, 2012, 5: p. 6111. Copyright 2012 The Royal Society of Chemistry, reproduced with permission.)

When fcc-NiNPs were used as H_2 evolution catalysts instead of hcp-NiNPs in the photocatalytic H_2 evolution, no stoichiometric amount of H_2 evolved by photoirradiation for 20 min as shown in Figure 2.27b [8]. As indicated in Figure 2.27c, hcp-NiNPs show much higher catalytic activity than fcc-NiNPs independent of the size of particles.

The crystal structure-dependent catalysis of NiNPs has been reported for various reaction systems [65–67]. For example, hcp-NiNPs show high product selectivity to H_2 in steam reforming of glycerol compared with that of fcc-NiNPs [66]. Also, high catalytic activity of hcp-Ni compared with fcc-Ni has been reported in propene hydrogenation [67]. The fcc-Ni structure is a thermally stable phase, whereas the hcp-Ni structure is known as metastable phase [65]. The excellent catalytic activity of hcp-NiNPs may be ascribed to the high surface energy of the hcp structure, in which the surfaces readily interact with substrates.

2.4 CONCLUSION

Photocatalytic H_2 evolution systems utilizing electron donor–acceptor linked dyads (Acr$^+$–Mes and QuPh$^+$–NA) as photosensitizers and MNPs (Pt, Ru, Ni) as H_2 evolution catalysts were reviewed. First, intramolecular ET in an electron donor–acceptor linked dyad was clarified by utilizing nanosecond transient absorption spectroscopy. Then, ET from an electron donor (NADH or oxalate) to the photoexcited state of the electron donor–acceptor linked dyad with long lifetime was monitored. Finally, the ET from the reduced photoexcited state to MNPs employed as H_2 evolution catalysts was scrutinized. These observations clarified that MNPs are involved in the rate-determining step. Although Pt is known to be the most active catalyst for proton reduction, reduction of its use or its replacement with earth-abundant metals is necessary. Thus, catalytic activity of MNPs was investigated by employing alternative metals. Further developments in catalysis are still demanded, however, Ni and Ru nanoparticles appear to be promising candidates for the optimization of size, shape, and crystal structure.

REFERENCES

1. J.M. Lehn, J.P. Sauvage, *New J. Chem.* 1 (1977) 449.
2. A. Moradpour, E. Amouyal, P. Keller, H. Kagan, *New J. Chem.* 2 (1978) 547.
3. K. Kalyanasundaram, J. Kiwi, M. Grätzel, *Helv. Chim. Acta* 61 (1978) 2720.
4. M. Grätzel, *Acc. Chem. Res.* 14 (1981) 376.
5. F. Teplý, *Collect. Czech. Chem. Commun.* 76 (2011) 859.
6. H. Kotani, R. Hanazaki, K. Ohkubo, Y. Yamada, S. Fukuzumi, *Chem.–Eur. J.* 17 (2011) 2777.
7. H. Kotani, T. Ono, K. Ohkubo, S. Fukuzumi, *Phys. Chem. Chem. Phys.* 9 (2007) 1487.
8. Y. Yamada, T. Miyahigashi, H. Kotani, K. Ohkubo, S. Fukuzumi, *Energy Environ. Sci.* 5 (2012) 6111.
9. Y. Yamada, T. Miyahigashi, H. Kotani, K. Ohkubo, S. Fukuzumi, *J. Am. Chem. Soc.* 133 (2011) 16136.
10. R.A. Marcus, *Annu. Rev. Phys. Chem.* 15 (1964) 155.
11. R.A. Marcus, N. Sutin, *Biochim. Biophys. Acta* 811 (1985) 265.
12. S. Fukuzumi, K. Ohkubo, T. Suenobu, K. Kato, M. Fujitsuka, O. Ito, *J. Am. Chem. Soc.* 123 (2001) 8459.
13. S. Fukuzumi, H. Kotani, K. Ohkubo, S. Ogo, N.V. Tkachenko, H. Lemmetyinen, *J. Am. Chem. Soc.* 126 (2004) 1600.
14. S.M. Hubig, J.K. Kochi, *J. Am. Chem. Soc.* 122 (2000) 8279.
15. S. Fukuzumi, I. Nakanishi, K. Tanaka, *J. Phys. Chem. A* 103 (1999) 11212.
16. S. Fukuzumi, K. Doi, A. Itoh, T. Suenobu, K. Ohkubo, Y. Yamada, K.D. Karlin, *Proc. Natl. Acad. Sci. U. S. A.* 109 (2012) 15572.
17. H. Kotani, K. Ohkubo, S. Fukuzumi, *Faraday Discuss.* 155 (2012) 89.
18. L. Biczók, H. Linschitz, *J. Phys. Chem. A* 105 (2001) 11051.
19. J.K. Kochi, R. Rathore, P.L. Maguères, *J. Org. Chem.* 65 (2000) 6826.
20. S. Fukuzumi, H. Kotani, K. Ohkubo, *Phys. Chem. Chem. Phys.* 10 (2008) 5159.
21. A.C. Benniston, A. Harriman, P.Y. Li, J.P. Rostron, H.J. van Ramesdonk, M.M. Groeneveld, H. Zhang, J.W. Verhoeven, *J. Am. Chem. Soc.* 127 (2005) 16054.
22. A.C. Benniston, A. Harriman, P.Y. Li, J.P. Rostron, J.W. Verhoeven, *Chem. Commun.* (2005) 2701.

23. X.Q. Zhu, Y. Yang, M. Zhang, J.P. Cheng, *J. Am. Chem. Soc.* 125 (2003) 15298.
24. S. Fukuzumi, T. Tanaka, in: M.A. Fox, M. Chanon (Eds.), *Photoinduced Electron Transfer, Part C*, Elsevier, Amsterdam, The Netherland, 1988.
25. S. Fukuzumi, D.M. Guldi, in: V. Balzani (Ed.), *Electron Transfer in Chemistry*, Wiley-VCH, Weinheim, Germany, 2001, p. 270.
26. S. Fukuzumi, H. Miyao, K. Ohkubo, T. Suenobu, *J. Phys. Chem. A* 109 (2005) 3285.
27. K. Ohkubo, H. Kotani, S. Fukuzumi, *Chem. Commun.* (2005) 4520.
28. Y. Yamada, T. Miyahigashi, K. Ohkubo, S. Fukuzumi, *Phys. Chem. Chem. Phys.* 14 (2012) 10564.
29. Q.G. Mulazzani, M. D'Angelantonio, M. Venturi, M.Z. Hoffman, M.A.J. Rodgers, *J. Phys. Chem.* 90 (1986) 5347.
30. I. Rubinstein, A.J. Bard, *J. Am. Chem. Soc.* 103 (1981) 512.
31. F. Kanoufi, A.J. Bard, *J. Phys. Chem. B* 103 (1999) 10469.
32. A.I. Krasna, *Photochem. Photobiol.* 31 (1980) 75.
33. F. Pina, Q.G. Mulazzani, M. Venturi, M. Ciano, V. Balzani, *Inorg. Chem.* 24 (1985) 848.
34. H. Kotani, K. Ohkubo, Y. Takai, S. Fukuzumi, *J. Phys. Chem. B* 110 (2006) 24047.
35. J.R. Darwent, P. Douglas, A. Harriman, G. Porter, M.C. Richoux, *Coord. Chem. Rev.* 44 (1982) 83.
36. S. Fukuzumi, Y. Yamada, T. Suenobu, K. Ohkubo, H. Kotani, *Energy Environ. Sci.* 4 (2011) 2754.
37. S. Fukuzumi, Y. Yamada, *J. Mater. Chem.* 22 (2012) 24284.
38. J.R. Anderson, *Structure of Metallic Catalysts*, Academic Press, London, UK, 1975.
39. Y. Sun, Y. Xia, *Science* 298 (2002) 2176.
40. N. Tian, Z.Y. Zhou, S.G. Sun, Y. Ding, Z.L. Wang, *Science* 316 (2007) 732.
41. K. An, N. Musselwhite, G. Kennedy, V.V. Pushkarev, L.R. Baker, G.A. Somorjai, *J. Colloid Interface Sci.* 392 (2013) 122.
42. Y. Li, Q. Liu, W. Shen, *Dalton Trans.* 40 (2011) 5811.
43. F. Zaera, *Catal. Lett.* 142 (2012) 501.
44. Y. Kang, J.B. Pyo, X. Ye, R.E. Diaz, T.R. Gordon, E.A. Stach, C.B. Murray, *ACS Nano* 7 (2013) 645.
45. M. Wolfsberg, *Acc. Chem. Res.* 5 (1972) 225.
46. M.J. Tanner, M. Brookhart, J.M. DeSimone, *J. Am. Chem. Soc.* 119 (1997) 7617.
47. E.A. Gelder, S.D. Jackson, C.M. Lok, *Chem. Commun.* (2005) 522.
48. O.W. Howarth, C.H. McAteer, P. Moore, G.E. Morris, *J. Chem. Soc., Chem. Commun.* (1982) 745.
49. V.C. Srivastava, P. Raghunathan, S. Gupta, *Int. J. Hydrogen Energy* 17 (1992) 551.
50. M.G. Basallote, S. Bernal, J.M. Gatica, M. Pozo, *Appl. Catal., A* 232 (2002) 39.
51. N. Ozawa, T.A. Roman, H. Nakanishi, H. Kasai, *Surf. Sci.* 600 (2006) 3550.
52. M. Conte, K. Wilson, V. Chechik, *Org. Biomol. Chem.* 7 (2009) 1361.
53. P. Lara, K. Philippot, B. Chaudret, *ChemCatChem* 5 (2013) 28.
54. J.M. Kleijn, G.K. Boschloo, *J. Electroanal. Chem.* 300 (1991) 595.
55. J.M. Kleijn, J. Lyklema, *Colloid Polym. Sci.* 265 (1987) 1105.
56. M. Kleijn, H.P. van Leeuwen, *J. Electroanal. Chem.* 247 (1988) 253.
57. E. Amouyal, P. Keller, A. Moradpour, *J. Chem. Soc., Chem. Commun.* (1980) 1019.
58. Y. Yamada, K. Yano, S. Fukuzumi, *Aust. J. Chem.* 65 (2012) 1573.
59. M.W. Breiter, *J. Electroanal. Chem.* 178 (1984) 53.
60. Y. Yamada, S. Shikano, S. Fukuzumi, *J. Phys. Chem. C* 117 (2013) 13143.
61. M.Y. Darensbourg, E.J. Lyon, J.J. Smee, *Coord. Chem. Rev.* 206 (2000) 533.
62. G.J. Kubas, *Chem. Rev.* 107 (2007) 4152.
63. J. Handman, A. Harriman, G. Porter, *Nature* 307 (1984) 534.
64. S. Ogo, R. Kabe, K. Uehara, B. Kure, T. Nishimura, S.C. Menon, R. Harada et al., *Science* 316 (2007) 585.

65. H.S. Bengaard, J.K. Nørskov, J. Sehested, B.S. Clausen, L.P. Nielsen, A.M. Molenbroek, J.R. Rostrup-Nielsen, *J. Catal.* 209 (2002) 365.
66. Y. Guo, M.U. Azmat, X. Liu, J. Ren, Y. Wang, G. Lu, *J. Mater. Sci.* 46 (2011) 4606.
67. G. Carturan, S. Enzo, R. Ganzeria, M. Lenrda, R. Zanoni, *J. Chem. Soc. Faraday Trans.* 86 (1990) 739.

3 CO$_2$ to Fuels

Atsushi Urakawa and Jacinto Sá

CONTENTS

3.1 Introduction 93
3.2 CO$_2$ Hydrogenation 94
    3.2.1 Carbon Monoxide 96
    3.2.2 Methane 98
    3.2.3 Higher Hydrocarbons 99
    3.2.4 Methanol and Dimethyl Ether 100
    3.2.5 Higher Alcohols 103
    3.2.6 Formic Acid 103
    3.2.7 CO$_2$ Hydrogenation: Perspective and Challenges 104
3.3 Photocatalytic Conversion of CO$_2$ to Fuel 105
    3.3.1 Photocatalytic Systems Based on TiO$_2$ 105
        3.3.1.1 Pristine TiO$_2$ Photocatalysis 106
        3.3.1.2 Metal-Modified TiO$_2$ Photocatalysis 110
    3.3.2 Alternative Solid Semiconductors 112
    3.3.3 Phosphides and Sulfides 115
    3.3.4 Photoreduction of CO$_2$: Perspective and Challenges 116
References 119

This chapter compiles the most recent technologies aimed at converting one of the most abundant and harmful greenhouse gases to fuels. The technologies range from classical CO$_2$ hydrogenation to photocatalytic conversion. Each section contains a historical summary of the processes, recent developments, and a future perspective.

3.1 INTRODUCTION

Hydrocarbon fuels are currently the most important source of energy due to their availability and high energy density [1]. However, their usage has led to a dramatic increase of atmospheric greenhouse gas concentration, in particular CO$_2$ [2], to values considered menacing to human life, urging the development of strategies to remedy the problem. One of the strategies to lower CO$_2$ emissions is to capture and store it in a geological formation; however, the process is energetically demanding and it is not leakage free. Potential risks associated with CO$_2$ leakage should be alerted because a rapid release of CO$_2$ can result in serious accidents as witnessed by the death of 1700 people, by asphyxiation, in Lake Nyos, Cameroon, in 1986 [3].

Conversion of thermodynamically stable CO_2 into useful chemicals and fuel requires energy inputs. This can be provided in the form of light, electricity, and chemical energy, and the reactions are often facilitated by the function of enabling catalysts [4]. For sustainable development of mankind and natural environment, our dependency on fossil fuels should be reduced and utilization of renewable energies is mandatory for the activation of CO_2 or to produce energetic molecules for CO_2 conversion. There are some approaches attractive from a practical point of view. Renewable energy sources such as wind, geothermal, solar, and hydropower are converted to electric energy as the most convenient form for distribution, sharing, and exchange. This versatility of electric energy renders the electrocatalytic conversion promising for CO_2 conversion [5]. A very popular electrocatalytic process is the production of H_2 by water electrolysis by feeding only water and electricity into the system. The technology is well established, being further developed, and available on a large industrial scale. The H_2 thus generated can be directly used in fuel cells, but it also offers a means to produce valuable chemicals and fuels by CO_2 hydrogenation. Fundamental chemicals like alcohols and hydrocarbon including olefins can be produced by or via CO_2 hydrogenation, giving a path for CO_2 to be the building block of commodity chemicals and fuel. Considering the enormous amount of CO_2 emitted worldwide, approximately 34.5 billion tons CO_2 per year in 2012 [6], and increasing capacity for CO_2 capture for further utilization at emitting sites, technologies capable of converting large quantities of CO_2 should be exploited. Compared to other CO_2 conversion technologies, the high reaction rates, and thus the possibility to convert a large quantity of CO_2 within a shorter time and a smaller space, make the technologies practically attractive. Catalytic CO_2 hydrogenation using heterogeneous catalysts will be discussed in Section 3.2 because of the practical importance of the processes for large-scale, continuous operation.

Sunlight, the most abundant renewable energy source on earth, can be directly used in CO_2 conversion, mimicking the nature's process by converting CO_2 and water into chemicals with the aid of photocatalytic reactions. Utilization of photons for CO_2 activation, often accompanied by photocatalytic water-splitting reaction, is an ultimate solution for CO_2 mitigation and from the carbon footprints viewpoints because little CO_2 release is associated with the operation of such processes. The current state of the art and perspectives for future developments will be discussed in detail in Section 3.3.

3.2 CO_2 HYDROGENATION

Continuing efforts in the development and installation of CO_2 capture and storage technologies at emission sites are enabling us to readily access abundant CO_2 at a low or possibly zero cost. For the transport and space requirements, captured CO_2 is typically stored in the compressed liquid form. This fact can be advantageous from chemical reaction engineering aspects, because CO_2 is available at the vapor pressure (5.73 MPa at 293 K) and ease further compression work if necessary, facilitating reactions requiring high pressures. The reactions of CO_2 with the lightest energetic diatomic molecule, H_2, offer various pathways to produce hydrocarbon fuels and also fundamental chemicals that can be used as future energy vector. Scheme 3.1

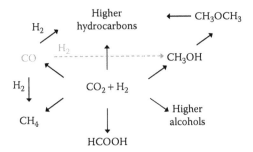

SCHEME 3.1 Hydrogenation of CO$_2$ to chemicals that can be used as energy vectors.

illustrates the major products that are and will be potentially demanded in a very large quantity in the next decades.

Among them, methane [synthetic natural gas (SNG)] and hydrocarbons used for transport fuels like gasoline and diesel oil can be synthesized by methanation and Fischer–Tropsch (FT) synthesis, respectively, occurring via CO or directly from CO$_2$ hydrogenation. These products are readily useful at present due to their compatibility with the existing infrastructure for transport and distribution. Also, methanol, dimethyl ether (DME), and some higher alcohols like ethanol can be stored in the liquid form and compatible with the existing transport and storage infrastructure. These chemicals can be used directly as fuel in combustion engines in either pure or blended form and directly in fuel cells [7]. Importantly, they can serve as a building block of fundamental chemicals like olefins and gasoline by methanol-to-olefin (MTO) and methanol-to-gasoline (MTG) processes [7]. Formic acid is the product produced from one CO$_2$ and one H$_2$ molecule. Its utilization as a chemical building block and fuel is more limited compared to other products shown in Scheme 3.1. However, formic acid can be considered as one of the most promising transportable media for hydrogen storage due to its high hydrogen density in the molecule [7].

Although it is out of the scope of this chapter, CO$_2$ is also used in dry reforming of methane, in which methane and CO$_2$ are used to produce syngas, according to Equation 3.1,

$$CH_4 + CO_2 \rightarrow 2CO + 2H_2 \qquad \Delta H_{298\,K} = 247.5 \text{ kJ mol}^{-1} \qquad (3.1)$$

instead of steam reforming of methane to produce syngas.

$$CH_4 + H_2O \rightarrow CO + 3H_2 \qquad \Delta H_{298\,K} = 205.9 \text{ kJ mol}^{-1} \qquad (3.2)$$

Recent work by the Nobel laureate, George A. Olah, and coworkers combines the two reactions to produce syngas with specific H$_2$ to CO ratio of 2.

$$3CH_4 + 2H_2O + CO_2 \rightarrow 4CO + 8H_2 \qquad \Delta H_{298\,K} = 659.3 \text{ kcal mol}^{-1} \qquad (3.3)$$

This process has been named bi-reforming, and the syngas with the specific composition is called metgas because of the stoichiometric ratio of CO hydrogenation to methanol [8].

$$CO + 2H_2 \rightarrow CH_3OH \qquad \Delta H_{298\,K} = -90.7 \text{ kJ mol}^{-1} \qquad (3.4)$$

Therefore, in the whole process, a mixture of methane, water, and carbon dioxide in the ratio of 3:2:1 produces 4 equivalents of methanol. The hydrogen required for methanol synthesis is taken from CH_4 and H_2O by reforming reactions, which require a large amount of thermal energy input. However, this process converts CO_2 using methane which is recognized to be abundant thanks to the recent discovery of shale gas, tight gas, coal bed methane, and even methane hydrates, besides the possible future availability of methane produced by CO_2 methanation. Therefore despite the dependency on the fossil fuel, it is economically and also energetically viable by using very inexpensive and abundant reactant molecules and by combining the endothermic reforming reactions Equations 3.1 and 3.2 with the exothermic methanol synthesis Equation 3.4, offering an intermediate and bridging solution to greener technologies based on renewable energies. These strategies are not the subject of this chapter, although they are definitely important as a feasible technology until more sustainable, less CO_2 emitting become available.

It is also important to evaluate the source of hydrogen for CO_2 hydrogenation because the conventional H_2 production by the steam reforming reaction Equation 3.2 should not be used due to its high-energy requirements and thus CO_2 emission associated with the plant operation. As a promising and greener alternative, hydrogen should be produced by water electrolysis with the electricity being produced from renewable sources such as wind and sunlight [9] or directly from sunlight using photocatalytic water splitting reaction [10]. These paths offer means to accomplish the challenging activation of energetically stable CO_2 using hydrogen derived from renewable-energy sources.

In the following sections, brief summaries of the reactions in Scheme 3.1 with selected state-of-the-art examples are presented with a strong focus on the syntheses and usage as chemical energy vector, that is, production in a large quantity, followed by a perspective view on CO_2 hydrogenation in the near future.

3.2.1 CARBON MONOXIDE

Carbon monoxide is an important molecule in industry and serves as the carbon source and C1 building block for the production of hydrocarbon fuels and lubricants by FT synthesis and of oxygenates by methanol synthesis and hydroformylation reactions. CO can be produced by hydrogenation of CO_2 by so-called reverse water–gas shift (RWGS) reaction.

$$CO_2 + H_2 \rightarrow CO + H_2O \qquad \Delta H = 41.2 \text{ kJ mol}^{-1} \qquad (3.5)$$

The reaction is endothermic and high temperature is highly beneficial for the reaction, while the reaction pressure is known to have minor impacts on the catalytic performance as shown by a high-pressure study up to 950 bar [11]. Typically, the reaction is limited by thermodynamics and it is of critical importance to be aware of the conversion limit set by the equilibrium. The limitations of this equilibrium are of critical importance for this reaction because the catalysts active for RWGS reaction are also active in

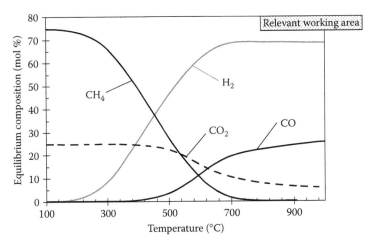

FIGURE 3.1 Thermodynamic equilibrium composition of the product gas of RWGS reaction at 0.1 MPa for a molar H_2/CO_2 inlet ratio of 3. (Reproduced from Jess, A. et al. *Chem. Ing. Tech.*, 85, 489, 2013. With permission.)

the backward reaction of Equation 3.5, namely water–gas shift (WGS) reaction. RWGS reaction Equation 3.5 often competes with methanation reaction (Section 3.2.2), and the latter is preferred at lower temperatures as depicted in Figure 3.1 [12]. According to the figure, very high temperatures are required to achieve good CO_2 conversion and high selectivity to CO. However, this does not take into account the reaction pathways and also kinetic factors; higher selectivity to CO can be attained because further hydrogenation of CO to methane is limited by kinetics [11] and also by addition of promoters it is possible to suppress methanation reaction.

Possible strategies to unlimited equilibrium conversion is to combine RWGS reaction with secondary reaction(s) producing additional product(s), even inducing phase separation, so that the equilibrium is drastically shifted to the product side. This approach is practical especially when a target product instead of CO is directly obtained, because CO is always an intermediate and never the desired final product. One such example is methanol synthesis via RWGS reaction and this will be discussed in Section 3.2.4.

There are a wide variety of catalysts active in both WGS and RWGS reactions. Recent reports mainly deal with low-temperature shift reactions below 600 K [13,14] and different catalysts are known to impose distinct pathways for RWGS reaction.

Copper-based catalysts, whose compositions are often similar to those of methanol synthesis catalysts, are widely used for RWGS reaction. A clear correlation has been found between Cu metal surface area and RWGS activity (Figure 3.2) [15]. An identical linear correlation has been reported for methanol synthesis [16], indicating possible decisive role of RWGS in methanol synthesis by CO_2 hydrogenation. Also, potassium promoters are known to boost the RWGS performance [17] and lower pressure is favorable for RWGS activity [18]. The high effectiveness of potassium promotion was explained by the high dispersion through the formation of nanocrystallites or even thin layers covering both support materials

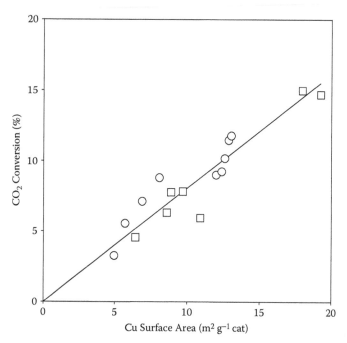

FIGURE 3.2 Activity of Cu/ZnO (circle) and Cu/ZnO/Al$_2$O$_3$ (square) catalysts in RWGS reaction as a function of Cu metal surface area. (Reproduced from Stone, F. S. and Waller, D. *Top. Catal.*, 22, 305, 2003 With permission.)

as well as Cu surface as evidenced by the absence of formate species adsorbed on Cu surface [18].

Another popular family of WGS/RWGS catalysts is precious metals (e.g., Pt, Au) supported on reducible oxide supports (e.g., TiO$_2$, CeO$_2$). Active role of support in the reaction has been proven by spectroscopic and temporal analysis of products (TAP) studies [19]. The major advantages of these materials are non–air sensitivity, no need for activation, and a wider operating temperature window compared to Cu-based catalysts [20]. One issue of these catalysts is the deactivation caused by the carbonaceous deposits on the catalyst surface [21], although CeO$_2$ modified with economically attractive Ni showed excellent performance with a long-term stability in RWGS reaction [22].

3.2.2 METHANE

Methane is the major constituent of natural gas and widely used for power generation and residential heating. In industry, catalytic methanation reaction is used to reduce CO and CO$_2$ concentrations in ammonia synthesis. When it comes to its use as fuel produced by hydrogenation of CO$_2$, existing gas grid for its distribution is probably the major driving force to produce methane in the light of counteracting low cost of methane in comparison to that of hydrogen.

Catalytic hydrogenation of CO_2 to CH_4 is also known as Sabatier reaction and is an exothermic reaction, requiring four H_2 molecules to convert one CO_2 molecule.

$$CO_2 + 4H_2 \rightarrow CH_4 + 2H_2O \qquad \Delta H_{298K} = -252.9 \, \text{kJ mol}^{-1} \qquad (3.6)$$

Although the reaction is thermodynamically favored, the reaction requires catalysts promoting the demanding four hydrogenation steps at the carbon atom to overcome kinetic limitations. This is important because conversion and selectivity would not be sufficient to evaluate the process performance and a value representing yield such as weight time yield (gram of methane produced per gram of catalyst per hour) should be compared.

Active catalysts typically contain Ni or precious metals such as Ru, Rh, and Pt [14, 23]. These metals are supported on metal oxides including mesoporous materials, such as SBA-15, to improve mass transfer characteristics [24]. Commercial catalysts often contain nickel, which is supported on alumina, and are operative at relatively low temperatures, approximately at or below 470 K. Recent works demonstrated that Rh/γ-Al₂O₃ could catalyze the reaction even at room temperature and atmospheric pressure, although deactivation by oxidation of Rh surface by CO_2 was observed [25]. Ru/TiO₂ catalysts exhibit excellent performance and stability when the catalyst was prepared in a special manner by a barrel sputtering method to increase the dispersion and 100% methane yield at approximately 433 K was reported [26]. Thermodynamically, lower temperature is preferred to achieve higher yield of methane; therefore, further catalyst optimization and design for highly active catalyst at low temperatures will be the main stream of R&D in CO_2 methanation.

Compared to RWGS reaction, proposed reaction mechanisms involve active or exclusive participation of metal components. Suggested reaction mechanisms of CO_2 methanation involve the presence of surface intermediates:

1. Dissociation of CO_2 into CO and oxygen as the first step and follow CO methanation path [21]
2. Reaction with support surface to produce (bi)carbonates which lead to CO adsorbed on metal surface prior to hydrogenation [27]
3. Formation of active surface carbon, which is subsequently reduced to form C–H bonds [28]

There are surface formates, (bi)carbonates, and also carbon species detected during the reaction, but so far there is no clear proof about how the carbon atom is hydrogenated and if adsorbed hydrogen or gaseous hydrogen is reacting to form C–H bonds.

3.2.3 HIGHER HYDROCARBONS

The global demands for gasoline and diesel oils are expected to remain high. Fischer–Tropsch (FT) synthesis is one of the key conversion technologies to produce liquid hydrocarbon fuels from syngas, which can be conveniently produced from steam reforming of natural gas Equation 3.2 or coal gasification. Syngas can be converted further to straight-chain hydrocarbons over a catalyst [12].

$$CO + 2H_2 \rightarrow (-CH_2-) + H_2O \qquad \Delta H_{298\,K} = -152 \text{ kJ mol}^{-1} \qquad (3.7)$$

FT synthesis is commonly performed at around 400–600 K and at elevated pressures that are favorable to enhance the reaction rate and also increase the chain length. Higher temperature is advantageous to increase the reaction rate but results in more prominent undesired methane formation. Therefore, lower temperature is more favored to increase the chain length, although a very long chain (wax) needs to be cracked to make it usable. Obviously, optimization of reaction conditions and catalyst plays decisive roles in the final product distribution and productivity. Co-based FT catalysts, typically containing either Fe or Co supported on alumina, are widely studied for analogous CO_2-based processes. Current trends and state of the art of the catalyst and process developments have been summarized in recent reviews [12,14,29].

The advantages of Fe-based catalysts in traditional FT are lower costs compared to Co-based ones, a wider window of operation temperature, and good activity at low H_2/CO ratio [12]. The major drawback commonly reported for Fe-based catalysts is its WGS activity, producing a large amount of CO_2 in the process [12,29]. However, this WGS activity turns into advantageous RWGS activity when the CO_2-based syngas is used [29]. Indeed, most of the beneficial promoters used for Fe catalysts, such as K [30,31], Cu [32], and CeO_2 [33], show good activities in RWGS reaction and minimize methanation reaction. Mn is also reported as a promoter for Fe catalysts working as modifier of structural and electronic properties of Fe to suppress methanation and increase alkene to alkane ratio [31]. Substitution of CO by CO_2 in syngas has minor influence on the product distribution [34]. The reaction mechanism sets the limit given by the equilibrium of RWGS reaction and innovative strategies like the use of membranes for water removal [35] are important for further development.

On the other hand, cobalt catalysts perform excellently in CO-based FT synthesis because of its low WGS activity, minimizing the formation of undesired CO_2. Promoters such as Pt, Ru, or Re are added to enhance the reducibility of catalyst and to maintain cobalt in the metallic state. However, traditional Co catalysts preferentially catalyze undesired methanation when CO_2-based syngas is used [36] and the product distribution does not follow the Anderson–Schulz–Flory product distribution probability, which is obeyed when CO-based syngas is used [29]. These observations suggest the impact of CO_2 on alternation of the reaction pathways. There is so far no clear strategy to improve Co-based catalysts for higher hydrocarbon syntheses by CO_2 hydrogenation.

3.2.4 Methanol and Dimethyl Ether

Methanol is one of the most celebrated and widely suggested alternatives for chemical energy carrier produced by CO_2 hydrogenation. Methanol is a very flexible molecule; it is an excellent fuel and a key starting material for important industrial reactions such as the syntheses of formaldehyde and acetic acid among many others. Methanol can be used as fuel by blending with gasoline or directly in fuel cell. DME, a potential substitute of diesel oil for its better combustion performance, can be synthesized from methanol via dehydration reaction. In addition, fundamental chemical products like olefins can be produced by a methanol-to-olefin (MTO) process. This versatility and increasing pivotal roles of methanol, and the derived DME, in the modern world have led George A. Olah to propose a shift from oil and gas economy toward "Methanol Economy" [37]. Indeed, there exist already a few commercial processes [e.g., Carbon Recycling International (CRI) in Iceland and Blue Energy Fuels in Canada] using captured CO$_2$ and H$_2$ generated by electrolysis, which is powered by renewable sources. The history, state of the art, and challenges related to the catalysts and processes used in methanol synthesis are discussed in detail in Chapter 6. Here, very recent progresses going beyond the state of the art are briefly discussed.

In methanol synthesis by CO$_2$ hydrogenation, generally three competing reactions are reported. The first is the direct methanol synthesis from CO$_2$.

$$CO_2 + 3H_2 \rightarrow CH_3OH + H_2O \qquad \Delta H_{298\,K,\,5\,MPa} = -40.9 \text{ kJ mol}^{-1} \qquad (3.8)$$

The second one is the hydrogenation of CO to methanol.

$$CO + 2H_2 \rightarrow CH_3OH \qquad \Delta H_{298\,K,\,5\,MPa} = -90.7 \text{ kJ mol}^{-1} \qquad (3.9)$$

Both methanol synthesis reactions are exothermic. The third one is the endothermic RWGS reaction Equation 3.5. Both kinetic and thermodynamic aspects have to be examined to fully understand the catalytic performance. The former is largely influenced by the choice of the catalyst as well as residence time of reactants in reactors, while the latter by reaction conditions such as temperature and pressure. Typically per-pass CO$_2$ conversion through a fixed-bed reactor is in the range of 20%–40%, while a rather high value of 35%–45% has been reported for a pilot plant scale operation [38]. It is also important to note that in most reports significant formation of CO is observed by RWGS reaction. These low CO$_2$ conversion and low methanol selectivity can be solved in practice by recycling the unreacted CO$_2$, H$_2$, and CO to the feed stream, thus achieving nearly full overall CO$_2$ conversion and high methanol selectivity. It is, however, desirable to omit the recycling process by achieving high CO$_2$ conversion at a high reactant space velocity and high methanol productivity. As obvious from Equations 3.8 and 3.9 and Le Châtelier's principle, lower temperature and higher pressure are the favorable conditions for methanol synthesis, while higher temperature is the favorable condition for RWGS reaction. This thermodynamics-based prediction, however, neglects important contributions from kinetic factors, that is, how fast the reactions take place. At low temperatures, the reaction does not proceed sufficiently fast, leading to low CO$_2$ conversion and/or high CO selectivity when the methanol formation occurs via RWGS reaction.

High-pressure processes have been recently reported to be highly advantageous to reach nearly full carbon oxides conversion. Heeres and coworkers have elegantly demonstrated using a batch reactor (see Figure 3.3) the advantage of phase separation of the liquid products (methanol and water) from the reactants to overcome the one-phase thermodynamic limitation [39]. The range where the phase separation takes place was predicted by detailed thermodynamic calculations including dew point and liquid phase formation [40]. The phase separation could be beneficially exploited to overcome the one-phase equilibrium conversion to methanol for both CO- and

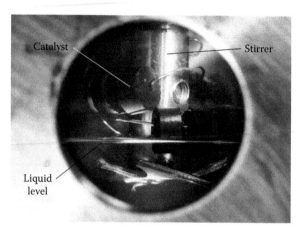

FIGURE 3.3 Visual inspection of liquid product formation during methanol synthesis from syngas ($H_2/CO/CO_2$ = 0.70/0.28/0.02) over a commercial $Cu/ZnO/Al_2O_3$ catalyst. P = 20 MPa, T = 473 K. (Reproduced from van Bennekom, J. G. et al., *Chem. Eng. Sci.*, 87, 204, 2013. With permission.)

CO_2-rich syngas, achieving carbon oxides conversion of 99.5% at 468 K and 92.5% at 484 K, respectively [39]. For the CO_2-rich syngas case, 46.9% higher conversion of carbon oxides could be attained by the phase separation compared to the reaction at more conventional pressure of 7.5 MPa at the same temperature.

Another study demonstrated the great kinetic as well as thermodynamic advantages unique under high-pressure conditions [41]. Figure 3.4a shows the remarkable kinetic advantage of the reaction with high H_2 partial pressure. The study consistently indicated that methanol is produced via RWGS reaction and the subsequent CO hydrogenation step is accelerated and reaches equilibrium conversion at CO_2:H_2 ratio of 1:≥10 at 260°C with almost full methanol selectivity. The reaction was too slow at the lower reaction temperatures, thus resulting in lower CO_2 conversion as

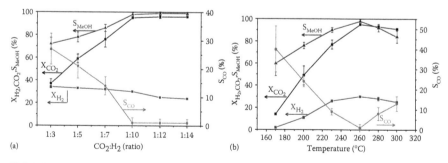

FIGURE 3.4 (a) Effects of CO_2:H_2 ratio at 260°C and (b) reaction temperature at CO_2:H_2 = 1:10 on CO_2 conversion (X_{CO_2}), H_2 conversion (X_{H_2}), and selectivity to methanol (S_{MeOH}) and CO (S_{CO}) over a $Cu/ZnO/Al_2O_3$ catalyst. Reaction conditions: 360 bar (including 10% of Ar with respect to 90% of H_2 as the internal standard for GC analysis), GHSV = 10,471 h^{-1}, CO_2:H_2 = 1:10. (Reproduced from Bansode, A. and Urakawa, A. *J. Catal.*, 309, 66, 2014. With permission.)

well as lower methanol selectivity as shown in Figure 3.4b. Furthermore, above the optimum temperature, both CO_2 conversion and methanol selectivity dropped and this observation was in accordance with thermodynamic calculation [41]. These two studies clearly present that playing with thermodynamics, for example, by high-pressure approach, opens up new opportunities to design efficient processes for CO_2 hydrogenation to methanol.

DME can be synthesized by dehydration reaction of methanol in a continuous operation using solid acids like γ-Al_2O_3 and zeolites. The reaction can be performed as a post-processing of methanol synthesis, but it can be combined with methanol synthesis to directly yield DME. This combined approach generally utilizes a mixture of methanol synthesis catalyst (e.g., $Cu/ZnO/Al_2O_3$) and dehydration catalyst (e.g., H-ZSM-5) [14]. Two important properties should be highlighted for the dehydrating catalyst in the combined approach. The first point is the stability. Serious deactivation can be caused by the presence of water and this was prominently observed for γ-Al_2O_3 [42]. This is detrimental for CO_2 hydrogenation to methanol because it produces H_2O as by-product. Probably, the most common and effective catalyst for this reaction is H-ZSM-5, because it does not deactivate and even promotion effects by water have been reported [42]. The second point is the vicinity of active centers for methanol synthesis and dehydration reaction. It has been reported [43] and also noticed by the author's experience that mixing (grinding) the two catalysts prior to pelletizing to improve the proximity of the two active sites surprisingly lowers the catalytic performance. Rather a physical mixture of two catalysts pelletized separately yields better or even outstanding results when combined with highly performing methanol synthesis process [41]. This may be caused by the water formation in the dehydration reaction. This water may induce WGS reaction to inhibit the methanol synthesis, although further mechanistic investigation and development of efficient catalysts and processes are necessary.

3.2.5 HIGHER ALCOHOLS

Higher alcohols (>C2), particularly ethanol, are highly demanded globally. They can be used in a variety of applications including fuel. The major challenge of CO_2 hydrogenation to these products is the selectivity, that is, how to stop the C–C bond formation at a desired chain number. Unlike the products of FT reaction, alcohol products generally require high purity and selective catalysts are of pivotal importance [44]. When they are produced from CO_2 by hydrogenation reaction, the most probable first step is RWGS and subsequent C–C bond formation for which FT type of catalysts can be employed. Catalysts and equilibrium limitations of ethanol synthesis are discussed in detail in Chapter 6.

3.2.6 FORMIC ACID

Formic acid is one of fundamental commodity chemicals used widely as preservative and antibacterial agent. Besides its chemical use, formic acid has gained considerable attention as a storage medium of H_2, which can be used as a fuel [7]. Formic acid itself can also be injected as fuel in direct formic acid fuel cells

(DFAFCs), generating protons at anode [45]. The synthesis of formic acid by CO_2 hydrogenation has been challenging and a breakthrough was reported by Noyori and coworkers using Ru-based homogeneous catalysts and supercritical CO_2 as solvent and reactant with very high turnover numbers [46].

$$CO_2 + H_2 \rightarrow HCOOH \qquad \Delta H_{298K} = -31.0 \, kJ \, mol^{-1} \qquad (3.10)$$

The reaction is not thermodynamically favored in apolar media, and high pressure conditions are necessary to enhance reaction rate, especially by increasing H_2 partial pressure because of the H_2 insertion step to the Ru–formate complex is the rate-limiting step [47]. However, the major problem is in the backward reaction; good catalysts are also active in the decomposition of formic acid into CO_2 and H_2. Therefore, stabilization of the formic acid product in aqueous media [48], by addition of base molecules, such as amines, or further transformation with alcohol by esterification reactions is generally reported.

Another big challenge of formic acid synthesis by CO_2 hydrogenation is to use heterogeneous catalysts. There are known conventional catalysts, for example, metals supported on metal oxides, active in formic acid decomposition. Supposedly they may show some activity in the forward reaction. However, the difficulty is the thermodynamics; formic acid is likely decomposed before product collection. Some attempts are made to immobilize active homogeneous Ru catalysts onto metal oxide support material. The major problems are the low stability of the immobilized complexes and leaching of the complex into the liquid phase [49], rendering the approach difficult to use in continuous processes, at present. Further transformation of formic acid into other chemicals makes its use difficult in fuel applications. A completely new strategy for large-scale formic acid synthesis is still longed for.

3.2.7 CO_2 HYDROGENATION: PERSPECTIVE AND CHALLENGES

The currently available catalytic processes technologically permit production of fuel by CO_2 hydrogenation. Advanced options are the syntheses of methanol, DME, methane, as well as higher hydrocarbons by FT reaction, while technologies for continuous large-scale production of formic acid are lacking at present. Flexibility and versatility of methanol in its use as fuel and C1 building block render methanol particularly attractive for large-scale production among the chemicals produced by CO_2 hydrogenation. There is still much scope for further technological innovation and advancement in CO_2 hydrogenation. As described earlier, several reactions are limited by both kinetics and thermodynamics, whose limitations are generally tackled by efficient catalyst design and optimization of reaction conditions and reactor types. High-pressure approach was found to go beyond the widely reported productivity in methanol synthesis. Also, the fundamental reaction occurring in most of CO_2 hydrogenation reactions and RWGS reactions may be drastically benefited from a use of membrane reactors or unsteady-state operations because of the serious thermodynamic limitations. A variety of catalysts have been employed in CO_2 hydrogenation and achieving and retaining high active site dispersion, for example, using mesostructured materials, are the key factors to enhance kinetic advantages allowing lower temperature and pressure operations. Furthermore, it is important to warn about the energetic requirements of CO_2

conversion processes because these processes should not emit more CO_2 than its consumption. Life cycle assessments avoiding common pitfalls [50] would be beneficial to understand the carbon footprints of the conversion and associated processes.

It is definitely worth highlighting the primary role of H_2 and its costs in the practical viability of these processes. In order to reduce CO_2 emission, H_2 has to be produced from renewable energy sources instead of the traditional H_2 production method by the steam reforming of natural gas. Technologies of water electrolysis are probably the most advanced in this respect. Economic and scalable technologies for water electrolysis including novel electrode catalyst design will indirectly but greatly contribute to the feasibility of CO_2 conversion processes.

3.3 PHOTOCATALYTIC CONVERSION OF CO₂ TO FUEL

Plants are able to convert CO_2 and H_2O into sugars and oxygen, a process called photosynthesis. Conversion of fast-growing crops and microalgae biomass is another strategy to recycle carbon. The sunlight-to-fuel energy conversion efficiency of photosynthesis is roughly 1% [51] and the biomass reforming requires extra energy input, leading to overall efficiencies way below 1%. Nature's process is extremely selective and efficient in performing its designed task, namely to sustain plants growth. However, in order to reduce CO_2 concentration in the atmosphere, the artificial photosynthesis (mankind mimic) process must outperform nature's process. The most sustainable option is the direct conversion of solar energy into chemical energy, that is, to photoconvert CO_2 to hydrocarbons, such as methane and methanol.

Halmann [52] reported in 1978 the photoelectrochemical reduction of CO_2 to formic acid, formaldehyde, and methanol, with a single crystal p-type GaP cathode and carbon anode. The work followed the initial findings of Fujishima and Honda, in which they reported the photoelectrochemical splitting of water with TiO_2 [53]. In 1979, Inoue et al. [54] demonstrated that the photoreduction of CO_2 to several hydrocarbons could be achieved with powder semiconductors. Hydrogen formation from water has a ΔG of 237 kJ/mol and a ΔH of 285 kJ/mol; the corresponding values for the formation of CO from CO_2 are 257 and 283 kJ/mol at 25°C (1 atm). Hence, the minimum energy per photon required for H_2O and CO_2 processes are 1.229 and 1.33 eV, respectively. Conceptually, the bandgap of a semiconductor for co-splitting CO_2 and H_2O should be at least 1.33 eV, which equates to absorption of photons with wavelengths below 930 nm. Furthermore, one must consider the energy losses associated with entropy changes [approximately 87 J/(mol K)] and other losses with the formation of $CO + O_2$ from CO_2. The optimal semiconductor bandgap is between 2 and 2.4 eV, which implies that photons with wavelengths below 600 nm are absorbed, limiting the maximum attainable efficiency to about 17% [55].

3.3.1 PHOTOCATALYTIC SYSTEMS BASED ON TiO₂

TiO_2, the most widely investigated photocatalyst, has a bandgap of around 3.0–3.2 eV, which implies the use of UV-A light to overcome the bandgap and separate the charges required to perform artificial photosynthesis. However, the wide bandgap means that TiO_2 is able to perform most of the desired reactions since their redox potential follows within its bandgap. The only step that is difficult to achieve with pristine TiO_2 is the reduction of CO_2 to CO [54], unless the material has high electron degeneracy, which is the case when nanoparticles are used [56].

3.3.1.1 Pristine TiO₂ Photocatalysis

As aforementioned, Inoue et al. [54] were the first to report the ability of TiO_2 to photocatalyze CO_2 and H_2O to formic acid, formaldehyde, and methanol. They found that methanol production increased as the conduction band becomes more negative with respect to the redox potential of H_2CO_3/CH_3OH, relating to the fact that the reactions of interest start following more and more within the bandgap (Figure 3.5).

Anpo and Chiba [57] anchored TiO_2 nanoparticles to transparent Vycol glass, and with it they were able to photoreduce CO_2 in water to CH_4 and CH_3OH. Interestingly, large amounts of products were desorbed when the temperature was increased from 275 to 673 K, suggesting that the products are strongly chemisorbed to the photocatalyst surface. They also demonstrated that the type of products formed depends on the ratios between water and CO_2. Lower ratio of water to CO_2 resulted in a higher production of methanol, whereas at high ratio the production of alkanes, including C_2H_6, and alkenes, dominated.

Yamashita et al. [59] demonstrated that photocatalytic performance is strongly influenced by TiO_2 crystal termination since with TiO_2 (100) they achieved yields of 3.5 and 2.4 nmol/(g_{cat} h) for CH_4 and CH_3OH, respectively, which dropped significantly when TiO_2 (110) was used instead; however, no explanation was given to justify the difference in performance.

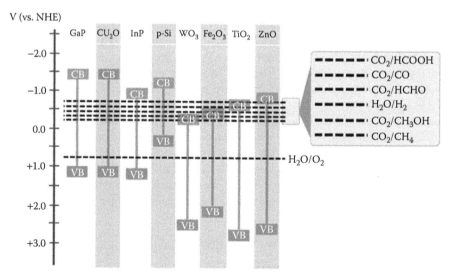

FIGURE 3.5 Illustration of the bandgap of semiconductor and thermodynamic reduction potentials of various compounds measured at pH = 7. (Reproduced from Rongé, J. et al., *Chem. Soc. Rev.*, DOI:10.1039/C3CS60424A, 2014. With permission.)

The reaction yields increased considerably if the reaction was carried out under CO_2 pressure. In respect to the formation of alkanes and alkenes, by increasing the reaction pressure from 1 to 2.5 MPa, as demonstrated by Mizumo et al. [60], a four-fold increase in products formation was reported. The shape of the reaction product formation curve revealed an exponential increase in product formation with the increase of reaction pressure. The increase of pressure also led to the formation of other oxygenated species, namely acids and aldehydes. However methanol formation reached its maximum at around 1 MPa. The formation of larger alcohols, for example, ethanol, at higher pressures indicated the occurrence of methanol methanation. The same authors, shortly after, performed an electron spin resonance (ESR) study, in which they claimed that Ti^{3+} centers are the initiators of the photoreduction of CO_2 in water [61]. Furthermore, the presence of holes scavenger, such as 2-propanol, dramatically increases the photocatalytic yield [62].

Another significant improvement in the photocatalysis was achieved when isolated Ti were loaded into zeolites, in particular to MCM-48. The authors attributed the increase in yield to the high dispersion of Ti sites and the presence of large pores in a three-dimensional channel structure of MCM-48 [63]. Green et al. [64] and Anpo et al. [65] demonstrated that carbon dioxide adsorbed preferentially on uncoordinated titanium surface sites, namely, on five- and four-coordinated, respectively.

A recent attempt to improve TiO_2 photocatalytic performance was to design hybrid materials containing another semiconductor, which is believed to improve charge separation, and to change surface chemistry. An example is TiO_2/ZnO systems that were found to increase significantly the rate of methane formation in comparison to TiO_2. Xi et al. [66] reported a methane formation rate for TiO_2/ZnO of 55 µ mol/(g_{cat}·h), which was 5–6 times higher than the one measured with TiO_2 P25 [9.3 µ mol/(g_{cat}·h)] under UV irradiation.

Composite and hybrid materials can be also used to improve the photocatalytic performance under visible light irradiation. Liang et al. [67] reported a two- to four-fold increase in methane production under visible light illumination when graphene–TiO_2 nanocomposite were used instead of P25, making it a promising strategy to shift TiO_2 activity into the visible range. Direct doping of TiO_2 with light elements, such as N, S, and C, is less effective because the doping also changes the surface chemistry leading to drastic decrease in activity under UV illumination that was not compensated by the gains in the visible light.

Despite the efforts, the photoefficiency of CO_2 reduction to methane and methanol over TiO_2 is still less than 0.1%, when water is used as electron donor. The low yield over TiO_2 is associated with two factors:

1. The strong oxidizing power of valence holes ($E_{VB} = +2.70$ V vs. NHE at pH $= 7$), resulting in the reaction of holes with molecular intermediates and products
2. The thermodynamically unfavorable one-electron reduction of CO_2, based on a very negative redox potential for this process in an aqueous homogeneous system [$E^0(CO_{2aq}/CO_{2aq}^-) = -1.9$ V]. However, the presence of proton donors and the chemical adsorption of CO_2 to the catalyst can decrease the redox potential by almost an order of magnitude [68].

The majority of research efforts are focused on the improvement of photocatalytic performance and light harvesting by design of novel and/or improved TiO_2-based materials; however, the reaction mechanism is still unclear. If known, this can help the rational design of materials with improved overall performances. The reduction of CO_2 to methane and methanol with water as electron donor is a multistep process governed by the following chemical equations [62,69]. The potentials are also mentioned in reference to NHE at pH = 7.

$$CO_2 + 2H^+ + 2e_{CB}^- \rightarrow HCOOH \qquad [E^0 = -0.61 \text{ V}] \qquad (3.11)$$

$$HCOOH + 2H^+ + 2e_{CB}^- \rightarrow HCOH + H_2O \qquad [E^0 = -0.48 \text{ V}] \qquad (3.12)$$

$$HCOH + 2H^+ + 2e_{CB}^- \rightarrow CH_3OH \qquad [E^0 = -0.38 \text{ V}] \qquad (3.13)$$

$$CH_3OH + 2H^+ + 2e_{CB}^- \rightarrow CH_4 + H_2O \qquad [E^0 = -0.24 \text{ V}] \qquad (3.14)$$

$$H_2O + 2h_{VB}^+ \rightarrow OH^{\bullet-} + H^+ \qquad [E^0 = +2.32 \text{ V}] \qquad (3.15)$$

$$2H_2O + 2h_{VB}^+ \rightarrow H_2O_2 + 2H^+ \qquad [E^0 = +1.35 \text{ V}] \qquad (3.16)$$

$$2H_2O + 4h_{VB}^+ \rightarrow O_2 + 4H^+ \qquad [E^0 = +0.82 \text{ V}] \qquad (3.17)$$

Conduction band electrons of TiO_2 are sufficiently energetic (E_{CB} = −0.50 V at pH = 7) to reduce CO_2 to methane [$E^0(CO_2/CH_4)$ = −0.24 V]. The process is believed to start with the formation of CO_2^- radical anion bound to the oxide surface [70]. However, a recent study from Yang et al. [71] questioned the feasibility of the process altogether. They offered an alternative explanation for the formation of CO (a possible intermediate), which did not rely on the use of photogenerated electrons. Using infrared and labeled CO_2, they suggested that CO was formed from one of two possible reactions:

$$^{13}CO_2 + {}^{12}C \rightarrow {}^{13}CO + {}^{12}CO \qquad (3.18)$$

$$H_2O + {}^{12}C \rightarrow {}^{12}CO + H_2 \qquad (3.19)$$

The reaction represented by Equation 3.18 is known as reverse Boudouard reaction, and the one summarized by Equation 3.19 is known as H_2O-induced photocatalytic surface carbon gasification. Since they measured a higher formation of ^{12}CO, they proposed Equation 3.19 to be the predominant contributor for the formation of CO. The presented result suggests that despite CO being formed from photoactivated water (steam), the carbon source is not CO_2 but carbon residues in the semiconductor structure, most likely left over from synthesis. This carbon deposit should deplete over time, and the reaction should cease; however, the experiment was not performed long enough to verify this, and categorically confirm that CO_2 photosplitting does not occur over TiO_2.

Dimitrijevic et al. [72] reported recently the initial steps in CO_2 reduction. They observed the formation of H atoms and CH_3 radicals upon illumination, suggestive of competitive electron transfer to adsorbed CO_2 and protons on TiO_2 surface (schematically represented in Figure 3.6a). The initial electron transfer leads to the

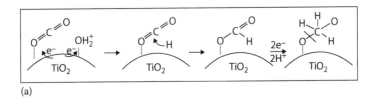

(a)

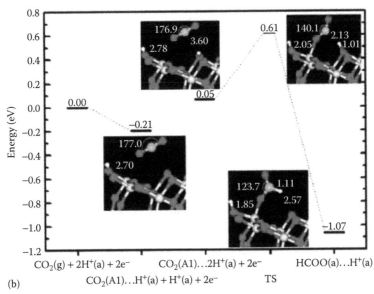

(b)

FIGURE 3.6 (a) Schematic representation of photocatalytic transformation of CO₂ to methoxy radical over TiO₂ in the presence of water. (b) Theoretical calculation of reaction pathway from CO₂ and hydrogen to formate on anatase (101) surface. Atoms colors are Ti in gray, O in red, C in blue, and H in white. (Reproduced from Dimitrijevic, N. M. et al., *J. Am. Chem. Soc.*, 133, 3964, 2011. With permission.)

break of a $C = O(O = C = O)$ bond, and the attachment of a H atom, leading to the formation of a formate species. The initial step of competitive electron transfer thus corresponds to two-electrons, one-proton transfer, according to Equation 3.20:

$$CO_2 + H^+ + 2e^-_{CB} \rightarrow HCOO^- \tag{3.20}$$

Consecutive electron/proton addition leads to the formation of methoxy radical. Calculations (Figure 3.6b) suggest that CO₂ prefers a linear vertical adsorption onto five-coordinated Ti sites on anatase (101) surface [73]. Dissolved CO₂ forms carbonate and bicarbonate species that compete with water for photogenerated holes.

The authors assigned three roles to water:

1. Charge stabilization (prevention of electron–hole recombination)
2. Electron donor (reaction of water with holes to form OH radicals)
3. Electron acceptor (formation of H atoms, a reaction involving electrons and protons on TiO₂ surface)

The study revealed the importance of hydrogen and water management on the surface.

3.3.1.2 Metal-Modified TiO_2 Photocatalysis

TiO_2 photocatalytic performance can be significantly affected with the addition of a metal cocatalyst, such as Ag, Cu, Pt, or Ru. Generally, the most noticeable changes are the increase in reaction rate and the change in product formation and distribution, which depend on parameters such as metal identity, amount, and preparation procedures, to name only a few. There are numerous reviews on the topic, for example, by Roy et al. [74] and Kubacka et al. [75]. In the following section, we report the most well-known cases.

One of the first metals to be added to TiO_2 was ruthenium in the form of RuO_2 [76]. The addition of RuO_2 led to a small increase in reaction efficiency. RuO_2 is believed to trap valence band holes (holes sink), which are involved in water oxidation, a key step when water is used as electron donor [77]. The main reaction product was formic acid; however, formaldehyde and methanol were also detected.

RuO_2/TiO_2 is also able to carry out the photomethanation of CO_2 in the presence of H_2. The methanation of CO_2 follows the Sabatier reaction:

$$CO_2 + H_2 \rightarrow CH_4 + H_2O \qquad \Delta G° = -27 \text{ kcal/mol} \qquad (3.21)$$

Rhampi et al. [78] demonstrated that under UV irradiation methane was produced exclusively at a rate of 116 μL/h, which was 4–5 times higher than the rate in the dark at 46°C. The authors confirmed that the enhancement was due to the involvement of electrons and holes generated during the photoexcitation of TiO_2. They also suggested that CO_2 photomethanation led to the formation of Ru carbidic surface species (Ru–C), which were subsequently hydrogenated to form methane, in a manner similar to the thermal process of CO_2 methanation. However, a subsequent study from another group ruled out the formation of Ru–C, since surface carbon on the sample was only found either in the form of inert graphite or partially hydrogenated C–O species [79].

Hirano et al. [80] reported an enhancement in the formation of methanol, formaldehyde, and formic acid when copper powder was added to a suspension of TiO_2, under UV irradiation. However under CO_2-pressurized water, Adachi et al. [81] only detected the formation of methane, ethylene, ethane, and hydrogen. Their best result was obtained with 5 wt% Cu on TiO_2, where they formed 15 μL/g$_{cat}$ [0.65 μL/(g$_{cat}$ h)] of methane, 22 μL/g$_{cat}$ [0.92 μL/(g$_{cat}$ h)] of ethylene, and <3 μL/g$_{cat}$ [0.13 μL/(g$_{cat}$ h)] of ethane. Several studies confirmed that the addition of Cu or Pt generally leads to the formation of methane with a low concentration of methanol and other hydrocarbons, such as C_2H_4 and C_2H_6 [82].

Varghese et al. extended the reactivity of Cu- and Pt-doped TiO_2 into the solar range by doping TiO_2 nanotubes with N [83]. The product generation rates are reported on Figure 3.7 (b and c). Both metals significantly improved the formation of products, which was believed to be due to the transference of photoexcited electrons from TiO_2 to the metal, and subsequently to the surface species [84], that is, promotion of water and CO_2 reduction. However, Pt and Cu affected the product formation in a

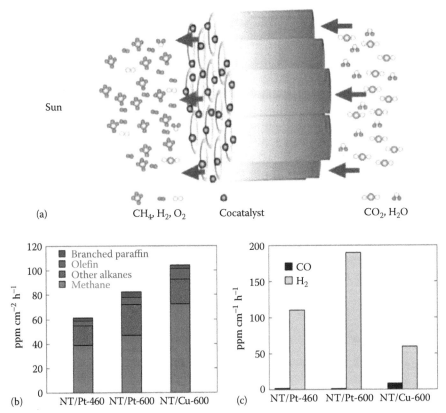

FIGURE 3.7 (a) Schematic representation of photocatalytic transformation of $CO_2 + H_2O$ to $CH_4 + O_2$ over N-doped TiO_2 nanotubes with cocatalysts under solar irradiation. (b and c) Product formation over N-doped TiO_2 nanotubes array with Pt (NT/Pt) and Cu (NT-Cu) cocatalysts annealed at 460°C or 600°C, under sunlight illumination. (b) Hydrocarbon generation rate; (c) H_2 and CO generation rates. (Reproduced from Varghese, O. K. et al., *Nano Lett.*, 9, 731, 2009. With permission.)

different way. In the case of Cu, high concentrations of CO were detected suggesting preferential reduction of CO_2, whereas in the case of Pt, high concentrations of H_2 were measured suggesting preferential reduction of water [85]. In fact, when Pt and Cu were combined, no CO was detected, and the rate of formation of hydrocarbons increased from 104 (only with Cu) to 111 ppm/(cm² h).

The addition of Pd as cocatalyst resulted in the higher formation of hydrocarbons in detriment of CO and/or alcohols (Figure 3.8) [86], contrasting with the case when Ag was added, where a significant production of alcohols, in particular methanol, was observed [87]. They reported a maximum yield of methanol with 1 wt% Ag on TiO_2 of 4.12 μmol/(g_{cat} h). Ag acts as an electron sink as the previous metals [88]; however, its catalytic properties favor the formation of partially hydrogenated CO products instead.

AbouAsi et al. [89] showed that photocatalytic reduction of CO_2 was possible under visible light if AgBr was used. As with previous Ag systems, the AgBr/TiO_2

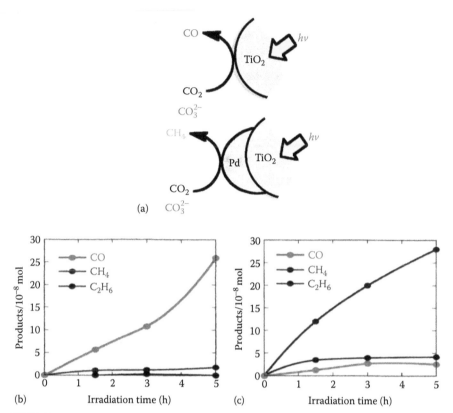

FIGURE 3.8 (a) Schematic representation of photocatalytic transformation of CO_2 to CH_4 over Pd/TiO_2 under UV irradiation. Photocatalytic reduction of CO_2 (650 Torr) by (b) TiO_2 and (c) 1 wt% Pd/TiO_2. (Reproduced from Yui, T. et al., *ACS Appl. Mater. Interfaces*, 3, 2594, 2011. With permission.)

system produced significant amounts of methanol, ethanol, and CO, together with methane (major product) (Figure 3.9). However, contrary to the other Ag systems on TiO_2, the role of Ag was to harvest the solar light, and inject the electrons into TiO_2, where, in their opinion, all the reduction steps took place.

3.3.2 ALTERNATIVE SOLID SEMICONDUCTORS

An initial study by Inoue et al. [54] revealed that SiC is one of the best semiconductors for the photoreduction of CO_2, in part due to the position of its bandgap in respect to the redox potentials associated with the transformation of CO_2 into useful chemicals, such as formic acid, methanol, and methane. In their study, SiC produced formaldehyde and methanol with a yield of 1.0×10^{-3} and 5.35×10^{-3} M, respectively. The yield of methanol production was 25 times higher than that obtained with TiO_2, and fivefold of what was measured with CdS or GaP. However, this never became the prime choice of material, in part due to its rather low energy conversion, almost two

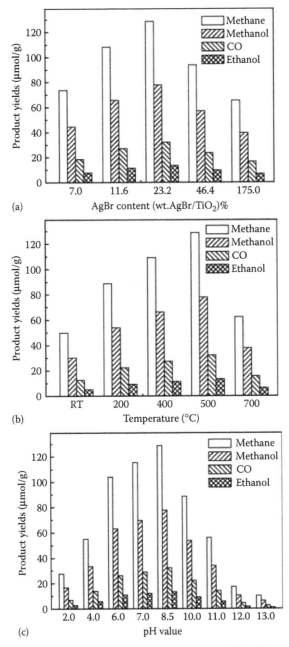

FIGURE 3.9 Yield of photocatalytic products under 5 h visible light irradiation on (a) AgBr content using AgBr/TiO$_2$ sintered at 500°C in pH 8.5 aqueous media; (b) calcination temperature using 22.3 wt%AgBr/TiO$_2$ in pH 8.5 aqueous media; and (c) pH value using 22.3 wt%AgBr/TiO$_2$ sintered at 500°C. (Reproduced from Abou Asi, M. et al., *Catal. Today*, 175, 256, 2011. With permission.)

orders of magnitude lower than the ones obtained with TiO_2 [90]. Another reason is a low scope for improvement, which is a major advantage of the TiO_2-based systems.

Titanates, in particular the perovskite-$SrTiO_3$ [90], produced methanol at a rate three times higher than TiO_2, resulting in a significantly higher energy conversion in respect to TiO_2. The prime reason for this enhancement seems to be related to a better position of $SrTiO_3$ bandgap in respect to the desired reactions' redox potential since the bandgap itself is very similar to that of TiO_2. The discovery that $SrTiO_3$ is active for the photoreduction of CO_2 was significant because it opened the prospect of using mixed oxide semiconductors, such as $NaNbO_3$ [91], $FeCaO_4$ [56], $InTaO_4$ [92], and $BiVO_4$ [93], as photocatalysts. This class of materials is highly flexible, cheap, and easy to tune and optimize. This can be achieved by changing parameters, such as elemental composition and metals identity.

Bi_2WO_6 is one of the simplest members of the Aurivillius family, comprising accumulated layers of corner-sharing WO_6 octahedral sheets and bismuth oxide sheets, able to absorb visible light ($\lambda > 420$ nm). Standard Bi_2WO_6 exhibits very low photocatalytic performance in CO_2 reduction. However, when shaped into nanoplates, its activity increased. Methane formation rate of standard Bi_2WO_6 is 0.045 $\mu mol/(g_{cat} h)$, which increased to 1.1 $\mu mol/(g_{cat} h)$ when nanoplates were used. More significantly, it photoreduces CO_2 to CH_4, exclusively [94]. The authors assigned this enhancement to three possible factors:

1. Higher surface area
2. Faster charge mobility due to the ultrathin geometry of nanoplates, which should reduce charge recombination
3. Exposed (001) facets, which are believed to be the active sites for the photocatalytic reduction of CO_2

Similar result was obtained with Zn_2GeO_4 nanoribbons [95]. The authors report a significant enhancement in methane formation [1.5 $\mu mol/(g_{cat} h)$] when compared to bulk material (only trace amounts after 1 h). The reasons for photocatalytic performance enhancement are similar to the ones suggested for Bi_2WO_6. The formation rate of methane improved significantly when Pt and RuO_2 were added to the material as cocatalysts. When 1 wt% of each cocatalyst was added, the methane formation jumped from 1.5 to 25 $\mu mol/(g_{cat} h)$.

However, the experiments had to be performed under UV irradiation due to Zn_2GeO_4's large bandgap ($Eg = 3.8$ eV). The photocatalytic performance of the material could be extended to the visible range when the material underwent nitridation, resulting in the formation of $Zn_{1.7}GeN_{1.8}O$ structure [96]. The resultant material was found active under visible irradiation ($\lambda > 420$ nm). As before, the addition of 1 wt% of Pt and RuO_2 increased significantly the rate of methane formation, reaching a value close to 10 $\mu mol/(g_{cat} h)$. The results are significant because they imply that nanoscale compounds behave significantly different from bulk materials, which opens a plethora of opportunities.

Ahmed et al. [97] proposed recently the use of double layer hydroxide (LDH) structures, which have the benefit of increasing CO_2 uptake due to intercalation of carbonate species between layers during reaction, which can be at posteriori converted to methanol (Figure 3.10). Carbonates result from the interaction of CO_2 with

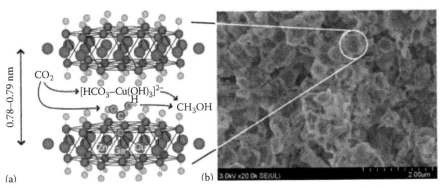

FIGURE 3.10 (a) LDH structure and schematic representation of photoreduction of CO$_2$ to methanol. (b) Scanning electron microscopy picture of LDH structures. (Reproduced from Ahmed, N. et al., *Catal. Today*, 185, 263, 2012. With permission.)

TABLE 3.1

Rates of Photocatalytic Conversion of CO$_2$ with H$_2$ into CH$_3$OH and CO over LDH Photocatalysts[a] under UV–Vis Irradiation with a 500 W Xe Arc Lamp

		Formation Rate (nmol h^{-1}gcat^{-1})			Conversion	Selectivity to CH$_3$OH
Entry	Photocatalyst	CH$_3$OH	CO	Σ	(%, C-base)	(mol%)
A	Zn$_3$Ga|CO$_3$	51(±4)	80(±6)	130	0.02	39(±4)
B	Zn$_{1.5}$Cu$_{1.5}$Ga|CO$_3$	170(±14)	79(±6)	250	0.03	68(±4)
C	Zn$_3$Ga|CO$_3$[b]	50(±4)	74(±6)	120	0.02	40(±4)
D	Zn$_{1.5}$Cu$_{1.5}$Ga|CO$_3$[b]	310(±9)	18D(±2)	500	0.07	63(±1)
E	Zn$_3$Ga|Cu(OH)$_4$	300(±9)	130(±10)	430	0.04	71(±2)
F	Zn$_{1.5}$Cu$_{1.5}$Ga|Cu(OH)$_4$	490(±15)	70(±6)	560	0.05	88(±2)
G	Zn$_3$Ga|Cu(OH)$_4$–3×ex	280(±8)	120(±9)	390	0.04	71(±3)
H	Zn$_{1.5}$Cu$_{1.5}$Ga|Cu(OH)$_4$–3×ex	430(±13)	48(±4)	480	0.05	90(±1)

Source: Reproduced from Ahmed, N. et al., *Catal. Today*, 185, 263, 2012. With permission.

[a] The catalyst amount was 100 mg. Values in the parentheses are the experimental errors for evaluation.

[b] Preheated at 423 K for 1 h under vacuum.

water. Table 3.1 summarizes the photocatalytic performance of their LDH systems. The LDH structures' light absorption occurs in the UV range and their photocatalytic activity increased if Cu was added to the structure.

3.3.3 PHOSPHIDES AND SULFIDES

In 1978, Halmann demonstrated that nonbiological reduction of CO$_2$ to organic raw materials and fuel was possible under solar irradiation with a p-type gallium phosphide semiconductor. After 18 h of irradiation (λ = 365 nm), the concentrations of produced formic acid, formaldehyde, and methanol were 1.2×10^{-2}, 3.2×10^{-2}, and 1.1×10^{-4} M, respectively [52]. Halmann estimated an optical conversion efficiency, according to Equation 3.22, of 5.6% for formaldehyde, and of 3.6% for methanol, determined between the bias voltages of −0.8 and −0.9 V. Considering that only 17.3% of the solar light can promote charge separation of p-GaP ($Eg = 2.3$ eV), the maximal solar energy conversion was estimated to be 0.97% and 0.61% for formaldehyde and methanol, respectively.

$$\text{Optical conversion efficiency} = \frac{100\,I_c\;[(\Delta H/Z) - V_B]}{I_a} \qquad (3.22)$$

where:

I_c is the current density (mA/cm²)

I_a is the incident light intensity (mW/cm²)

ΔH is the heat of combustion (2.962, 2.639, 5.915, and 7.259 eV for hydrogen, formic acid, formaldehyde, and methanol, respectively)

Z is the number of electrons required in the reduction of a single CO_2 molecule to a molecule of product (2, 2, 4, and 6, for hydrogen, formic acid, formaldehyde, and methanol, respectively)

V_B is the electrical bias (V)

Shortly after, Irvine and coworkers followed Hamman's strategy. They used CdS and reported the formation of 7.1×10^{-4} and 9.1×10^{-6} M of formic acid and formaldehyde, respectively, using light with wavelengths ranging from 320 to 580 nm [98]. CdS follows the so-called protonation route, which leads to the formation of aldehydes, acids, and alcohols despite the plethora of molecular species determined during reaction.

ZnS has also been tested for photoreduction of CO_2 either on its own [99] or in combination with other metals [100]. Due to ZnS large bandgap ($Eg = 3.66$ eV), illumination needs to be in the UV range. Quantum yields close to 0.001% were measured with bulk ZnS [99], equating to a production of 1.8 mmol formate/mmol ZnS after 72 h under illumination with $\lambda \geq 290$ nm. The mechanism is believed to involve the transference of two electrons from ZnS to adsorbed hydrogen carbonate (key step of CO_2 fixation). Significant enhancement in the production of formic acid was observed when metals such as Zn, Ag, Pb, Ni, and Cd were used in combination. In the case of Cd, the quantum yield for the production of formic acid in the presence of 2-propanol as hole scavenger jumped from 0 to 32.5%, which was double from the second best metal, namely Zn.

Despite the potential of phosphide and sulfide systems, they contain metals such as Cd that are highly toxic and can be bioaccumulated, which hampers significantly their commercial application, at least prior to studies on their leaching into the environment are carried out.

3.3.4 Photoreduction of CO_2: Perspective and Challenges

Converting solar energy into energy contained in chemical bonds is the ultimate objective of scientists working in the field of energy storage and production because this process does not involve emission of greenhouse gases, and energy can be stored and used whenever desired. Photosynthesis is nature's way of converting sunlight, water, and CO$_2$ into sugars and oxygen. The energy-conversion efficiency can reach 7% in optimum conditions, although an efficiency of less than 1% is commonly expected for agricultural crops over their entire lifecycle [101]. Artificial photosynthesis is the man-made chemical process related to the capturing and storing of sunlight energy in chemical bonds. The most well-known artificial photosynthesis processes are water splitting and carbon dioxide reduction. Water splitting forms H$_2$ and O$_2$, whereas the CO$_2$ reduction forms hydrocarbons such as CH$_4$, CH$_3$OH, and O$_2$. The yields of artificial photosynthesis, in its present state, are an order of magnitude lower than those obtained in nature [102].

Three elementary processes determine the outcome of any photocatalytic reaction, namely, light harvesting, charge generation and separation, and the catalytic reaction itself. In order to improve the current efficiency (~0.1%), all the three elemental processes need to be understood and improved.

Concerning light harvesting, as aforementioned, TiO$_2$ is the most widely used photocatalyst; however, its wide bandgap also means that in order to promote charge separation we need UV photons, which account for only 4% of the solar spectrum. Asahi et al. [103] tried to circumvent this difficulty by doping TiO$_2$ with light elements in order to shift the absorption to the visible range. They reported a significant improvement in the visible light absorption when TiO$_2$ was doped with nitrogen in substitutional places, making it a popular strategy to improve photocatalytic activity in the visible range. However, the strategy very often leads to a drastic decrease in the photocatalytic performance in the UV range, which is not compensated by the gains in the visible range [104]. Furthermore, by narrowing the bandgap one decreases the breath of reactions that can be photocatalyzed.

An alternative strategy is to use sensitizers, which can harvest solar light, and inject electrons into TiO$_2$. The most famous is the Grätzel cell [105] consisting of a dye sensitizer coated on TiO$_2$ surface. The success of the Grätzel cell relates to the fast electron injection from dye to TiO$_2$ (<10 fs) [106], and slow back transfer of electrons to the oxidized dye (up to ms) [107], enabling the transference of the electrons to a counter-electrode (increased lifetime). The problems with organic dyes are low light cross sections, instability, and unreactive hole. As mentioned previously, photoreduction of CO$_2$ requires the involvement of holes as well as electrons. Therefore, the strategy to be followed is the development of solid sensitizers that have large light cross sections, are stable under reaction conditions, and possess reactive holes able to carry out photooxidations.

Metallic nanoparticles are interesting sensitizer candidates because of their localized surface plasmons, which have large optical cross sections. Gold group metals exhibit plasmonic resonances, which can be tuned by changing their shape, size, and/ or composition, enabling a good match with the solar spectrum [108]. Furthermore, due to their d10 configuration they are chemically stable. Recently, the excitation of Au and Ag localized surface plasmons nanostructures was shown to improve solar cell charge transfer from sensitizer to semiconductor [109], increase the photocurrents under solar irradiation [110], and improve photoinitiated catalytic oxidations [111]. However, there is still a lot that is unknown about localized surface plasmons, including the mechanism that leads to charge separation. We have recently carried out a high-resolution X-ray absorption study, which demonstrated that Au density-of-states changes when the plasmon is excited, demonstrating that charge separation occurs, and the electrons are energetic enough to be injected into TiO_2 [112]. This is the first study reporting on the mechanism; however, one forecasts many more studies to be published in the future due to the increasing popularity of these systems.

The second aspect relates to charge separation and mobility, which is a crucial part of the photocatalytic process. As aforementioned, the preparation of nanosize materials can effectively increase charge lifetime and mobility. It is expected that this approach will gain prominence over the next decade, primarily due to the advances in nanoscience and materials synthesis, which enable researchers and industry to prepare materials unthinkable a decade ago. The nanomaterials not only have high performance but in many cases exhibit noble properties absent in bulk materials.

Concerning catalytic performance, the field will continue to be dominated by the metal cocatalysts; however, photocatalyst surface modification aiming at increasing the adsorption of CO_2 is expected to gain importance. This can be achieved by adding chemicals that make the catalytic surface more basic, which would then promote CO_2 adsorption since CO_2 is a mild acid. Concerning the cocatalysts able to promote photoreductions, a methodical study on loading, metal identity, and shape and size is well overdue. The future trend seems to be the use of multielements that are able to promote specific steps of the reaction.

Another important issue that has been largely neglected is the oxidation side, in particular if one excludes the RuO studies. Nocera's group reported a significant enhancement in the formation of oxidation products when Co^{2+} was added to the catalyst composition [113]. Cobalt modification uses a cobalt alkoxide or with a terminal thiol group that coordinates to the free metal in the porous structure. After coordination, the organic precursor will be removed via oxidation, resulting in the formation of CoO centers. Therefore, we expect this area to become an area of intense research in the future.

The reaction mechanism and active sites, even on TiO_2, are still largely unknown. Significant progress on artificial photosynthesis cannot be realistically achieved before elucidation of the reaction mechanism and identification of the active sites. Latest developments in operando spectroscopy made scientists hopeful that in the next decade a consistent reaction mechanism will be proposed at least for reactions carried out on TiO_2. It is hoped that the methodologies developed to attain this objective can be adapted to the study of other photocatalysts, enabling for rational development of photocatalytic system with new or improved properties.

Finally, we forecast that reactor design will start to play a bigger role since some reports have demonstrated that wired reactors are more efficient than the wireless ones [114]. Furthermore, a wired reactor prevents the problem of gas separation since the gases are generated in different compartments, and each process can be optimized independently. The two compartments are separated by a proton exchange membrane and connected with a wire, which carries the electrons from the light harvesting side to the dark side. Recent developments in 3D printing technology enable manufacture and optimization of the reactor design by the researchers carrying out the development of the photocatalytic systems [115]. This does not mean that engineering solutions have no role to play in the development of this area; on the contrary, scientists and engineers should work in tandem to ensure rational and incremental progress.

As a final remark, scientists working in this area should agree on a standardized way to present results. This is absent and in our view damages the field and its progress.

REFERENCES

1. Energy Information Administration, Annual Energy Review, US Department of Energy (2008).
2. R. Lal, *Energy Environ. Sci.* 1 (2008) 86.
3. M. Halbwachs, J. C. Sabroux, *Science* 292 (2001) 438.
4. (a) X. Xiaoding, J. A. Moulijn, *Energy Fuels* 10 (1996) 305; (b) E. V. Kondratenko, G. Mul, J. Baltrusaitis, G. O. Larrazábal, J. Pérez-Ramírez, *Energy Environ. Sci.* 6 (2013) 3112.
5. C. Costentin, M. Robert, J.-M. Savéant, *Chem. Soc. Rev.* 42 (2013) 2423.
6. PBL Netherlands Environmental Assessment Agency, "Trends in Global CO$_2$ Emissions," http://edgar.jrc.ec.europa.eu/news_docs/pbl-2013-trends-in-global-co2-emissions-2013-report-1148.pdf.
7. M. Grasemann, G. Laurency, *Energy Environ. Sci.* 5 (2012) 8171.
8. (a) G. A. Olah, A. Goeppert, M. Czaun, G. K. S. Prakash, *J. Am. Chem. Soc.* 135 (2013) 648; (b) G. A. Olah, *Catal. Lett.* 143 (2013) 983.
9. M. Carmo, D. L. Fritz, J. Mergel, D. Stolten, *Int. J. Hydrogen Energy.* 38 (2013) 4901.
10. (a) A. Kudo, Y. Miseki, *Chem. Soc. Rev.* 38 (2009) 253; (b) K. Maeda, K. Domen, *J. Phys. Chem. Lett.* 1 (2010) 2655.
11. B. Tidona, A. Urakawa, P. Rudolf von Rohr, *Chem. Eng. Process.* 65 (2013) 53.
12. P. Kaiser, R. B. Unde, C. Kern, A. Jess, *Chem. Ing. Tech.* 85 (2013) 489.
13. C. Ratnasamy, J. P. Wagner, *Catal. Rev.* 51 (2009) 325.
14. W. Wang, S. Wang, X. Ma, J. Gong, *Chem. Soc. Rev.* 40 (2011) 3707.
15. F. S. Stone, D. Waller, *Top. Catal.* 22 (2003) 305.
16. T. Fujitani, J. Nakamura, *Catal. Lett.* 56 (1998) 119.
17. C. S. Chen, W. H. Cheng, S. S. Lin, *Appl. Cat. A.* 238 (2003) 55.
18. A. Bansode, B. Tidona, P. Rudolf von Rohr, A. Urakawa, *Catal. Sci. Technol.* 3 (2013) 767.
19. (a) A. Goguet, F. C. Meunier, D. Tibiletti, J. P. Breen, R. Burch, *J. Phys. Chem. B* 108 (2004) 20240; (b) L. C. Wang, M. Tahvildar Khazaneh, D. Widmann, R. J. Behm, *J. Catal.* 264 (2009) 67.
20. Q. Fu, H. Saltsburg, M. Flytzani-Stephanopoulos, *Science* 301 (2003) 935.
21. A. Goguet, F. Meunier, J. P. Breen, R. Burch, M. I. Petch, A. Faur Ghenciu, *J. Catal.* 226 (2004) 382.
22. L. Wang, S. Zhang, Y. Liu, *J. Rare Earths* 26 (2008) 66
23. W. Wang, J. Gong, *Front. Chem. Sci. Eng.* 5 (2011) 2.
24. B. Lu, K. Kawamoto, *RSC Adv.* 2 (2012) 6800.
25. (a) M. Jacquemin, A. Beuls, P. Ruiz, *Catal. Today* 157 (2010) 462; (b) A. Beuls, C. Swalus, M. Jacquemin, G. Heyen, A. Karelovic, P. Ruiza, *Appl. Cat. B.* 113–114 (2012) 2.
26. T. Abe, M. Tanizawa, K. Watanabe, A. Taguchi, *Energy Environ. Sci.* 2 (2009) 315.
27. M. Marwood, R. Doepper, A. Renken, *Appl. Cat. A.* 151 (1997) 223.
28. N. M. Gupta, V. S. Kamble, K. Annaji Rao, R. M. Iyer, *J. Catal.* 60 (1979) 57.
29. R. W. Dorner, D. R. Hardy, F. W. Williams, H. D. Willauer, *Energy Environ. Sci.* 3 (2010) 884.

30. T. Herranz, S. Rojas, F. J. Pérez-Alonso, M. Ojeda, P. Terreros, J. L. G. Fierro, *Appl. Cat. A.* 311 (2006) 66.
31. R. W. Dorner, D. R. Hardy, F. W. Williams, H. D. Willauer, *Appl. Cat. A.* 373 (2010) 112.
32. S. Li, S. Krishnamoorthy, A. Li, G. D. Meitzner, E. Iglesia, *J. Catal.* 20 (2002) 202.
33. F. J. Pérez-Alonso, M. Ojeda, T. Herranz, S. Rojas, J. M. González-Carballo, P. Terreros, J. L. G. Fierro, *Catal. Commun.* 9 (2008) 1945.
34. T. Riedel, M. Claeys, H. Schulz, G. Schaub, S. S. Nam, K. W. Jun, M. J. Choi, G. Kishan, K. W. Lee, *Appl. Cat. A.* 186 (1999) 201.
35. M. P. Rohde, D. Unruh, G. Schaub, *Ind. Eng. Chem. Res.* 44 (2005) 9653.
36. R. W. Dorner, D. R. Hardy, F. W. Williams, B. H. Davis, H. D. Willauer, *Energy Fuels* 23 (2009) 4190.
37. (a) G. A. Olah, A. Goeppert, G. K. S. Prakash, *J. Org. Chem.* 74 (2009) 487; (b) G. A. Olah, A. Goeppert, G. K. S. Prakash, "Beyond Oil and Gas: The Methanol Economy," Wiley-VCH Verlag Gmbh & Co. KGaA, Weinheim (2009).
38. F. Pontzen, W. Liebner, V. Gronemann, M. Rothaemel, B. Ahlers, *Catal. Today* 171 (2011) 242.
39. J. G. van Bennekom, R. H. Venderbosch, J. G. M. Winkelman, E. Wilbers, D. Assink, K. P. J. Lemmens, H. J. Heeres, *Chem. Eng. Sci.* 87 (2013) 204.
40. J. G. van Bennekom, J. G. M. Winkelman, R. H. Venderbosch, S. D. G. B. Nieland, H. J. Heeres, *Ind. Eng. Chem. Res.* 51 (2012) 12233
41. A. Bansode, A. Urakawa, *J. Catal.* 309 (2014) 66.
42. K. W. Jun, H. S. Lee, H. S. Roh, S. E. Park, *Bull. Korean Chem. Soc.* 23 (2002) 803.
43. A. García-Trenco, A. Martínez, *Appl. Cat. A.* 411–412 (2012) 170.
44. V. Subramani, S. K. Gangwal, *Energy Fuels* 22 (2008) 22, 814.
45. X. Yu, P. G. Pickup, *J. Power Sources* 182 (2008) 124.
46. (a) P. G. Jessop, T. Ikariya, R. Noyori, *Nature* 368 (1994) 231; (b) P. G. Jessop, T. Ikariya, R. Noyori, *Science* 269 (1996) 1065.
47. (a) P. G. Jessop, Y. Hsiao, T. Ikariya, R. Noyori, *J. Am. Chem. Soc.* 118 (1996) 344; (b) A. Urakawa, F. Jutz, G. Laurenczy, A. Baiker, *Chem. Eur. J.* 13 (2007) 3886; (c) A. Urakawa, M. Iannuzzi, J. Hutter, A. Baiker, *Chem. Eur. J.* 13 (2007) 6828.
48. C. Federsel, R. Jackstell, A. Boddien, G. Laurenczy, M. Beller, *ChemSusChem* 3 (2010) 1048.
49. (a) L. Schmid, O. Kröcher, R. A. Köppel, A. Baiker, *Micropor. Mesopor. Mater.* 35–36 (2000) 181; (b) M. Rohr, M. Günther, F. Jutz, J.-D. Grunwaldt, H. Emerich, W. van Beek, A. Baiker, *Appl. Cat. A.* 296 (2005) 238; (c) M. Baffert, T. K. Maishal, L. Mathey, C. Copéret, C. Thieuleux, *ChemSusChem* 4 (2011) 1762.
50. N. von der Assen, J. Jung, A. Bardow, *Energy Environ. Sci.* (2013) 2721.
51. (a) I. Zelitch, *Science* 188 (1975) 626; (b) N. S. Lewis, *MRS Bull.* 32 (2007) 808; (c) O. Morton, "Eating the Sun: How Plants Power the Planet," Harper, New York (2008).
52. M. Halmann, *Nature* 275 (1978) 115.
53. A. Fujishima, K. Honda, *Nature* 238 (1972) 37.
54. T. Inoue, A. Fujishima, S. Konishi, K. Honda, *Nature* 277 (1979) 637.
55. (a) K. Tanaka, K. Miyahara, I. Toyoshima, *J. Phys. Chem.* 88 (1984) 3504; (b) O. K. Varghese, C. A. Grimes, *Sol. Energy Mater. Sol. Cells* 92 (2008) 374.
56. Y. Matsumoto, M. Obata, J. Hombo, *J. Phys. Chem.* 98 (1994) 2950.
57. M. Anpo, K. Chiba, *J. Mol. Catal.* 74 (1992) 207.
58. J. Rongé, T. Bosserez, D. Martel, C. Nervi, L. Boarino, F. Taulelle, G. Decher, S. Bordiga, J. A. Martens, *Chem. Soc. Rev.* (2014) doi: 10.1039/C3CS60424A.
59. H. Yamashita, N. Kamada, H. He, K.-I. Tanaka, S. Ehara, M. Anpo, *Chem. Lett.* (1994) 855.
60. T. Mizuno, K. Adachi, K. Ohta, A. Saji, *J. Photochem. Photobiol. A: Chem.* 98 (1996) 87.

61. S. Kaneco, H. Kurimoto, K. Ohta, T. Mizuno, A. Saji, *J. Photochem. Photobiol. A: Chem.* 109 (1997) 59.

62. S. Kaneco, Y. Shimizu, K. Ohta, T. Mizuno, *J. Photochem. Photobiol. A: Chem.* 115 (1998) 223.

63. M. Anpo, H. Yamashita, K. Ikeue, Y. Fujii, S. G. Zhang, Y. Ichihashi, D. R. Park, Y. Suzuki, K. Koyano, T. Tatsumi, *Catal. Today* 44 (1998) 327.

64. J. Green, E. Carter, D. M. Murphy, *Chem. Phys. Lett.* 477 (2009) 340.

65. (a) M. Anpo, H. Yamashita, Y. Ichihashi, S. Ehara, *J. Electroanal. Chem.* 396 (1995) 21; (b) M. Anpo, T. Takeuchi, *J. Catal.* 216 (2003) 505; (c) M. Anpo, J. M. Thomas, *Chem. Commun.* (2006) 3273.

66. G. Xi, S. Ouyang, J. Ye, *Chem. Eur. J.* 17 (2011) 9057.

67. Y. T. Liang, B. K. Vijayan, K. A. Gray, M. C. Hersam, *Nano Lett.* 11 (2011) 2865.

68. (a) H. Yoneyama, *Catal. Today* 39 (1997) 169; (b) J. C. Hemminger, R. Carr, G. A. Somorjai, *Chem. Phys. Lett.* 57 (1978) 100.

69. (a) H. I. Tseng, W. C. Chang, J. C. S. Wu, *Appl. Cat. B.* 37 (2002) 37; (b) A. D. Belapurkar, K. Kishore, *Photochem. Photobiol. A* 163 (2004) 503; (c) G. R. Dey, K. K. Pushpa, *Res. Chem. Intermed.* 33 (2007) 631.

70. G. Centi, S. Perathoner, G. Wine, M. Gangeria, *Green Chem.* 9 (2007) 671, and references cited therein.

71. C.-C. Yang, Y.-H. Yu, B. van der Linden, J. C. S. Wu, G. Mul, *J. Am. Chem. Soc.* 132 (2010) 8398.

72. N. M. Dimitrijevic, B. K. Vijayan, O. G. Poluektov, T. Rajh, K. A. Gray, H. He, P. Zapol, *J. Am. Chem. Soc.* 133 (2011) 3964.

73. H. He, P. Zapol, L. Curtiss, *J. Phys. Chem.* 114 (2010) 21474.

74. S. C. Roy, O. K. Varghese, M. Paulose, C. A. Grimes, *ACSNano* 4 (2010) 1259.

75. A. Kubacka, M. Fernández-García, G. Colón, *Chem. Rev.* (2011) doi: 10.1021/cr100454n.

76. M. Halmann, V. Katzir, E. Borgarello, J. Kiwi, *Sol. Energy Mater.* 10 (1984) 85.

77. J. Kiwi, M. Grätzel, *Chimia* 33 (1979) 289.

78. K. R. Thampi, J. Kiwi, M. Grätzel, *Nature* 327 (1987) 506.

79. J. Melsheimer, W. Guo, D. Ziegler, M. Wesemann, R. Schlögl, *Catal. Lett.* 11 (1991) 157.

80. K. Hirano, K. Inoue, T. Yatsu, *J. Photochem. Photobiol. A: Chem.* 64 (1992) 255.

81. K. Adachi, K. Ohta, T. Mizuno, *Sol. Energy* 53 (1994) 187.

82. For example (a) R. Cook, R. C. Macduff, A. F. Sammells, *J. Electrochem. Soc.* 135 (1988) 429; (b) M. Anpo, K. Chiba, *J. Mol. Catal.* 74 (1992) 207; (c) M. Anpo, H. Yamashita, I. Ichihashi, Y. Fujii, M. Honda, *J. Phys. Chem. B* 101 (1997) 2632.

83. O. K. Varghese, M. Paulose, T. J. LaTempa, C. A. Grimes, *Nano Lett.* 9 (2009) 731.

84. A. L. Linsebigler, G. Lu, J. T. Yater Jr, *Chem. Rev.* 95 (1995) 735.

85. X. Feng, J. D. Sloppy, T. J. LaTempa, M. Paulose, S. Komarneni, N. Bao, C. A. Grimes, *J. Mater. Chem.* 21 (2011) 13429.

86. T. Yui, A. Kan, C. Saitoh, K. Koike, T. Ibusuki, O. Ishitani, *ACS Appl. Mater. Interfaces* 3 (2011) 2594

87. J. C. S. Wu, T.-H. Wu, T. Chu, H. Huang, D. Tsai, *Top. Catal.* 47 (2008) 131.

88. J. Sá, M. Fernádez-Garcia, J. A. Anderson, *Catal. Commun.* 9 (2008) 1991.

89. M. Abou Asi, C. He, M. Su, D. Xia, L. Lin, H. Deng, Y. Xiong, R. Qiu, X.-Z. Li, *Catal. Today* 175 (2011) 256.

90. B. Aurian-Blajeni, M. Halmann, J. Manassen, *Sol. Energy* 25 (1980) 165.

91. H. Shi, T. Wang, J, Chen, C. Zhu, J. Ye, Z. Zou, *Catal. Lett.* 141 (2011) 525.

92. P. W. Pan, Y. W. Chen, *Catal. Commun.* 8 (2007) 1546.

93. Y. Liu, B. Huang, Y. Dai, X. Zhang, X. Qin, M. Jiang, M. H. Whangbo, *Catal. Commun.* 11 (2009) 210.

94. Y. Zhou, Z. Tian, Z. Zhao, Q. Liu, J. Kou, X. Chen, J. Gao, S. Yan, Z. Zou, *ACS Appl. Mater. Interfaces* 3 (2011) 3594.
95. Q. Liu, Y. Zhou, J. Kou, Z. Tian, J. Gao, S. Yan, Z. Zou, *J. Am. Chem. Soc.* 132 (2010) 14385.
96. Q. Liu, Y. Zhou, Z. Tian, X. Chen, J. Gao, Z. Zou, *J. Mater. Chem.* 22 (2012) 2033.
97. N. Ahmed, M. Morikawa, Y. Izumi, *Catal. Today* 185 (2012) 263.
98. J. T. S. Irvine, B. R. Eggins, J. Grimshaw, *Sol. Energy* 45 (1990) 27.
99. H. Kisch, G. Twardzik, *Chem. Ber.* 124 (1991) 1161.
100. H. Inoue, H. Moriwaki, K. Maeda, H. Yoneyama, *J. Photochem. Photobiol. A* 86 (1995) 191.
101. J. Barber, *Chem. Soc. Rev.* 38 (2009) 185.
102. A. Nishimura, N. Sugiura, S. Kato, *N. Proc. Int. Energy Convers. Eng. Conf. Rhode Island* (2004) 824.
103. R. Asahi, T. Morikawa, T. Ohwaki, K. Aoki, Y. Taga, *Science* 293 (2001) 269.
104. S. Hoang, S. Guo, N. T. Hahn, A. J. Bard, C. B. Mullins, *Nano Lett.* 12 (2012) 26.
105. (a) B. O'Regan, M. Grätzel, *Nature* 353 (1991) 737; (b) M. Grätzel, *Nature* 414 (2001) 338.
106. O. Bräm, A. Cannizzo, M. Chergui, *Phys. Chem. Chem. Phys.* 14 (2012) 7934.
107. A. Hagfeldt, M. Grätzel, *Chem. Rev.* 95 (1995) 49.
108. For example: http://nanocomposix.com/products.
109. (a) K. R. Catchpole, A. Polman, *Opt. Exp.* 16 (2008) 21793; (b) D. Duche, P. Torchio, L. Escoubas, F. Monestier, J.-J. Simon, F. Flory, G. Mathian, *Sol. Energ. Mat. Sol.* 93 (2009) 1377; (c) M. D. Brown, T. Suteewong, R. S. S. Kumar, V. D'Innocenzo, A. Petrozza, M. M. Lee, U. Wiesner, H. J. Snaith, *Nano Lett.* 11 (2011) 438; (d) L. M. Peter, *J. Phys. Chem. Lett.* 2 (2011) 1861.
110. (a) Y. Nishijima, K. Uno, Y. Yokota, K. Murakoshi, H. Misawa, *J. Phys. Chem. Lett.* 1 (2010) 2031; (b) F. Wang, N. A. Melosh, *Nano Lett.* 11 (2011) 5426; (c) Y. Tian, T. Tatsuma, *J. Am. Chem. Soc.* 127 (2005) 7632.
111. (a) P. Christopher, H. Xin, S. Linic, *Nature Chem.* 3 (2011) 467; (b) D. Tsukamoto, Y. Shiraishi, Y. Sugano, S. Ichikawa, T. Sanaka, T. Hirai, *J. Am. Chem. Soc.* 134 (2012) 6309; (c) P. Christopher, H. Xin, A. Marimuthu, S. Linic, *Nature Mater.* 11 (2012) 1044; (d) S. Linic, P. Christopher, D. B. Ingram, *Nature Mater.* 10 (2011) 911; (e) W. Hou, S. B. Cronin, *Adv. Funct. Mater.* 23 (2013) 1612; (f) Z. Zhang, L. Zhang, M. N. Hedhili, H. Zhang, P. Wang, *Nano Lett.* 13 (2013) 14.
112. J. Sá, G. Tagliabue, P. Friedli, J. Szlachetko, M. H. Rittmann-Frank, F. G. Santomauro, C. J. Milne, H. Sigg, *Energy Environ. Sci.* 6 (2013) 3584.
113. (a) M. W. Kanan, D. G. Nocera, *Science* 321 (2008) 1072; (b) Y. Surendranath, M. Dincă, D. G. Nocera, *J. Am. Chem. Soc.* 131 (2009) 2615.
114. S. Y. Reece, J. A. Hamel, K. Sung, T. D. Jarvi, A. J. Esswein, J. J. H. Pijpers, D. G. Nocera, *Science* 334 (2011) 645.
115. M. D. Symes, P. J. Kitson, J. Yan, C. J. Richmond, G. J. T. Cooper, R. W. Bowman, T. Vilbrandt, L. Cronin, *Nature Chem.* 4 (2012) 349.

4 Methane Activation and Transformation over Nanocatalysts

Rajaram Bal and Ankur Bordoloi

CONTENTS

4.1 Introduction 123
4.2 Reforming of Methane... 125
    4.2.1 Oxidative Coupling of Methane 127
4.3 Direct Conversion to Aromatics and Hydrocarbon in Absence
    of Oxygen.. 136
    4.3.1 Coupling of Methane 137
    4.3.2 Methane to Methanol... 139
4.4 Conclusions.. 141
References.. 142

4.1 INTRODUCTION

Nanocatalysis is one of the most stimulating fields to have emerged from current science and technology as sustainable alternatives to conventional materials. The main objectives of nanocatalysis research are to produce catalysts with 100% selectivity, extremely high activity, low energy consumption, and long lifetime by changing the size, dimensionality, chemical composition, and morphology of the reaction center [1–3]. This approach initiates the new avenues for an atom-by-atom catalyst design. The development of nanoparticle (NP) catalysis actually began in the 1950s when research labs, in an attempt to reduce the cost for large commercial applications, developed supported metal catalysts with particle size of less than 100 nm.

Natural gas is an abundant resource in various parts of the world. Methane is the main constituent of natural gas, coal-bed gas, and biogas. Methane is produced in large quantities during oil production as associated gas and also produced in the petrochemical refining and petrochemical processes. It is also available in large quantities as methane hydrate at seabeds. By 2009, the proven natural gas reserves were 6219.25 Tcf. According to this 2009 report, advances in technology could bring natural gas resources to 7775 Tcf (an increase of around 25%) by 2017. This increase may additionally be attributed to extractable methane from shale rock. Most of these reserves are located in the Middle East with 2658 Tcf, or 40% of the world total, and Europe and the former USSR with 2331 Tcf, or 35% of total world reserves. With the present

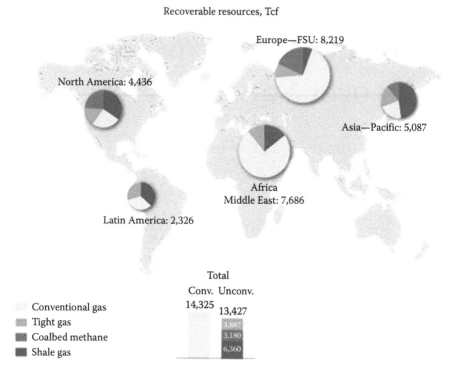

FIGURE 4.1 World gas scenario. (World Energy Outlook 2009.)

usage rate of methane at 115.554 Tcf, the present reserves can sustain the energy needs for at least next 50 years. Based on the data from BP (British Petroleum) at the end of 2009 proved gas reserves were dominated by three countries: Russia (24%), Iran (16%), and Qatar (14%) (see Figure 4.1 for methane gas global distribution).

In many respects, methane is an ideal fuel for these purposes because of its availability in most populated centers, its ease of purification to remove sulfur compounds, and the fact that among the hydrocarbons, it has the largest heat of combustion relative to the amount of CO_2 formed. On the other hand, methane is a greatly underutilized resource for chemicals and liquid fuels. Much of the methane is found in regions that are far removed from industrial complexes and often it is produced offshore. Pipelines may not be available for transporting this remote gas to potential markets and liquefaction for shipping by oceangoing vessels is expensive. Approximately, 11% of this gas is reinjected, and unfortunately, another 4% is flared or vented.

Methane (natural gas) is frequently produced along with liquid petroleum and some contracts require that the producer takes the gas along with the oil. Because of this arrangement, the methane may actually have "negative value," that is, it is a liability to the producer. For safety reasons, gassy underground coal mines must drain methane from their coal seam. Most coal mines vent this methane to the atmosphere, which not only represents the loss of a valuable fuel source but also contributes to global warming, as methane is a potent greenhouse gas. Coal mines do not produce enough methane to fuel large methanol plants, but one or more very gassy mines typically produce enough methane to fuel a small (25–30 million gallons per year) methanol plant. Alternatively, smaller (3–5 million gallons per year) mobile methanol plants currently used at offshore oil rigs may be a potential option for use at coal mines.

The mastery of methane chemistry would provide chemicals and liquid fuels, presenting an alternative to petroleum in these applications and enabling the use of a plentiful, though often remote, natural gas that is currently uneconomical to transport to target markets. In addition, it could also reduce the severe greenhouse effect of CH_4 (21 times higher than the equivalent volume of CO_2) or the gas flaring associated with petroleum production, and it might even provide a way of upgrading landfill gas.

Because of the increasing importance of the CO_2 sequestration for avoiding global warming, CO_2 reforming of methane to syngas (synthesis gas $H_2 + CO$) has been considered as an important process for the utilization of CO_2. Apart from the high energy requirement, this process also suffers badly from a very rapid carbon deposition on the catalyst, particularly, Ni-based catalysts. It is therefore of great practical importance to develop a non-noble metal-based catalyst, which allows only a little or no carbon formation in the CO_2 reforming reaction. In a free market economy, the profitability of these facilities will depend on the selling price of the product and the cost of alternative technologies. Methane is the most inert among hydrocarbons and hence its activation at low temperature (<600°C) is very challenging.

The current ways to activate methane (Figure 4.2) are as follows:

1. Steam and carbon dioxide reforming or partial oxidation of methane (POM) to form carbon monoxide and hydrogen, followed by Fischer–Tropsch chemistry
2. Direct oxidation of methane to methanol and formaldehyde
3. Oxidative coupling of methane to ethylene
4. Direct conversion to aromatics and hydrogen in the absence of oxygen

Syngas (a mixture of CO and H_2) is a versatile feedstock for methanol and ammonia syntheses and oxo-processes and also for a number of Fischer–Tropsch syntheses for the production of liquid fuels, olefins, and oxygenates. Hence, the conversion of methane to value-added products via its conversion to syngas is highly promising one, provided the syngas from methane is produced economically. Presently, syngas is produced by the steam reforming of methane (SRM) or higher hydrocarbons. Because of its several limitations/drawbacks (high energy requirement, high H_2/CO ratio, low selectivity/yield for CO, high capital cost, and low space-time yield), the steam reforming process is uneconomical for its use in methane conversion.

4.2 REFORMING OF METHANE

Syngas is also used for ammonia/urea production and methanol synthesis; after removal of carbon monoxide, hydrogen is extensively used by the oil industry in the hydrotreating and in hydrocracking processes. Today, the main process for producing syngas is based on the *steam reforming of methane*, which is the main component of natural gas.

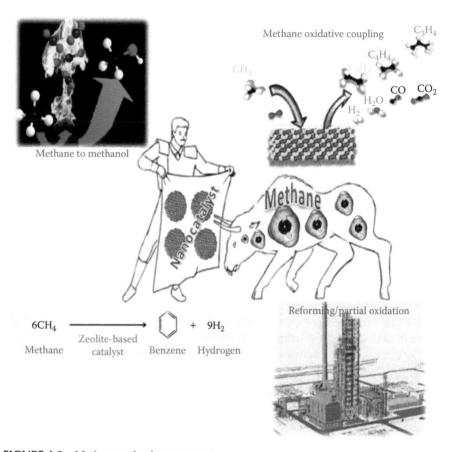

FIGURE 4.2 Methane activation processes.

$$CH_4 + H_2O \rightarrow CO + 3H_2 \qquad \Delta H_{298\,K} = 206 \text{ kJ/mol} \qquad (4.1)$$

$$CO + H_2O \rightarrow CO_2 + 3H_2 \qquad \Delta H_{298\,K} = -41.2 \text{ kJ/mol} \qquad (4.2)$$

This is a highly endothermic reaction having some disadvantages: high temperatures and pressures are required for methane conversions exceeding 95%; high temperature- and high pressure-resistant materials are needed; and a H_2/CO mole ratio of 3.0 is obtained, which is not the desired stoichiometry of some chemical reactions such as methanol and Fischer–Tropsch synthesis (which require H_2/CO mole ratios of 2.0).

POM to synthesis gas is a mildly exothermic reaction that is free from the limitations of the aforementioned reaction and would be more energetically efficient. In fact POM, having a H_2/CO mole ratio of 2.0, would be a viable alternative reaction to the methane steam reforming reaction for syngas generation.

$$CH_4 + \frac{1}{2}O_2 \rightarrow CO + 2H_2 \qquad \Delta H_{298\,K} = -36 \text{ kJ/mol} \qquad (4.3)$$

Although *dry reforming* (Equation 4.4) seems to be a potent technique to utilize and sequester carbon dioxide, however its high endothermicity along with coke forming propensity are some of its drawbacks which are detrimental to its prevalent usage compared to the above two reactions.

$$CH_4 + CO_2 \rightarrow 2CO + 2H_2 \qquad \Delta H_{298\,K} = 247.3 \text{ kJ/mol} \qquad (4.4)$$

4.2.1 OXIDATIVE COUPLING OF METHANE

The oxidative coupling of methane (OCM) involves the reaction of CH_4 and O_2 over a catalyst at high temperatures to form C_2H_6 as a primary product and C_2H_4 as a secondary product. Unfortunately, both the CH_4 and the C_2H_4 may be converted to CO_2, and the single-pass combined yield of C_2H_4 and C_2H_6 (C2 products) is limited to about 25%.

In the OCM process at around 800°C, the following selective (Equations 4.5 through 4.7) and nonselective (Equation 4.8) reactions occur simultaneously:

$$2CH_4 + \tfrac{1}{2}O_2 \rightarrow C_2H_6 + H_2O \qquad (\Delta H = -174.2 \text{ kJ/mol}) \qquad (4.5)$$

$$C_2H_6 + \tfrac{1}{2}O_2 \rightarrow C_2H_4 + H_2O \qquad (\Delta H = -103.9 \text{ kJ/mol}) \qquad (4.6)$$

$$C_2H_6 \rightarrow C_2H_4 + H_2 \qquad (\Delta H = +114.6 \text{ kJ/mol}) \qquad (4.7)$$

$$C_xH_y + O_2 \rightarrow CO_2 + H_2O + \text{large amount of heat} \qquad (4.8)$$

Today, the annual world energy demand amounts to some 350×10^{18} J, which is projected to be 2–5 times higher by the year 2050. The use of crude oil will decrease in the next century mainly because of global climate change, depletion of petroleum resources, and concomitant decline in crude oil quality. In addition, current energy systems are not closed cycles and therefore are ecologically harmful. An increased use of coal and natural gas as well as the development of alternative energy sources has become significantly important in recent years for fulfillment of our future energy needs [4]. The effective exploitation of natural gas, which is mainly composed of methane, also normally contains small quantities of higher hydrocarbons (e.g., ethane, propane), and sometimes other gases such as hydrogen sulfide, carbon dioxide, and nitrogen are crucial to the protection of new energy resources and environment. However, natural gas reserves are located in remote locations and are hence difficult to bring to energy market due to the expensive cost of compression, transportation, and storage. On a contrary, petroleum products are relatively cheap in current market. In last few decades, several researchers and companies are putting tremendous effort to investigate an economically viable pathway to convert natural gases to liquids or higher hydrocarbons. Nevertheless, a good process with high productive rate is yet to arrive.

Converting natural gas to syngas is an important process for natural exploitation of methane, because syngas could be exploited for (1) ammonia/urea synthesis, (2) oil refining operations, (3) methanol synthesis, (4) Fischer–Tropsch processes, and (5) other applications such as fine chemical synthesis and electricity generations. Steam reforming is the usual route for syngas generation. The process of SRM produces syngas ($H_2 + CO$) with a ratio $H_2/CO = 3$. In this catalytic process, methane reacts with water steam in the presence of a catalyst. The product of this reaction is the syngas [5]. Since the process of SRM leads to the formation of syngas with the major H_2/CO ratio, this type of reforming process is considered ideal to obtain hydrogen flow of high purity. The SRM is an endothermic process and, therefore, requires very high temperatures, which makes this process very expensive. The concern with the economic viability issue led to the development of alternative processes to reforming of methane, such as dry reforming, autothermal reforming, and partial oxidation [5,6]. Conventional catalysts for the SRM reaction are Ni on various supports, such as Al_2O_3, MgO, $MgAl_2O_4$, or their mixtures [7]. Selection of support material is a very crucial issue as it has been evident that metal catalysts are not very active for the SRM when supported on inert oxides [8]. Watanabe et al. [9] reported nanosized Ni particles supported on hollow Al_2O_3 ball by spraying a mixed solution of nickel and aluminum nitrates. They used a fixed-bed quartz tubular reactor for SRM. They reported a 92% methane conversion for their nanocatalyst. Sadykov et al. [8] also reported nanocomposite catalysts comprised of Ni particles embedded into the complex oxide matrix comprised of Y or Sc-stabilized Zr (YSZ, ScSZ) combined with doped ceria–zirconia oxides or La–Pr–Mn–Cr–O perovskite and promoted by Pt, Pd, or Ru were synthesized via different routes. Nano-NiO/SiO_2 with crystallite, size 9–15 nm supported on alumina is prepared by sol–gel technique and used successfully for SRM [10].

The dry reforming of natural gas is a process where methane reacts with carbon dioxide in the presence of a catalyst, and syngas at a $H_2/CO = 1$ ratio is obtained as a product of this reaction [5,11]. Due to the value of the H_2/CO ratio shown by the syngas obtained in the dry reforming of methane, this process is considered the ideal when it comes to the use of syngas produced as a raw material for the synthesis of important fuel liquids, which require H_2 and CO as raw materials. On the other hand, this type of reforming process is considered very expensive because, being an endothermic process, it consumes a great amount of energy.

The main disadvantage of dry reforming of methane is the significant formation of coke that is subsequently deposited on the surface of the catalyst, which is active in the reaction. The deposition of coke on the surface of the catalyst contributes to the reduction of its useful life. The large formation of coke occurred in this process is explained by the presence of the CO_2 reagent introduced in the catalytic process input which increases the production of coke. Thus, dry reforming is the unique process for reforming of methane powered by two greenhouse gases that contain carbon (CH_4 and CO_2) [5,11].

The main challenge for the industrial application of the reforming of methane with CO_2 is related to the development of active catalytic materials, but with a very low coke formation rate, either on the catalysts or in the cold zones of the reactor. The carbon formation in this process can be controlled by using a support that favors the dissociation reaction of CO_2 into CO and O, the last species being responsible for the cleaning of the metallic surface [12]. Although research into the CO_2 reforming of natural gas had been initiated in the 1920s, it gained renewed interest in the 1990s because of its potential applications in the greenhouse chemistry [13]. It is a promising means of disposing and recycling two important greenhouse gasses, CH_4 and CO_2, and a route to producing valuable synthesis gas [14].

Compared with SRM or POM, carbon dioxide reforming of methane provides synthesis gas with a relatively low H_2/CO ratio, which is more desirable for the direct use as feedstock for the synthesis of oxygenates, hydroformylation, oxo synthesis, and so on [14,15]. Also, this reaction is usually considered as chemical energy transmission system (CETS) due to its strong endothermic characteristic, in which a power source generated from solar or nuclear energy drives this intensively endothermic reforming reaction and converts these inexpensive energies into valuable chemical energy [16].

However, the major drawback of this catalytic process till now remains the rapid deactivation of the catalysts originating from the sintering of the metal active sites as well as the carbon deposition [14–16]. Therefore, the recent research focus in this field has been mainly concentrated on developing catalysts with favorable capacity of anticoke and antisintering. Many kinds of catalysts using Ni or noble metals such as Ru, Rh, Pd, Ir, and Pt have been reported to be active in this reaction. Noble metals have promising catalytic properties and low sensitivities to carbon deposits compared to nickel, but the unavailability and high cost of noble metals limit their application in large-scale processes [16]. Hence, Ni catalysts have been extensively investigated because of the metal availability and economic reasons [17–20].

Although Ni-based catalysts exhibit high activity and selectivity and are cheap, the major drawback of this reaction, however, is the rapid deactivation of catalysts as a result of carbon deposition and sintering of Ni metal particles [17,18]. Thus, in recent years, much effort has been devoted to developing Ni-based catalysts with improved performance, that is, lower coke deposition and higher stability against metal sintering [19,21]. Former experimental and theoretical studies have confirmed that size of the Ni particle has a crucial role in suppressing coke [18]. It has been reported that carbon deposition can occur only when the metal cluster is greater than a critical size. Therefore, to inhibit carbon deposition, it should be ensured that the size of the metal cluster is smaller than the critical size needed for coke formation. Hence, many methods have been explored recently in order to obtain mesoporous nanocrystalline powders with high surface areas for catalytic applications [19]. They provide catalysts with more edges and corners, which can lead to higher performance.

Among the catalyst supports, magnesium aluminate spinel, $MgAl_2O_4$, has been widely used in industrial applications [16]. This material has unique properties, such as high melting temperature ($2135°C$), high mechanical strength at elevated temperature, high chemical inertness, good thermal shock resistance, and catalytic properties [14]. Several synthesis methods for the preparation of $MgAl_2O_4$ spinel powders have been employed, such as, sol–gel, hydrothermal, combustion, and coprecipitation. For many of its applications especially as catalyst support, a high surface area, small crystalline size, high porosity, and more active sites are more desired. Due to the low density and good thermal stability, it has long been used as catalyst support for catalytic reforming [19].

Also, the studies report that ceria is an effective promoter to prevent the metallic sintering and to favor the activity as well as the resistance to coke formation [21]. It is known for its high oxygen storage/transport capacity (OSC), that is, its ability to use its lattice oxygen under an oxygen-poor environment and quickly reoxidize under an oxygen-rich environment. It increases nickel dispersion, thus enhancing resistance toward sintering and coke formation [22]. Table 4.1 presents a review of dry reforming catalysts that have been used recently.

Gonzalez-Delacruz et al. [23] reported the effect of reduction procedures using CO or H_2 on the size of nickel particles in Ni/ZrO_2 dry reforming of methane (DRM) catalysts. An increase in the dispersion of the nickel metallic phase has been observed at high temperature treatment using CO as reducing agent. Their X-ray absorption spectroscopy analysis reveals a lower coordination number of Ni present in the sample treated with CO than that reduced by H_2. The report also stated that using CO will lead to the formation of $Ni(CO)_4$ complexes, promoting the decrease in nickel particles size maintaining the same nickel content. They experimentally evidenced and demonstrated the whole redispersion phenomenon using several experiments of CO on nickel catalysts. Qu et al. [24] studied the CO_2 reforming of methane using Ni NPs immobilized at the tips of single-walled carbon nanotubes (SWNTs). They observed that (1) SWNTs area better support for nanocatalysts at high temperatures; (2) NPs at the tips of SWNTs are small enough to reduce or even eliminate carbon deposition on them; and (3) NPs at the tips of SWNTs do not sinter during DRM. Rezaei et al. [1] studied CO_2 reforming over nanocrystalline zirconia-supported nickel catalysts with methane conversion around 70%. The author also correlate the performance of catalysts with the surface area of the catalyst, the reaction temperature, and the feed portions. Various reports have been found [25–27] stating that preparation of a catalyst with high surface area has a great effect on methane activation via DRM. Nanocatalyst has several advantages in contrast to the ordinary catalysts with similar composition, and their effects can alter the reaction conditions and results. $Ni_xMg_{1-x}O$ nanoplatelet has been fabricated using topotactic decomposition methodology reported for DMR. Both nickel and magnesium were found to be very homogeneously dispersed with tunable composition; however, the surface orientation is preferential and independent of composition variations. The reduced $Ni_xMg_{1-x}O$ solid solutions are found to be stable toward methane dry reforming, due to homogeneous distribution of Ni [28].

POM may be advantageous compared to other syngas generation processes due to the following reasons. Although, the mechanism of partial oxidation is not clear till date, however, according to the literature reports, it can be broadly divided into two categories: One is the indirect oxidation mechanism involving total combustion of methane and steam and dry reforming reactions, often referred to as the combustion and reforming reactions (CRR) mechanism, and the other one is the direct oxidation mechanism in which surface carbon and oxygen species react to form primary products, known as the direct partial oxidation (DPO) mechanism. According to the CRR mechanism, synthesis gas is the secondary product, however, synthesis gas is also the primary product in the DPO mechanism.

The POM schematically and thermodynamically gives some information. This process is likely to become more important in the future of methane conversion due to the thermodynamic advantages of this process over other reforming processes. Possible pathways via which methane is converted are shown in Figure 4.3. At high temperature, the main reaction products between methane and oxygen are, however, limited to CO, CO_2, H_2O, and H_2 [29,30], apart from some intermediates.

TABLE 4.1
Review of Dry Reforming Catalysts

Catalyst	Space Velocity (ml/h/g)	Temperature (°C)	Conversion (%)		Reference
			CH_4	CO_2	
Ni–CeO$_2$ (26% Ni)	3,00,000	750	50	NA	[22]
Ni–CeO$_2$ (13% Ni)	3,00,000	750	40	NA	[22]
Ni–CeO$_2$ (7% Ni)	3,00,000	750	35	NA	[22]
Ni/CeO$_2$–ZrO$_2$	30,000	973	39	39	[21]
Ce-Zr1.51Ni0.49Rh0.03	30,000	800	90	90	[20]
Ni/Al	2,00,000	750	68	NA	[19]
Ni/Ce (3%) Al	2,00,000	750	75	NA	[19]
Rh–Ni(3%)/Ce Al	2,00,000	750	85	NA	[19]
Ni/Al$_2$O$_3$	30,000	800	92	95	[15]
Ni/MgO–Al$_2$O$_3$	30,000	800	92.5	91.8	[15]
NiO–MgO–Al$_2$O$_3$ (15% Ni, 2% Mg/83% Al)	15,000	700	94.64	95.9	[16]
(0.5%) Mo–(1%) Ni/SBA-15	4,000	800	96	NA	[15]
7 wt% Ni/MgAl$_2$O$_4$	18,000	700	70	74	[18]
NiFe$_2$O$_4$ SG	54,000	800	80	95	[20]
La$_2$NiO$_4$/α-Al$_2$O$_3$	1,500	800	98	NA	[17]

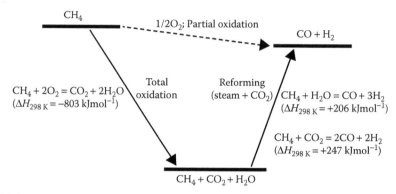

FIGURE 4.3   Partial oxidation of methane.

1. Partial oxidation is slightly exothermic; hence, reactor could be more economical in terms of heat and can make the processes more energy efficient.
2. Partial oxidation process may be hydrogen efficient because the H_2/CO ratio produced in the process is around 2, and this ratio is ideal for downstream processes, in the synthesis of methanol, which is a bridge between gaseous and liquids fuels.
3. POM can produce very low amount of CO_2, which can be easily removed before the use of syngas.
4. Partial oxidation process can avoid the expensive superheated steam in large scale.

Catalysis is becoming a strategic field for generation of syngas via partial oxidation. Heterogeneous catalysts for partial oxidation are classified into two large categories:

1. Noble metal-based catalysts
2. Non-noble metal-based catalysts

Noble metal-based catalyst such as rhodium, ruthenium, iridium, platinum, and palladium based in heterogeneous form showed excellent catalytic performance in partial oxidation. Owing to huge importance in industries, various supports have been exploited for noble and non-noble catalyst in partial oxidation process [31]. The supports may be classified into the following categories: (1) ceramic monoliths, (2) hydrotalcite type, (3) alkaline/rare earth doped perovskite type and other type of oxides, (4) alumina/CaO, and (5) others.

The catalytic partial oxidation has been first reported by Liander [32], Padovani, and Franchetti [33], and Prettre et al. [34] with a H_2/CO ratio of 2 around 727°C–927°C and 1 atm over supported nickel catalysts. Various evidences were also found in the literature justifying that the nickel is very active for the production of syngas with the necessary H_2/CO ratio; however, it also facilitates carbon formation. A basic work from Lunsford and coworkers demonstrates that CO selectivity was 95% and total conversion of methane has been obtained at temperatures higher than 700°C. However, the higher stoichiometric oxygen is a necessary requirement (i.e., $O_2/CH_4 > 0.5$) for the stable operation of the process. Three catalyst species were observed by them in the catalyst bed: (1) $NiAl_2O_4$, (2) NiO/Al_2O_3, and (3) Ni/support. The first two species are responsible for complete combustion of methane to CO_2 and H_2O; and the third phage is active for reforming of methane with CO_2 and H_2O to produce syngas [35]. Nickel is a wonderful catalyst for partial oxidation; nevertheless, it is also responsible for the carbon deposition. Much effort has been made to modify the support, to limit the carbon formation, to increase the stability of the catalyst system, and further to extend the catalyst life.

Supported nickel catalyst over ytterbium oxide, CaO, TiO_2, ZrO_2, ThO_2, UO_2, and rare earth oxide-modified alumina supports has been reported for POM by Choudhary et al. [36–44]. They found quite promising results on NiO containing MgO, CaO, rare earth oxides, or alumina catalysts with extremely low contact time. The researchers found the following order based on performance: $NiO/ThO_2 > Ni/UO_2 > NiO/ZrO_2$. However, SiO_2 and TiO_2 were found not good supports for the reaction, due to the sintering of Ni and formation of inactive binary metal oxide phases under the reaction conditions. The catalyst stability can be improved to some extent with suitable modification of support, conversely, catalyst deactivation is still hard to avoid due to both the loss of nickel metallic surface area and unwanted carbon formation. Ruckenstein [45,46] and Santos et al. [47] studied Ni/MgO catalyst system in detail and observed high stability of the catalyst with substantial amount of carbon formation during reaction. They conclude the high catalyst stability due to the formation of solid solution, for example, Ni occupies sites in the MgO lattice and is distributed evenly in the catalyst. The Ni is present as very small particles and the weak basicity of the MgO somewhat suppresses carbon deposition [48–50]. Addition of rare earth metal oxide or alkaline metal oxide to alumina or the use of rare earth metal oxide as support can restrict carbon deposition [51–57]. Nickel catalysts supported with rare earth metal oxide-modified and alkaline metal oxide-modified alumina have been tested for 500 hours without observable decrease of CH_4 conversion and H_2/CO ratio [58]. The promotion of the rare earth oxide addition on the catalyst support is probably due to its capability for oxygen storage, which can help by oxidizing the surface carbon deposited. It is also believed that the presence of a rare earth oxide such as CeO_2 can stabilize the support and prevent it from sintering during the high-temperature reaction.

Various modified alumina supports have been explored in lieu to eliminate the carbon deposition. Supports like $CaAl_2O_4$ have a high resistivity toward sintering and carbon deposition, hence giving good CH_4 conversion and CO and H_2 selectivity. The poor stability has been observed in case of supports like $AlPO_4$. In case of Ni-support, perovskites ($Ni/Ca0:8Sr0:2TiO_3$) are also quite resistive toward carbon deposition [59,60]. The reports proposed that the support was able to control the size of the metal crystallites, maintaining them below the threshold size at which carbon formation becomes a problem [58]. They also suggested that oxygen atoms in the support might be able to react with surface carbon, thereby keeping the nickel surface carbon-free. Another way to control nickel metal crystallite particles in a supported catalyst is to prepare a catalyst by reduction of Ni containing Ni/Mg/Al hydrotalcite-type precursors [61]. The catalyst activity and selectivity are related to the reductivity of nickel metal in the catalyst, and the nickel metal content and contact time of the reactants have a significant influence on the catalyst performance. Further studies on this catalyst system are currently ongoing.

Metals like Co and Fe are also been applied along with Ni to improve the stability and decrease the carbon deposition of nickel catalyst. Provendier et al. [42] observed that addition of iron could stabilize nickel catalysts via control of the reversible migration of nickel from the structure to the surface. Choudhary [43] observed that addition of cobalt to NiO/Yb_2O_3, NiO/ZrO_2, and NiO/ThO_2 catalysts causes a drastic reduction in the rate of carbon formation and also resulted in a huge decrease in the catalyst activation temperature of the oxidative conversion of methane to synthesis gas because the presence of Co promotes the reduction of Ni and thus improves the catalyst activity. Independent studies have also been made for Co and Fe supported studies; however, comparatively low activity has been observed since CoO and Fe_2O_3 have higher activity for complete oxidation of methane [56,62,63]. The order for POM activity of the supported catalysts is Ni/Co > Fe. It was noticed that the cobalt catalysts are active for partial oxidation to syngas only when promoters have been used to favor Co reducibility [40,64,65]. Similarly, metallic cobalt is the active species for POM reaction of cobalt-based catalyst; moreover, the activity also depends on the method of preparation and choice of support [66–68].

In conclusion, supported Ni and Co catalysts have been widely studied for methane partial oxidation to synthesis gas. Very limited attention has been paid to iron catalysts. The deactivation of nickel catalyst for POM is due to carbon deposition, and the loss of nickel due to the high flow rate. Co and Fe have higher melting and vaporizing points than nickel, which may be alternatives for nickel catalysts, if a higher performance can be obtained. Further clues to techniques for modifying nickel catalysts to suppress carbon formation may be obtained from the literature on steam reforming, where methods for controlling crystallite size or the addition of dopants are well established [69].

Green et al. [70,71] summarized the production of syngas over all noble metal catalysts. Palladium catalyst behaves almost like nickel, with substantial amount of carbon formation, however, iridium and rhodium catalyst show high resistivity toward carbon deposition [72,73]. Poirier and coworkers observed that the products of POM were basically governed by the kinetics while applying high gas hour space velocity (GHSV) [74]. Moreover, Rh was found to be more active than Ni despite its low loading in alumina support.

Hochmuth et al. [75–78] and Schwiedernoch et al. [79] observed that the activity of noble metals for syngas generation via POM not only depends upon noble metal itself but also relates to the method of preparation and support properties. Basile et al. [80] observed that syngas generation activity follow the order Rh > Ru ~ Ir = Pt > Pd on clay support. Yan et al. [81] indicated that the conversion and selectivity were relatively stable over the Rh-based catalyst, however, change has been observed in case of Ru-based catalyst. Furthermore, the pulse reaction studies showed that the reaction mechanisms over these two catalysts systems are different, CO was only the main carbonaceous product formed over Rh catalyst, on the other hand, CO_2 has formed along with CO in case of Ru catalyst. Ir showed activity order of supports as follows: $TiO_2 \le ZrO_2 \le Y_2O_3 > La_2O_3 > MgO \le Al_2O_3 > SiO_2$ [82] for POM. A series of rare earth supported noble metal catalysts was studied, among them Pt/Gd$_2$O$_3$ and Pd/Sm$_2$O$_3$ gave good catalytic performance [83] with high CO selectivity. They conclude that alkali earth or rare earth metal oxides not only improve the selectivity but also disperse the noble metal on the support.

It was observed that the presence of highly dispersed small metallic particles (below 10 nm) on support and its modification with basic oxides are required to avoid sintering and coking problems [84]. In order to overcome these problems, some reports have been found in which metal NPs were prepared via polyol process. Claridge et al. have shown that the relative order of carbon formation is Ni > Pd >> Rh, Ru, Ir, Pt [85,86]. A series of noble metal catalysts (Ru, Rh, Ir, Pt, and Pd) supported on alumina-stabilized magnesia were prepared and employed in POM. The following order of activity was observed for different catalysts in POM: Rh = Ru > Ir > Pt > Pd. Takenaka et al. [87,88] reported that silica-coated Ni catalysts showed high activity and improved stability in the steam reforming of propane and POM. Compared to conventional Ni/Al$_2$O$_3$, Ni/MgO, and Ni/SiO$_2$ catalysts, core–shell structured Ni catalysts show less carbon deposition. It was thought that the strong interaction of the Ni cores with the silica shell prevents Ni particles from sintering as well as hindering carbon deposition. NPs of narrow size distribution encapsulated inside meso- and microporous silica were prepared through *in situ* reduction of NiO NPs coated with silica. By varying the preparation parameters, the mean size of Ni NPs can be fine-tuned in the range of 4.5– 6 nm. It was found that with variation in core size, microcapsular cavity, and shell porosity, the as-obtained Ni@meso-SiO$_2$ catalysts for the POM to synthesis gas are notably different in catalytic activity and durability [89,90]. The catalyst activity and durability are essentially determined by the size of the Ni cores, and also somewhat by the porosity of SiO$_2$ shells, as well as the extent of core–shell interaction, which is influenced by the microcapsular cavity structure.

Overall, apart from the type of process, methane activation is an important chemical practice in the energy sector worldwide, because it is the first catalytic step to convert methane, and design the road maps for subsequent catalytic processes that are essential for fuel and chemicals. In general, the ultimate purpose of methane reforming is to obtain fuel and chemicals; however, the type of method used in methane conversion process to syngas influences on H$_2$/CO ratio obtained. The main type of reforming processes of methane is the process called steam reforming, because it generates syngas with the highest H$_2$/CO ratio. The product of the reforming process is a gas flow that is considered ideal for the development of catalytic processes to acquire high-purity gaseous hydrogen and to be used as feedstock in the petrochemical industries to produce liquid fuels and methanol. However, since the process of steam reforming is rather expensive, the other three types of catalytic chemical processes are considered as alternative processes for carrying out the reforming of methane, and they were developed with the aim of making savings in thermal energy consumption required by these catalytic processes. Thus, the choice of the most appropriate catalytic chemical process of methane reforming must account for the economic viability of the process with regard to the destination to be given to the syngas produced. In other words, it can be said in short that the choice of the type of catalytic chemical process of reforming to be used in the conversion of methane in syngas should be made based on the final application that will be given to syngas obtained. Dry reforming and partial oxidation can also be good choices to produce syngas, if the value of H_2/CO ratio is adequate, especially when it comes to reduce the consumption of thermal energy, one of the most important factors in the implementation of an industrial project.

The cost of syngas produced by means of the SRM process is acutely dependant on natural gas prices and is currently the least expensive among all bulk hydrogen production technologies. A well-developed natural gas infrastructure already exists in the United States, a key factor that makes syngas production from natural gas by means of steam reforming very attractive for hydrogen generation. SRM is widely used in industry today. Hydrogen is produced by the SRM process in large centralized industrial plants for use in numerous applications, including chemical manufacturing and petroleum refining. Research and development programs are currently concerned with the development of small-scale technologies for SRM to enable distribution of hydrogen and improve delivery infrastructure [91–99].

4.3 DIRECT CONVERSION TO AROMATICS AND HYDROCARBON IN ABSENCE OF OXYGEN

The direct nonoxidative conversion of CH_4 is thermodynamically unfavorable. Nevertheless, an alternative approach had been deployed, but still it attracted the attention of many researchers. In early 1970s, Olah and his group showed the possibility to convert CH_4 into higher hydrocarbons via the homologation of methane in a superacid medium [90]. Till today, extensive research in this area has been undertaken to direct production of aromatic and hydrogen via nonoxidative conversion of lighter alkanes. In heterogeneous catalysis, various metals have been discovered that can chemisorb CH_4 at moderate temperatures and that can decompose CH_4 to aromatic (mainly BTX) and H_2 at higher temperatures. Choudhary and coworkers independently walked on a "single step" process where 36.3% methane conversion with 93.8% aromatic selectivity was achieved through low temperature nonoxidative conversion of methane in combination with higher alkanes over H-GaAlMFI [91]. Dehydrogenation and aromatization of CH_4 on modified ZSM-5 zeolite have been studied under nonoxidative condition by Wang and his group, they found benzene is the only hydrocarbon product of the catalytic conversion of methane at 600°C with a fixed-bed continuous-flow reactor [92]. Aboul-Gheit found that 6% Mo/HZSM-5 exhibits higher aromatization (benzene and naphthalene) up to 6%–7%, whereas higher electronegativity of Fe, Co, and Ni was assumed to contribute to the lower dehydroaromatization activity to benzene [93]. Working on the same, Liua et al. prepared MoO_3/HZSM-5 by impregnation over the distorted octahedrally coordinated Mo_{6c} species, ~70% BTX selectivity was achieved with 2%–3% methane conversion [94].

In order to overcome the thermodynamic limit and to enhance high-end aromatics in direct conversion of CH_4 under nonoxidative conditions, plasma excitation has also been attempted. High yields of aromatics and hydrogen were obtained by Li et al. [95]

in anaerobic conversion of methane using a two-stage pulsed spark discharge plasma reactor over Ni/HZSM-5 catalyst at atmospheric pressure and low temperature. BTX were found as major aromatic products with a maximum of 35% methane conversion and 47% aromatic selectivity. The work was extended; they utilized stable kilohertz spark discharges, which give 72.1 vol% of hydrogen concentration in the product stream having 81.5% of methane conversion [96].

ZSM-5 and Ga/ZSM-5 were found to give consistently high (>95%) n-decane conversion over the temperature range of 300°C–460°C where the parent zeolite, ZSM-5, produce almost equal yields of cracked hydrocarbons and aromatics [97]. The Ga-modified ZSM-5 produced predominantly BTX and other heavier aromatics. TNU-9 and IM-5 showed high content of methylated aromatics and polyaromatics from methanol at atmospheric pressure at 350°C and weight hours space velocity (WHSV) of 9/hour but fast deactivation was observed in these two structures. However, the product streams of TNU-9 and IM-5 contained unstable penta- and hexamethyl benzenes with 100% methanol conversion [98]. Propane aromatization to high yield (33.5%) C_6^+ aromatics were studied over Zn/Na–ZSM-5 by Iglesia et al. [99].

4.3.1 COUPLING OF METHANE

La_2O_3 nanorods with large surface area and strong basic sites were found to be highly active toward oxidative coupling of methane and high selectivity for C2 hydrocarbons [100]. Computational fluid dynamics (CFD), which has been used for obtaining detailed rate and temperature profiles through the porous catalytic pellet, where reaction and diffusion compete, indicated that the temperature variation within the catalyst pellet is <–271°C, thereby taking part in oxidative coupling reaction, as observed by Maghrebi et al. [101]. Moreover, Fe- and Au-modified MgO have also been reported for oxidative coupling of methane [102]. Jeon et al. reported that, at low temperature, oxidative coupling of methane has occurred facilely over Mg–Ti mixed oxide supported on Na/W/Mn catalysts at low temperatures, due to the presence of more active surface lattice O atoms [103]. Low temperature and oxidative coupling of methane with CO_2, as oxidant, in the presence of electric field, have been reported by Oshima et al. with greatly enhanced catalytic activity [104]. Samarium NPs with and without alumina support have been prepared using microemulsion, organic matrix decomposition, and wet impregnation methods effectively evaluated for oxidative coupling of methane. Moreover, Elkins et al. observed that samaria NPs are very selective toward C2 hydrocarbons [105]. Visinescu et al. reported the preparation of nanosized nickel-substituted zinc aluminate oxides by the gradual insertion of nickel cations within the zinc aluminate lattice, using starch as active ingredient and the catalyst showed high catalytic activity in oxidative methane coupling reaction, owing to the synergistic effect in the system [106]. Solution combustion synthesis method has been applied for the preparation of nanostructured complex metal oxides such as (1) Sr–Al complex oxides, (2) La_2O_3, (3) La–Sr–Al complex oxides, and (4) Na_2WO_4–Mn/SiO$_2$ and capitalized for oxidative coupling of methane. It was observed that Sr–Al complex oxide is active for oxidative coupling reaction and the Na_2WO_4–Mn/SiO$_2$ couple showed a yield of 25% in that reaction [107].

Baidya et al. reported Sr/Al$_2$O$_3$ couple as highly active catalyst for the oxidative coupling of methane at higher temperatures and they also correlate the activity with Al-IV/Al-VI ratio and Sr content [108]. The WO$_4$ tetrahedron on the catalyst surface appears to play a crucial role in achieving high methane conversion and high C2 hydrocarbon selectivity in the oxidative coupling of methane of M–W–Mn/SiO$_2$ (where M = Li, Na, K, Ba, Ca, Fe, Co, Ni, and Al) catalyst system observed by Ji et al. [109]. Theoretical calculations also support that the WO$_4$ tetrahedron interacts with the CH$_4$, giving suitable geometry and similar energy configuration with CH$_4$, and this may be the reason for high methane activation. Another report has also been found for oxidative coupling of methane (OCM) with a yield of 18% at 825°C and 16.5% at 775°C over Mg–Ti mixed oxides supported on Na/W/Mn catalysts [110].

The silver-catalyzed, OCM to C2 hydrocarbons is found to be an exceptionally structure-sensitive reaction. Reaction-induced changes in the silver morphology lead to changes in the nature and extent of formation of various bulk and surface-terminating crystal structures. This, in turn, impacts the adsorption properties and diffusivity of oxygen in silver, which is necessary for the formation of subsurface oxygen. A strongly bound, Lewis-basic, oxygen species, which is intercalated in the silver crystal structure, is formed as a result of these diffusion processes. This species is referred to as O-gamma and acts as a catalytically active site for the direct dehydrogenation of a variety of organic reactants, which is observed by Nagy et al. [110]. About 20% of CH$_4$ conversion and 80% of C2$^+$ selectivity during the oxidative coupling of methane over Mn/Na$_2$WO$_4$/SiO$_2$ and Mn/Na$_2$WO$_4$/MgO catalysts at 800°C was achieved by Pak et al. [111]. A relationship between the state of surface oxygen and C2 selectivity of Nd$_2$O$_3$ and SrO (1 at%)/Nd$_2$O$_3$ catalysts was derived from contact potential differences (CPD) between a reference electrode and the catalyst surface as a function of oxygen partial pressure and temperature for OCM has been reported by Gayko et al. [112]. Jeon et al. reported oxidative coupling of methane with a yield of 18% at 825°C and 16.5% at 775°C over Mg–Ti mixed oxides supported on Na/W/Mn catalysts [103]. A nonoxidative methane coupling has been reported by Moya et al. using Pd on alpha-Al$_2$O$_3$ catalysts where Pd particles with low nanometer range, and the catalyst was found to be very efficient [113]. Szeto et al. performed nonoxidative coupling of methane in classical fixed-bed reactor tungsten hydride catalyst supported on SiO$_2$–Al$_2$O$_3$ or gamma-Al$_2$O$_3$ [114]. An excellent example of photoinduced nonoxidative methane coupling at room temperature has been demonstrated by Yoshida et al. using highly dispersed zirconium oxide species on silica [115]. Murata et al. reported that Li-doped sulfated zirconia catalysts are found to be effective for methane, with 80% C2 selectivity attained at 800°C with 43% CH$_4$ conversion [116]. BaCO$_3$-supported vanadium oxide catalysts, which consist of BaCO$_3$ and small amounts of a barium orthovanadate Ba-3(VO$_4$) phase, exhibit high catalytic activity for oxidative coupling of methane, as reported by Dang et al. [117]. Moreover, Silica-supported tantalum hydride proves to be the first single-site catalyst for the direct nonoxidative coupling transformation of methane into higher alkanes and hydrogen at moderate temperatures, with a high selectivity (>98%) [118].

Li/Ce/MgO catalyst has been reported by Goncalves et al. They observed comparatively higher methane conversion and C2 selectivity than the Na/Ce/MgO catalyst [119]. They correlate the activity to different types of involved active site promoted by ceria during the reaction. Chen et al. reported oxidative coupling of methane at the temperature of 750°C over the Na–W–Mn–Zr–S–P/SiO$_2$ catalyst, on which the C2 yield was 23.5% at the methane conversion of 43.8% [120]. One of the elegant examples of dehydrogenative coupling of methane to ethane was also found in the literature using Zn$^+$-modified Zeolite in the presence of sunlight [121]. Takanabe et al. reported that OH radicals are very much responsible for the rate and C2 selectivity in oxidative coupling of methane for Mn/Na$_2$WO$_4$/SiO$_2$ catalyst system [122].

4.3.2 METHANE TO METHANOL

Methanol is currently manufactured from natural gas through an energy-intensive, two-step process. Methane reforming step for the current process is a highly endo-thermic and thermodynamically equilibrated process operated at high temperature (850°C). The direct methane to methanol oxidation process at low temperature contin-ues to be an interesting alternative to the current process. If successful, this new process could lead to a decrease of methanol price in the current methanol market, as well as opening of new methanol markets in methanol-to-chemical and methanol-to-fuel areas.

More efficient methods for the oxidation of low value, light alkane feedstocks, such as natural gas, to the corresponding alcohols or other useful liquid products would accelerate the use of natural gas feedstocks as a complement to petroleum. Methane is an abundantly available fuel whose use is mainly limited to that of a primary energy source due to its low reactivity. Methanol, on the other hand, is a use-ful intermediate material in many chemical manufacturing processes as well as a safe-to-handle liquid fuel for transportation and storage. Therefore, there is a long-standing industrial interest in producing methanol from methane effectively. Current technologies for the conversion of natural gas to liquid products proceed by gen-eration of carbon monoxide and hydrogen (syngas) that is then converted to higher products through Fischer–Tropsch chemistry. The initial formation of syngas in these processes is energy intensive and proceeds at high temperatures, typically 850°C. In contrast, direct methods partially oxidize the alkane molecule, functionalizing one C–H bond, and in principle can proceed more efficiently and cost-effectively through lower temperature routes.

The direct oxidation of methane to methanol has received much attention as the next step in methanol production since it avoids the above multistep processes. However, this oxidation is regarded as a very difficult reaction, especially in the gas phase at low pressure, because of the need to operate at high temperatures (>400°C), where methanol is quickly oxidized to formaldehyde and CO$_x$. One approach for oxi-dizing methane at lower temperatures is to apply an electrochemical cell to the reac-tion system. Otsuka and Yamanaka et al., for example, have reported the selective oxidation of light alkanes to oxygenates by the electrochemically activated oxygen species that are generated at the cathode in polymer electrolyte fuel cells (PEFCs) and phosphoric acid fuel cells (PAFCs) [123,124]. The reaction pathway and ener-getics for methane-to-methanol conversion by first-row transition metal oxide ions (MO$^+$s) are discussed from density functional theory (DFT) B3LYP calculations, where M is Sc, Ti, V, Cr, Mn, Fe, Co, Ni, and Cu. The methane-to-methanol con-version by these MO$^+$ complexes is proposed to proceed in a two-step manner via two transition states: $MO^+ + CH_4 \rightarrow OM^+(CH_4) \rightarrow [TS] \rightarrow OH^-M^+\text{-}CH_3 \rightarrow [TS] \rightarrow M^+(CH_3OH) \rightarrow M^+ + CH_3OH$. Both high-spin and low-spin potential energy surfaces are characterized in detail. A crossing between the high-spin and the low-spin potential energy surfaces occurs once near the exit channel for ScO^+, TiO^+, VO^+, CrO^+, and MnO^+, but it occurs twice in the entrance and exit channels for FeO^+, CoO^+, and NiO^+. Our calculations strongly suggest that spin inversion can occur near a crossing region of potential energy surfaces and that it can play a significant role in decreasing the barrier heights of these transition states. The reaction pathway from methane to methanol is uphill in energy on the early MO^+ complexes (ScO^+, TiO^+, and VO^+); thus, these complexes are not good mediators for the formation of methanol. On the other hand, the late MO^+ complexes (FeO^+, NiO^+, and CuO^+) are expected from the general energy profiles of the reaction pathways to efficiently convert methane to methanol. Measured reaction efficiencies and methanol branching ratios for MnO^+, FeO^+, CoO^+, and NiO^+ are rationalized from the energetics of the high-spin and the low-spin potential energy surfaces. The energy diagram for the methane-to-methanol conversion by CuO^+ is downhill toward the product direction, and thus CuO^+ is likely to be an excellent mediator for methane hydroxylation [125].

Platinum catalysts are reported for the direct, low temperature, oxidative conversion of methane to a methanol derivative at greater than 70% one-pass yield based on methane. The catalysts are platinum complexes derived from the bidiazine ligand family that are stable, active, and selective for the oxidation of a carbon–hydrogen bond of methane to produce methyl esters. Mechanistic studies show that platinum (II) is the most active oxidation state of platinum for reaction with methane, and are consistent with reaction proceeding through carbon–hydrogen bond activation of methane to generate a platinum–methyl intermediate that is oxidized to generate the methyl ester product. Roy A. Periana showed that dichloro (η-2-{2,2′-bipyrimidyl}) platinum (II) is the most effective one. The reaction of methane (34 bar, 115 mmol) with 80 ml of 102% H_2SO_4 containing a 50 mM concentration of the catalyst at 220°C for 2.5 hours resulted in 90% methane conversion and the formation of 1 M solutions of methyl bisulfate at 81% selectivity [126]. A dynamic model of oxidation of methane to methanol has been developed in a fixed-bed reactor applying V_2O_5/SiO_2 as the reaction catalyst. The effects of temperature (450°C–500°C) and pressure (20–120 bar) with residence time of 3 seconds on methane conversion and methanol or formaldehyde selectivity have been examined. Oxygen was used as an oxidant and the amount of oxygen in feed was 5% mol/mol of methane amount. The results showed with increasing the conversion of methane 0.66% to 1.52%, and the selectivity to methanol decreased from 93.4% to 91.9% [127]. A direct selective oxidation of methane to a methanol derivative with the catalysis of palladium acetate/benzoquinone/molybdovanado–phosphoric acid using molecular oxygen in trifluoroacetic acid has been suggested. Methyl trifluoroacetate is the only liquid product, and its highest yield could be obtained at 80°C–100°C [128]. Comparatively high CH_3OH selectivity (60.0%) and yield (6.7%) were obtained on $MoO_x/(LaCoO_3 + Co_3O_4)$ catalysts in selective oxidation of methane to methanol using molecular oxygen as oxidant. The interaction between MoO_x and La-Co-oxide modified the molecular structure of molybdenum oxide and the ratio of O^-/O_2^- on the catalyst surface, which controlled the catalytic performance of $MoO_x/(LaCoO_3 + Co_3O_4)$ catalysts [129].

A laboratory-scale simulated countercurrent moving bed chromatographic reactor (SCMCR) for the direct, homogeneous partial oxidation of methane to methanol has been constructed and tested. Reaction conditions were evaluated from independent experiments with a single-pass tubular reactor. Separation was effected by gas–liquid partition chromatography with 10% Carbowax on Supelcoport. At the optimal reaction conditions of 477°C, 100 atm, and feed methane-to-oxygen ratio of 16, the methane conversion in the SCMCR was 50%, the methanol selectivity was 50%, and the methanol yield was 25% [130]. In strong acid solvents such as triflic or sulfuric acid, cations [generated by dissolution Au(III) of Au_2O_3] react with methane at 180°C to selectively generate methanol (as a mixture of the ester and methanol) in high yield. The irreversible formation of metallic gold is very evident after these reactions and, unlike reactions with Hg(II), Pt(II), and Pd(II) that are catalytic in 96% H_2SO_4, only stoichiometric reactions [turnover numbers (TONs) < 1] are observed with Au(III). Soluble cationic gold is essential for these reactions as no methanol is observed under identical conditions without added Au(III) ions, or in the presence of metallic gold which is not dissolved in hot H_2SO_4 [131]. Graham J. Hutchings reported methane to methanol synthesis over Au–Pd/TiO_2 at 30°C–90°C using hydrogen peroxide. The maximum TOF and the highest methanol selectivity under these conditions (19%) were achieved at 90°C. Remarkably, methanol was stable at 90°C under our reaction conditions, but we noted that methyl hydroperoxide was the major reaction product in all cases. Suss-Fink and coworkers have shown that methyl hydroperoxide is transformed to formaldehyde and formic acid at temperatures above 40°C in the absence of a catalyst. However, at the temperatures we employed (30°C–90°C) these particular products were not observed, suggesting that the formation of methyl hydroperoxide and methanol is due to the presence of the Au–Pd catalyst [132].

Given the potential for high payoff, the goal of direct, selective alkane oxidation has been the focus of substantial effort since the 1970s. Despite these extensive efforts, very few selective alkane oxidation processes are known. Except in a few special cases, the basic chemistry for the selective, low temperature, direct, oxidative conversion of alkane C–H bonds to useful functional groups in high one-pass yield has not yet been developed. Such development is challenging, because alkane C–H bonds are among the least reactive known and the desired products of oxidation are typically more reactive than the starting alkanes and are consumed before recovery. Consequently, only uneconomically low one-pass yields can be obtained with direct alkane oxidation chemistries available today without prohibitively expensive separations and recycle. Regina et al. have shown that highly active Pt-carbon-based solid catalysts for methane oxidation by SO_3, and that these systems are stable over at least five recycling steps with high stability [133]. Pieter et al. reported Cu–ZSM5 for the transformation of methane to methanol with detailed mechanistic insight [134]. Ramakrishnan et al. reported biological dicopper centers for methane activation. Pt clusters and tricopper clusters have also been reported for methane to methanol synthesis [135].

4.4 CONCLUSIONS

Catalytically, methane can be activated using different ways such as (1) steam and carbon dioxide reforming or partial oxidation of methane to form carbon monoxide and hydrogen, followed by Fischer–Tropsch chemistry; (2) the direct oxidation of methane to methanol and formaldehyde; (3) oxidative coupling of methane to ethylene; and (4) direct conversion to aromatics and hydrogen in the absence of oxygen. The activation of methane and its transformation at ambient conditions can also reduce the greenhouse gas at the atmosphere—a threat for global warming. Despite significant efforts by several researchers the development of catalysts for the activation and transformation of methane at the commercial level is still a challenge for the scientific community. The importance of methane as a critical energy resource is not expected to diminish in the future and there is plenty of room for further research to bring about efficient competitive methane conversion technologies.

REFERENCES

1. Q. Shu, B. Yang, H. Yuan, S. Qing, G. Zhu, *Catal. Commun.* 8, 2007, 2159.
2. A. Farsi, S. Ghader, A. Moradi, S. S. Mansouri, V. Shadravan, *J. Nat. Gas Chem.* 20, 2011, 325.
3. S. W. Guo, L. Konopny, R. Popovitz-Biro, *Adv. Mater.* 12, 2000, 302.
4. M. Absi-Halabi, A. Stanislaus, H. Qabazard, *Hydrocarb. Process* 76, 1997, 45.
5. J. R. Rostrup-Nielsen, *Catal. Sci. Technol.* 5, 1984, 1.
6. J. N. Armor, *Appl. Catal. A: Gen.* 21, 1999, 159.
7. S. S. Maluf, E. M. Assaf, *Fuel* 88, 2009, 1547.
8. V. Sadykov, N. Mezentseva, G. Alikina, R. Bunina, V. Pelipenko, A. Lukashevich, S. Tikhov et al., *Catal. Today* 146, 2009, 132.
9. M. Watanabe, H. Yamashita, X. Chen, J. Yamanaka, M. Kotobuki, H. Suzuki, H. Uchid, *Appl. Catal. B: Env.* 71, 2007, 237.
10. B. Bej, N. C. Pradhan, S. Neogi, *Catal. Today* 207, 2013, 28.
11. J. A. Lercher, J. H. Bitter, A. G. Steghuis, J. G. V. Ommen, K. Seshan, *Environ. Catal., Catal. Sci. Series* 1999, 1.
12. Z. X. Cheng, J. L. Zhao, J. L. Li, Q. M. Zhu, *Appl. Catal. A: Gen.* 205, 2001, 31.
13. S. M. Stagg, E. Romeo, C. Padro, D. E. Resasco, *J. Catal.* 178, 1998, 137.
14. J. R. H. Ross, A. N. J. Keulen Van, M. E. S. Hegarty, K. Seshan, *Catal. Today* 30, 1996, 193.
15. K. K. Moon, K. K. Won, S. I. Wun, K. Ho-Younge, *Fuel Process. Techn.* 92, 2011, 1236.
16. X. Leilei, S. Huanling, C. Lingjun, *Appl. Catal. B: Env.* 108–109, 2011, 177.
17. B. S. Barros, D. M. A. Melo, S. Libs, A. Kiennemann, *Appl. Catal. A: Gen.* 378, 2010, 69.
18. H. Narges, R. Mehran, M. Zeinab, M. Fereshteh, *J. Nat. Gas Chem.* 21, 2012, 200.
19. O. Marco, P. Francisco, G. Gloria, *Catal. Today* 172, 2011, 226.
20. B. Koubaissy, A. Pietraszek, A. C. Roger, A. Kiennemann, *Catal. Today* 157, 2010, 436.
21. A. Kambolis, H. Matralis, A. Trovarelli, C. Papadopoulou, *Appl. Catal. A: Gen.* 377, 2010, 16.
22. V. M. Gonzalez-Delacruz, R. Pereniguez, F. Ternero, J. P. Holgado, A. Caballero, *ACS Catal.* 1, 2011, 82.
23. Y. Qu, A. M. Sutherland, T. Guo, *Energ. Fuels* 22, 2008, 2183.
24. M. Rezaei, S. M. Alavi, S. Sahebdelfar, P. Bai, X. Liu, Z. F. Yan, *Appl. Catal. B: Env.* 77, 2008, 346.
25. J. D. Aiken, R. G. Finke, *J. Mol. Catal. A* 145, 1999, 1.
26. C. Trionfetti, I. V. Babich, K. Seshan, L. Lefferts, *Appl. Catal. A* 310, 2006, 105.
27. H. Xiao, Z. Liu, X. Zhou, K. Zhu, *Catal. Commun.* 34, 2013, 11.
28. J. J. Zhu, J. G. Van Ommen, L. Leerts, *J. Catal.* 255, 2004, 388.
29. P. Aghalayam, Y. K. Park, N. Fernandes, V. Papavassiliou, A. B. Mhadeshwar, D. G. Vlachos, *J. Catal.* 213, 2003, 23.

30. T. V. Choudhary, V. R. Choudhary, *Angew. Chem. Int. Ed.* 47, 2008, 1828.
31. H. Liander, *Trans. Faraday Soc.* 25, 1929, 462.
32. C. Padovani, P. Franchetti, *Giorn. Chem. Ind. Appl. Catal.* 15, 1933, 429.
33. M. Prettre, C. Eichner, M. Perrin, *Trans. Faraday Soc.* 42, 1946, 335.
34. D. Dissanyake, M. P. Rosynek, K. C. C. Kharas, J. H. Lunsford, *J. Catal.* 132, 1991, 117.
35. V. R. Choudhary, A. S. Mamman, S. D. Sansare, *Angew. Chem. Int. Ed.* 31, 1992, 1189.
36. V. R. Choudhary, R. M. Ramarjeet, V. H. Rane, *J. Phys. Chem.* 96, 1992, 8686.
37. V. R. Choudhary, A. M. Rajput, B. Prabhakar, *J. Catal.* 139, 1993, 326.
38. V. R. Choudhary, V. H. Rane, A. M. Rajput, *Catal. Lett.* 22, 1993, 289.
39. V. R. Choudhary, A. M. Rajput, B. Prabhakar, *Catal. Lett.* 15, 1992, 363.
40. V. R. Choudhary, S. D. Sansare, A. S. Mamman, *Appl. Catal.* 90, 1992, 1.
41. V. R. Choudhary, A. M. Rajput, V. H. Rane, *Catal. Lett.* 16, 1992, 269.
42. V. R. Choudhary, V. H. Rane, A. M. Rajput, *Appl. Catal. A: Gen.* 162, 1997, 235.
43. V. R. Choudhary, A. M. Rajput, B. Prabhakar, A. S. Mamman, *Fuel* 77, 1998, 1803.
44. E. Ruckenstein, Y. H. Hu, *Appl. Catal. A: Gen.* 183, 1999, 85.
45. Y. H. Hu, E. Ruchenstein, *Ind. Eng. Chem. Res.* 37, 1998, 2333.
46. A. Santos, M. Menendez, A. Monzon, J. Santamaria, E. E. Miro, E. A. Lombardo, *J. Catal.* 158, 1996, 83.
47. V. R. Choudhary, A. S. Mamman, *Appl. Energy* 66, 2000, 161.
48. V. R. Choudhary, A. S. Mamman, *Fuel Process. Technol.* 60, 1999, 203.
49. S. Tang, J. Lin, K. L. Tan, *Catal. Lett.* 51, 1998, 169.
50. T. Zhu, M. Flytzani-Stephanopoulos, *Appl. Catal. A: Gen.* 208, 2001, 403.
51. S. Liu, G. Xiong, H. Dong, W. Yang, S. Sheng, W. Chu, Z. Yu, *Stud. Surf. Sci. Catal.* 130, 2000, 3573.
52. W. Chu, Q. Yan, X. Liu, Q. Li, Z. Yu, G. Xiong, *Stud. Surf. Sci. Catal.* 119, 1998, 849.
53. V. A. Tsipouriari, X. E. Verykios, *Stud. Surf. Sci. Catal.* 119, 1998, 795.
54. Y. Lu, Y. Liu, S. Shen, *J. Catal.* 177, 1998, 386.
55. V. R. Choudhary, A. M. Rajput, A. S. Mamman, *J. Catal.* 178, 1998, 576.
56. A. Slagtern, U. Olsbye, *Appl. Catal. A: Gen.* 110, 1994, 99.
57. Y. Chu, S. Li, J. Lin, J. Gu, Y. Yang, *Appl. Catal. A: Gen.* 134, 1996, 67.
58. T. Hayakawa, H. Harihara, A. G. Andersen, A. P. E. York, K. Suzuki, H. Yasuda, K. Takehira, *Angew. Chem. Int. Ed.* 35, 1996, 192.
59. B. L. Basini, M. D'Amore, G. Fornasari, A. Guarinoni, D. Matteuzzi, G. Del Piero, F. Trifiro, A. Vaccari, *J. Catal.* 173, 1998, 247.
60. T. Borowiecki, *Appl. Catal.* 4, 1982, 223.
61. H. Provendier, H. C. Petit, C. Estournes, S. Libs, A. Kiennemann, *Appl. Catal. A: Gen.* 180, 1999, 163.
62. A. Slagtern, H. M. Swaan, U. Olsbye, I. M. Dahl, C. Mirodatos, *Catal. Today* 46, 1998, 107.
63. H. M. Swaan, R. Rouanet, P. Widyananda, C. Mirodatos, *Stud. Surf. Sci. Catal.* 107, 1997, 447.
64. Y. F. Chang, H. Heinemann, *Catal. Lett.* 21, 1993, 215.
65. H. Y. Wang, E. Ruckenstein, *J. Catal.* 199, 2001, 309.
66. V. D. Sokolovskii, N. J. Coville, A. Parmaliana, I. Eskendirov, M. Makoa, *Catal. Today* 42, 1998, 191.
67. X. Bi, P. Hong, S. Dai, *Fenzi Cuihua* 12, 1998, 342.
68. J. R. Rostrup-Nielsen, *Catalysis Science and Technology*, eds. J.R. Andersen and M. Boudart (Springer, Berlin, 1984), 5, 1.
69. A. P. E. York, T. Xiao, M. L. H. Green, *Top. Catal.* 22, 2003, 3.
70. A. P. E. York, T. C. Xiao, M. L. H. Green, J. B. Claridge, *Catal. Rev. Sci. En.* 49, 2007, 511.
71. R. A. Periana, D. J. Taube, E. R. Evitt, D. G. Loffler, P. R. Wentrcek, G. Voss, T. Masuda, *Science* 259, 1993, 340.

72. J. K. Dixon, J. E. Longfield, *Catalysis*, Vol. VII, ed. P. H. Emmett (Reinhold, New York, 1960), ch. 4, p. 281.

73. D. A. Hickmann, L. D. Schmidt, *J. Catal.* 138, 1992, 267.

74. D. A. Hickmann, L. D. Schmidt, *Science* 259, 1993, 343.

75. D. A. Hickmann, L. D. Schmidt, in: *Synthesis Gas Formation by Direct Oxidation of Methane over Monoliths*, eds. S. T. Oyama and J. W. Hightower, ACS, 523, 1993, 416.

76. D. A. Hickmann, E. A. Haupfear, L. D. Schmidt, *Catal. Lett.* 17, 1993, 223.

77. D. A. Hickmann, L. D. Schmidt, *AIChE J.* 39, 1993, 1164.

78. J. R. Rostrup-Nielsen, *J. Catal.* 31, 1973, 173.

79. J. R. Rostrup-Nielsen, J. H. Bak Hansen, *J. Catal.* 144, 1993, 38.

80. J. B. Claridge, S. C. Tsang, M. L. H. Green, *Catal. Today* 21, 1994, 455.

81. F. Basile, L. Basini, G. Fornasari, M. Gazzano, F. Trifiro, A. Vaccari, *Stud. Surf. Sci. Catal.* 118, 1998, 31.

82. K. Nakagawa, K. Anzai, N. Matsui, N. Ikenaga, T. Suzuki, Y. Teng, T. Kobayashi, M. Haruta, *Catal. Lett.* 51, 1998, 163.

83. S. Xu, R. Zhao, X. Wang, *Fuel Proces. Technol.* 86, 2004, 123.

84. J. B. Claridge, M. L. H. Green, S. C. Tsang, A. P. E. York, A. T. Ashcroft, P. D. Battle, *Catal. Lett.* 22, 1993, 299.

85. K. M. Khajenoori, M. Rezae, B. Nematollahi, *J. In. Eng. Chem.* 19, 2013, 981.

86. A. I. Tsyganok, M. Inaba, T. Tsunoda, K. Suzuki, K. Takehira, T. Hayakawa, *Appl. Catal. A: Gen.* 275, 2004, 149.

87. J. R. Nielsen, *Catalytic Steam Reforming*, Springer-Verlag, Berlin, 1984.

88. L. Li, P. Lu, Y. Yao, W. Ji, *Catal. Comm.* 26, 2012, 72.

89. L. Li, S. He, Y. Song, J. Zhao, W. Ji, C. T. Au, *J. Catal.* 288, 2012, 54.

90. G. A. Olah, J. Halpern, J. Shen, Y. K. Mo, *J. Am. Chem. Soc.* 93, 1971, 1251.

91. V. R. Choudhary, A. K. Kinage, T. V. Choudhary, *Science* 275, 1997, 1286.

92. L. Wang, L. Tao, M. Xie, G. Xu, J. Huang, Y. Xu, *Catal. Lett.* 21, 1993, 35.

93. A. K. A. Gheit, M. S. El-Masry, A. E. Awadallah, *Fuel Proces. Technol.* 102, 2012, 24.

94. H. Liua, X. Bao, Y. Xu, *J. Catal.* 239, 2006, 441.

95. X. S. Li, C. Shi, Y. Xu, K. J. Wang, A. M. Zhu, *Green Chem.* 9, 2007, 647.

96. X. S. Li, C. K. Lin, C. Shi, Y. Xu, Y. N. Wang, A. M. Zhu, *J. Phys. D: Appl. Phys.* 41, 2008, 175203.

97. S. Pradhan, R. Lloyd, J. K. Bartley, D. Bethell, S. Golunski, R. L. Jenkins, G. J. Hutchings, *Chem. Sci.* 3, 2012, 2958.

98. F. Bleken, W. Skistad, K. Barbera, M. Kustova, S. Bordiga, P. Beato, K. P. Lillerud, S. Svelle, U. Olsbye, *Phys. Chem. Chem. Phys.* 13, 2011, 2539.

99. J. A. Biscardi, E. Iglesia, *Phys. Chem. Chem. Phys.* 1, 1999, 5753.

100. P. Huang, Y. Zhao, J. Zhang, Y. Zhu, Y. Sun, *Nanoscale* 5, 2013, 10844.

101. P. Schwach, M. G. Willinger, A. Trunschke, R. Schlogl, *Angew. Chem. Int. Ed.* 52, 2013, 11381.

102. R. Maghrebi, N. Yaghobi, S. Seyednejadian, M. H. Tabatabaei, *Particulogy* 11, 2013, 506.

103. W. Jeon, J. Y. Lee, M. Lee, J. W. Choi, J. M. Ha, D. J. Suh, I. W. Kim, *Appl. Catal. A: Gen.* 464, 2013, 68.

104. K. Oshima, K. Tanaka, T. Yabe, E. Kikuchi, Y. Sekine, *Fuel* 107, 2013, 879.

105. T. W. Elkins, H. W. Hagelin-Weaver, *Appl. Catal. A: Gen.* 454, 2013, 100.

106. D. Visinescu, F. Papa, A. C. Ianculescu, I. Balint, O. Carp, *J. Nano. Res.* 15, 2013, 1456.

107. R. Ghose, H. T. Hwang, A. Varma, *Appl. Catal. A: Gen.* 452, 2013, 147.

108. T. Baidya, N. van Vegten, R. Verel, Y. J. Jiang, M. Yulikov, T. Kohn, G. Jeschke, A. Baiker, *J. Catal.* 281, 2011, 241.

109. S. F. Ji, T. C. Xiao, S. B. Li, L. J. Chou, B. Zhang, C. Z. Xu, R. L. Hou, A. P. E. York, M. L. H. Green, *J. Catal.* 220, 2003, 47.

110. A. J. Nagy, G. Mestl, R. Schlogl, *J. Catal.* 188, 1999, 58.

111. S. Pak, P. Qiu, J. H. Lunsford, *J. Catal.* 179, 1998, 222.
112. G. Gayko, D. Wolf, E. V. Kondratenko, M. Baerns, *J. Catal.* 178, 1998, 441.
113. S. F. Moya, R. L. Martins, A. Ota, E. L. Kunkes, M. Behrens, M. Schmal, *Appl. Catal. A: Gen.* 411, 2012, 105.
114. K. C. Szeto, S. Norsic, L. Hardou, E. LeRoux, S. Chakka, J. T. Cazat, A. Baudouin, C. Papaioannou, J. M. Basset, M. Taoufik, *Chem. Commun.* 46, 2010, 3985.
115. H. Yoshida, M. G. Chaskar, Y. Kato, T. Hattori, *Chem. Commun.* 2002, 2014.
116. K. Murata, T. Hayakawa, K. I. Hayakawa, K. I. Fujita, *Chem. Commun.* 1997, 221.
117. Z. Y. Dang, J. F. Gu, J. Z. Lin, D. X. Yang, *Chem. Commun.* 1996, 1901.
118. D. Soulivong, S. Norsic, M. Taoufik, C. Coperet, J. T. Cazat, S. Chakka, J. M. Basset, *J. Am. Chem. Soc.* 130, 2008, 5044.
119. R. L. P. Goncalves, F. C. Muniz, F. B. Passos, M. Schmal, *Catal. Lett.* 135, 2010, 26.
120. F. Q. Chen, W. Zheng, N. Zhu, D. G. Cheng, X. L. Zhan, *Catal. Lett.* 125, 2008, 348.
121. L. Li, G. D. Li, C. Yan, X. Y. Mu, X. L. Pan, X. X. Zou, K. X. Wang, J. S. Chen, *Angew. Chem. Int. Ed.* 50, 2011, 8299.
122. K. Takanabe, E. Iglesia, *Angew. Chem. Int. Ed.* 47, 2008, 7689.
123. A. Tomita, J. Nakajima, T. Hibino, *Angew. Chem. Int. Ed.* 47, 2008, 1462.
124. K. Otsuka, I. Yamanaka, *Catal. Today* 41, 1998, 311.
125. Y. Shiota, K. Yoshizawa, *J. Am. Chem. Soc.* 122, 2000, 12317.
126. R. A. Periana, D. J. Taube, S. Gamble, H. Taube, T. Satoh, H. Fujii, *Science* 280, 1998, 560.
127. L. Vafajoo, M. Sohrabi, M. Fattahi, *World Acad. Sci. Eng. Technol.* 73, 2011.
128. J. Yuan, L. Wang, Y. Wang, *Ind. Eng. Chem. Res.* 50, 2011, 6513.
129. X. Zhang, D. H. He, Q. J. Zhang, B. Q. Xu, Q. M. Zhu, *Chin. Chem. Lett.* 14, 2003, 1066.
130. M. C. Bjorklund, R. W. Carr, *Ind. Eng. Chem. Res.* 41, 2002, 6528.
131. C. J. Jones, D. Taube, V. R. Ziatdinov, R. A. Periana, R. J. Nielsen, J. Oxgaard, W. A. Goddard, *Angew. Chem. Int. Ed.* 43, 2004, 4626.
132. M. H. A. Rahim, M. M. Forde, R. L. Jenkins, C. Hammond, Q. He, N. Dimitratos, J. A. L. Sanchez et al., *Angew. Chem. Int. Ed.* 52, 2013, 1280.
133. R. Palkovits, M. Antonietti, P. Kuhn, A. Thomas, F. Schüth, *Angew. Chem. Int. Ed.* 48, 2009, 6909.
134. P. J. Smeets, R. G. Hadt, J. S. Woertink, P. Vanelderen, R. A. Schoonheydt, B. F. Sels, E. I. Solomon, *J. Am. Chem. Soc.* 127, 2005, 1394.
135. R. Balasubramanian, S. M. Smith, S. Rawat, L. A. Yatsunyk, T. L. Stemmler, A. C. Rosenzweig, *Nature* 465, 2010, 115.

5 Fischer–Tropsch
Fuel Production with Cobalt Catalysts

Cristina Paun, Jacinto Sá, and Kalala Jalama

CONTENTS

5.1   Introduction 147
5.2   Low-Temperature FT Cobalt Catalysts.. 152
     5.2.1   Support Effect.. 152
     5.2.2   Effect of Metal Dispersion... 153
     5.2.3   Catalyst Preparation... 154
     5.2.4   Effect of Noble Metal Promoters.. 156
          5.2.4.1   Effect of Pt Promotion 156
          5.2.4.2   Effect of Ru Promotion... 157
          5.2.4.3   Effect of Pd Promotion 159
          5.2.4.4   Effect of Re Promotion 160
          5.2.4.5   Effect of Au Promotion... 161
     5.2.5   Catalyst Pretreatment.. 163
5.3   Future Perspectives... 164
References.. 165

This chapter compiles the most recent technologies aiming at converting synthetic gas "syngas" ($CO + H_2$) into fuel with Co catalysts. It contains an historical summary, the recent developments, mainly concerning doping and high dispersion Co catalysts, and a future perspective.

5.1 INTRODUCTION

Currently, the world's energy demand is met primarily by the traditional fossil fuels, such as petroleum and natural gas. With a population exceeding 7 billion on March 2012, and on growing continuously, the global energy demands will continue to increase [1]. The regional energy demand growth is depicted in Figure 5.1.

However, the natural reserves are limited and expected to be depleted in the near future (Figure 5.2); therefore, research into alternative energy sources and production is a top priority in both academia and industry [3]. Additionally, research into alternative fuels should aim for a carbon neutral production-usage cycle, that is, avoiding

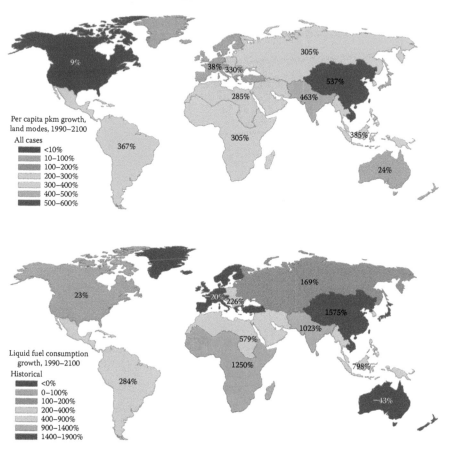

FIGURE 5.1 (Top) Regional demand growth in per capital and passenger km (pkm) traveled (all scenarios). (Bottom) Regional demand growth for all liquid fuels, including conventional oil and alternative liquid fuels (historical scenario). (Reproduced from Brandt, A. R. et al., *Environ. Sci. Technol.*, 47, 8031, 2013. With permission.)

increase of greenhouse gases in the atmosphere, which can cause severe environmental issues [4].

In order to meet the increasing global energy needs, ensure energy security, and help with environmental protection, many efforts have been made to develop new technologies for the production of alternative fuels and/or synthetic fuels. Some well-known alternative fuels include biodiesel, bioalcohol (methanol, ethanol, butanol), chemically stored electricity (batteries and fuel cells), hydrogen, nonfossil natural gas including methane, vegetable oil, and other biomass sources. Synthetic fuel is a liquid fuel produced from coal, natural gas, oil shale, or biomass [5]. However, more commonly the term "synthetic fuels" refers to the fuels obtained via Fischer–Tropsch synthesis (FTS), methanol to gasoline process, or direct coal liquefaction. The chapter aims to cover aspects of FTS with Co-based catalysts, leaving the other processes to be discovered by the reader from the published literature [6].

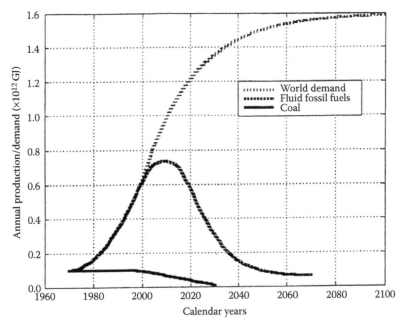

FIGURE 5.2 Estimated energy demand and fossil fuel production. (Reproduced from Veziroğlu, T. N. and Şahi'n, S., *Energy Convers. Manag.*, 49, 1820, 2008. With permission.)

The FTS is one of the primary technologies, together with the methanol to gasoline (Mobil process), which produces fuel from syngas. The synthesis is essentially a polymerization reaction in which carbon bonds are formed from carbon atoms derived from carbon monoxide, under the influence of hydrogen in the presence of a metal catalyst. The reaction leads to a range of products that depend on the reaction conditions and catalysts employed, that is, liquid hydrocarbons (fossil fuels like gasoline or kerosene) and potentially waxes [7].

Sabatier and Senderens [8] laid the fundamentals of this reaction, which has become of great industrial significance, when they reported direct CO hydrogenation over cobalt and nickel catalysts with the formation of methane in 1902. Few years later, Badische Anilin and Soda Fabrick (BASF) reported on the production of hydrocarbons, mainly oxygenated derivatives (synthol) from syngas, using alkali-promoted osmium and cobalt catalysts at high pressure and were awarded patents in 1913 and 1914 [9]. In the 1920s, after World War I, Fischer and Tropsch [10] also reported the formation of a product similar to the synthol product over alkalized iron shavings at 100 atm and 400°C. They also synthesized small amounts of ethane and higher hydrocarbons at atmospheric pressure and at 370°C over Fe_3O_4–ZnO catalysts [11,12]. Because of the rapid deactivation exhibited by iron-based catalysts, further studies focused on the use of cobalt and nickel catalysts. Fischer and Meyer developed Ni–ThO_2–Kieselguhr and Co–ThO_2–Kieselguhr catalysts in the early 1930s [13]. The limited supply of cobalt made the researchers concentrate their initial studies on nickel catalysts, but the high yields of methane over these catalysts shifted their attention back to cobalt. In 1937, Fischer and Pichler found improved product yields and longer catalyst lifetime when alkalized iron catalysts were used at medium pressures (5–30 atm) [7]. Pichler also reported the use of ruthenium-based catalysts, in 1938, which formed high boiling waxes [14].

Germany was first to use FTS at an industrial level with nine operating plants with a combined production capacity of about 660×10^3 t per year by 1938. The operating plants used cobalt catalysts at medium pressures [15] but were shut down after World War II. However, the fear of an impending shortage of petroleum kept the interest in the FT process alive. An FTS plant with a capacity of 360×10^3 t per year was built and operated in Brownsville, Texas, during the 1950s. This plant was based on syngas produced from methane, but a sharp increase in the price of methane caused the plant to be shut down [16,17]. To address the worldwide prediction of increasing crude oil prices, the South African Coal Oil and Gas Corporation (SASOL) commissioned an FT plant based on coal in Sasolburg in South Africa during the same time period.

Research on FTS has continued ever since at Sasol. Due to the oil crises of the mid-1970s, Sasol constructed two much larger FT plants, which came online in 1980 and 1982, respectively. These plants operated on coal-derived synthesis gas and increased Sasol's production to approximately 6000×10^3 t per year (three plants combined) [16]. Some commercial ventures in FTS by Shell International in Malaysia for the production of waxes [18], the Norway Statoil GMD slurry process [19], and the Mossgas project in South Africa have also been described. The Mossgas plant (PetroSA) in South Africa and the Shell plant at Bintulu, Malaysia, came on stream in 1992 and 1993, respectively [16]. These two plants use methane-derived syngas.

As mentioned, in the last few years the interest for FTS has significantly grown due to the increase in oil prices as well as the high demand for energy. Therefore, recent commercial ventures include the following:

1. The development of the world's largest gas-to-liquids (GTL) facility, Pearl GTL in Ras Laffan Industrial City in Qatar by Shell and the government of the Qatar State. The plant with a capacity of 260,000 barrels per day was fully completed in 2012 [20].
2. The development of a GTL plant, Oryx GTL, in a joint venture of Sasol with Qatar Petroleum at Ras Laffan in Qatar. The plant, with a capacity of 34,000 barrels per day, was fully commissioned in 2007.
3. The development of 33,000 barrels per day GTL plant at Escravos in Nigeria by Chevron Nigeria, National Petroleum Corporation, and Sasol. The plant is in an advanced stage of completion [21].
4. The development of 96,000 barrels per day GTL facility in North America by Sasol. The front-end engineering and design is expected to be completed in 2016 [22].

Many other small GTL projects are also taking place around the globe.

Traditionally, the feedstock for syngas production is coal, natural gas, and more recently biomass or other carbon sources (plastics). As such, the research for renewable fuels has been expanded toward the biomass-to-liquid via Fischer–Tropsch

(BTL-FT) synthesis. Renewable fuels from BTL-FT are usually much cleaner and environmentally friendly, and they contain little or even no sulfur and other contaminant compounds, satisfying the upcoming stricter environmental regulations in both Europe and the United States [23]. In the BTL-FT process, biomass (e.g., woodchips) is firstly gasified with air, oxygen, and/or steam to produce raw biosyngas (typically containing 22.16% CO, 17.55% H_2, 11.89% CO_2, 3.07% CH_4, with N_2 and other gases as the balance [24]). Then, a cleaning process is applied to the raw biosyngas to remove contaminants (e.g., small char particles, ash, and tar) followed by introduction into a catalytic reactor to perform FT synthesis to result in liquid fuels (green gasoline, diesel, and other clean biofuels) [25].

FT diesel is similar to fossil diesel with regards to its energy content, density, and viscosity, and can be produced directly from FTS; however, a higher yield can be achieved if the FT wax is produced first, followed by hydrocracking [26]. Regarding some fuel characteristics, FT diesel is even more favorable, that is, a higher cetane number (better autoignition qualities) and lower aromatic content, which results in lower NO_x and particle emissions.

In terms of operation mode, the FTS can be operated in two modes:

1. The high-temperature process (300°C–350°C) with iron-based catalysts used for the production of gasoline and linear low molecular mass olefins
2. The low-temperature process (200°C–240°C) with either iron or cobalt catalysts used for the production of high molecular mass linear waxes [16]

FT reaction is highly exothermic; therefore, it is very important to take into account the design of the reactor, the rapid and efficient removal of heat from the catalyst particles. An overheating of the catalyst would adversely affect product selectivity and catalyst lifetime. There are three main types of reactors used for FT reaction:

1. Fixed-bed reactors [Sasol—high-value linear waxes at low temperatures (225°C)]
2. Fluidized-bed reactors with either a fixed or a circulating bed (Sasol in Secunda)
3. Slurry-bed reactors in which gas is bubbled through a suspension of finely divided catalyst in a liquid which has a low vapor pressure at the temperature of operation (Sasol, Exxon—Co-based catalyst for wax production)

The critical features of FTS reactors are summarized in Table 5.1.

Despite the recent commercial applications of coal-to-liquids (CTL) and gas-to-liquids (GTL) processes by Sasol, PetroSA, and Shell [28], in most cases, the FT synthesis is still characterized by products with a wide range of molecular weights, whose distribution can be described by the Anderson–Schulz–Flory model [13]. A stepwise carbon chain growth process occurring on the catalyst surface explains the FT product distribution [16]. Therefore, the choice of catalyst plays an important role, and research (both ex situ and situ) into catalyst synthesis, metals and promoters, activity and reactivity, and mechanism represents a major part in FT synthesis research.

TABLE 5.1
Comparison of Selected FTS Reactors

Feature	Fixed Bed	Fluid Bed (Circulating)	Slurry
Temperature control	Poor	Good	Good
Heat exchanger surface	240 m² per 1,000 m³ feed	15–30 m² per 2,000 m³ feed	50 m² per 1,000 m³ feed
Max. reactor diameter	<0.08 m	Large	Large
CH₄ formation	Low	High	As fixed bed or lower
Flexibility	Intermediate	Little	High
Product	Full range	Low mol. weight	Full range
Space-time yield (C₂⁺)	>1,000 kg/m³ day	4,000–12,000 kg/m³ day	1,000 kg/m³ day
Catalyst affectivity	Lowest	Highest	Intermediate
Back-mixing	Little	Intermediate	Large
Minimum H₂/CO feed	As slurry or higher	Highest	Lowest
Construction			Simplest

Source: Reproduced from Davis, B. H., *Top. Catal.*, 32, 143, 2005. With permission.

5.2   LOW-TEMPERATURE FT COBALT CATALYSTS

Co-based catalysts are only used in the low-temperature FT (LTFT) process where they possess high activity and selectivity for middle distillates and waxes products and a lower water–gas shift reaction activity compared to Fe catalysts [16,29]. A high operating temperature results in production of excess methane. This means in order to produce gasoline (C3–C11) or diesel (C12–C18), the heavy waxes fraction is catalytically cracked with acid catalysts (usually silica–alumina and zeolites). Conventional cobalt LTFT catalyst contains a high surface area support (Al_2O_3, SiO_2, TiO_2), which accounts for 60%–80% of the catalyst, 15%–30% of Co, 1%–10% oxide promoter (ZrO_2, La_2O_3, CeO_2), and 0.05%–0.1% noble metal promoter (Pt, Ru, Rh, and Pd) [30].

For relatively large cobalt particles (d > 8 nm), FTS reaction rate is proportional to the overall number of cobalt surface sites. The number of cobalt surface sites in metal-supported catalysts depends on particle size, particle morphology, extent of metal reduction, and particle stability. Therefore, optimization of cobalt particle size and cobalt reducibility seems to be the most obvious goals in the design of any efficient cobalt catalysts for FT synthesis. These parameters can be affected by many factors such as type of support, metal dispersion, preparation method, type of promoters, pretreatment conditions, and so on.

5.2.1   Support Effect

High metal surface areas available for catalysis are achieved by dispersing metal particles on a support, which also stabilizes the dispersed metal crystallites [31,32].

Reuel and Bartholomew [33] reported on the effect that the supports have on the specific activity and selectivity properties of cobalt in CO hydrogenation. They tested the effect of support on low loading cobalt catalysts (3 wt%), with nearly the same dispersion and presumably presented a more intimate interaction of cobalt with the support. They observed a decrease in hydrogenation activity (1 atm and 225°C) in the following order: Co/TiO$_2$ > Co/SiO$_2$ > Co/Al$_2$O$_3$ > Co/C > Co/MgO. The higher activity of Co/TiO$_2$ was considered to be a result of strong metal–support interactions (SMSI) as found for the Ni/TiO$_2$ [34,35]. Subsequent studies from other research groups confirmed that SMSI Co/TiO$_2$ system is responsible for high CO hydrogenation activity [36]. Metal–support interaction post–pretreatment was also reported for Co/Al$_2$O$_3$ [37] and Co/SiO$_2$ [38] systems.

5.2.2 EFFECT OF METAL DISPERSION

The effect of cobalt particle size on CO hydrogenation was only reported in the literature in 1984, by Reuel and Bartholomew [33]. They observed a significant decrease in the specific activity cobalt catalysts that decreased with the increase of metal dispersion. Furthermore, high dispersion Co catalysts were found to produce lower molecular weight hydrocarbon products and are less reducible. This effect was attributed to the presence of stable oxides in the well-dispersed and poorly reduced catalysts, which catalyze the water–gas shift reaction leading to an increase of the H$_2$/CO ratio at the surface.

Fu and Bartholomew [39] extended the study to Co catalysts supported on Al$_2$O$_3$ catalysts. Changes in specific activity with dispersion were explained by variations in the distribution of low and high coordination sites and by changes in the nature of the adsorbed CO species available for reaction. High specific activity was believed to be favored on sites where CO is strongly coordinated. However, it should be mentioned that other studies did not observe a significant effect of cobalt particle size on specific activity [40–49].

In terms of selectivity, the decrease of cobalt particle size leads to an increase selectivity to methane production and consequent decrease selectivity to higher hydrocarbons size [33,39,45,46,48,50]. This effect was explained by the presence of cobalt oxides in the catalysts [33,39] or by olefin readsorption [45]. Some controversial effects of cobalt particle size on CO hydrogenation product selectivity have also been reported [51,52]. Kikuchi et al. [51] reported a decrease in methane selectivity and increase in C$_{5+}$ selectivity with a decrease in particle size. The apparent controversy on the effect of cobalt particle size on catalyst behavior for CO hydrogenation suggests that the reported observations are affected by variables like support, dispersion range, and so on.

Ho et al. [41] studied the effect of cobalt particle size on silica in order to minimize the metal–support interaction contribution. They selected a range of catalyst calcination conditions that allowed for complete reduction of the cobalt phase. The specific CO hydrogenation activity was found to be invariant with cobalt dispersion in the range of 6%–20% dispersion. Barbier et al. [53] showed that the intrinsic activity and chain growth probability on a series of silica-supported cobalt catalysts first increased with increasing the particle size and stabilized at a critical diameter of 6 nm. A similar observation was reported when Co was supported on graphitic carbon nanofibers [54]. The critical cobalt particle size over which the specific activity and product selectivity became invariant was 6 and 8 nm for reaction pressures between 1 and 35 bar. These effects were attributed to nonclassical structure sensitivity in combination with CO-induced surface reconstruction.

It should be mentioned that the reaction rate and selectivity can be greatly affected by intraparticle diffusion in larger grains of cobalt catalyst. Thus, uniform distribution of cobalt metal sites through the catalyst grain does not necessarily mean a better catalyst.

Related to Co component, an efficient LTFT Co-based catalysts necessitate [55]:

1. High density of cobalt surface metal sites
2. Cobalt metal particles larger than 6–8 nm
3. Low concentration of nonreducible Co sites
4. Optimal distribution
5. High stability under LTFT reaction conditions
6. Cost effective (optimized Co content)

5.2.3   Catalyst Preparation

FT cobalt catalysts are usually prepared by impregnation of the support with a cobalt solution (precursor) and subsequent drying and calcination to decompose the cobalt salt by forming cobalt oxide stabilized on the support. Cobalt nitrate is usually used; however, the need to control parameters affecting the behavior of catalyst, for example, cobalt dispersion, has prompted researchers to study the effect of various cobalt salts used as precursors. Some early studies showed that the use of cobalt carbonyl as a precursor leads to catalysts with higher activity for CO hydrogenation than the conventional, nitrate-derived catalysts [56]. Some of these studies [57] reported that catalysts prepared using cobalt carbonyl were more selective toward alcohols and also contended that the high activity measured for these catalysts is related to the high dispersion.

Niemela et al. [58] conducted a study where they distinguished between two types of carbonyl complexes, $Co_2(CO)_8$ and $Co_4(CO)_{12}$, and compared the effect on reduction extent, dispersion, and reactivity on a silica support. These effects were compared to the conventional silica-supported cobalt catalyst. The near surface reduction was found lower for $Co_2(CO)_8$-derived catalysts than for the $Co_4(CO)_{12}$-based ones. The cobalt dispersion was found to decrease in the precursor order $Co_2(CO)_8 > Co_4(CO)_{12} > Co(NO_3)_2$. Also the carbonyl-derived catalysts exhibited greater initial activity in CO hydrogenation than catalysts prepared from cobalt nitrate. The effect of organic cobalt precursors have been studied mostly on a silica support [59].

Matsuzaki et al. [56f] used cobalt acetate to prepare highly dispersed cobalt metal catalysts supported on silica. The resulting catalyst was hardly reduced in a hydrogen stream at 450°C compared to catalysts prepared using cobalt nitrate and cobalt chloride. The authors proposed that divalent cobalt is strongly connected to the Si of the SiO_2 support through the oxygen for the cobalt acetate–derived catalyst even after thermal treatment at 450°C in a hydrogen stream. Based on EXAFS results, the authors contended that the structure of cobalt oxide was similar to that of cobalt (II) acetate. They proposed a mechanism for oxide formation and suggested that coordinated ligands such as acetate are more strongly connected to a cobalt cation compared to counter anions such as nitrate or chloride and that the structure of the Co–O bonds is kept under reduction conditions. In the case of nitrate and chloride ions, the uncoordinated counter anions are more easily removed from the cobalt cation and thus the cation is easily reduced to metallic cobalt below 400°C. The cobalt acetate–derived catalyst was not active for CO hydrogenation.

Sun et al. [59c] prepared catalysts by mixed impregnation of cobalt (II) nitrate and cobalt (II) acetate and measured higher activity for the FT reaction than on catalysts prepared from mono-precursor. A nitrate/acetate ratio of 1 was shown to be the optimum ratio. This effect was explained assuming that readily reduced cobalt metal from cobalt nitrate promoted the reduction of highly dispersed Co^{2+} from cobalt acetate to metallic cobalt by a hydrogen spillover mechanism during reduction. Highly dispersed Co metal provided the main active sites. Some other studies [48,59d–f] have shown that the use of organic precursors such as cobalt acetate and/or cobalt acetylacetonate on silica results in formation of a catalyst with high dispersion and low reducibility and hence low activity for the CO hydrogenation. The high dispersion in the catalyst leads to formation of poorly reducible cobalt silicates [59d–f].

Van Steen et al. [60] reported on the dependence of cobalt silicate formation on the pH of the impregnating solution. At pH above 5, for example, when cobalt acetate is used as cobalt precursor, more cobalt silicates are formed. The surface cobalt silicate precursor is destroyed by drying or low-temperature calcination.

Ming et al. [61] have also reported that the pH of the impregnating solution influences cobalt dispersion on a support. They explained that below the point of zero charge for silica (between 2 and 3.5), the surface is positively charged and the adsorption of positively charged Co ions is slowed down leading to a low dispersion. At a pH > point of zero charge, deposition of Co is favored and improves the cobalt dispersion.

Girardon et al. [59f] have reported that Co silicate formation arises from thermal treatment and not from the pH of the impregnating solution as suggested by Ming et al. [61]. They explained that cobalt silicate formation depends on the exothermicity of the cobalt salt decomposition in air and the temperature of the oxidative pretreatment. The high exothermicity of cobalt acetate decomposition leads primarily to amorphous and low-reducible cobalt silicates. They suggested that a more efficient heat flow control at the stage of cobalt acetate decomposition significantly increases the concentration of cobalt oxide species, which is more easily reducible in the oxidized catalyst.

The effect of organic precursor on alumina- and titania-supported cobalt catalysts has also been reported in the literature. Van der Loosdrecht et al. [62] have reported that the use of cobalt Ethylenediaminetetraacetic acid (EDTA) and ammonium cobalt citrate in the preparation of low-loaded alumina-supported Co catalysts (2.5 wt%) gave very small oxide particles, which reacted with the support during thermal treatment in a reducing gas to form Co aluminates which were not active for the FT reaction. However, catalysts prepared from cobalt nitrate had larger particle sizes, which were easily reduced to Co metal and therefore active under FT conditions. Kraun et al. [63] have shown that the use of cobalt oxalate, cobalt acetate, or cobalt acetylacetonate as cobalt precursors for the preparation of titania-supported Co catalysts gives higher cobalt dispersions and higher activity for the FT reaction than catalysts prepared using cobalt nitrate. They suggested that various Co precursors seemed to influence the interaction of the active cobalt sites with the support that may be ascribed to the decomposition of the organocobalt compounds. They were unable to elucidate the mechanism of the decomposition process.

A review on recent developments on FT cobalt catalysts has been given by Khodakov et al. [64] and can be consulted for more details.

5.2.4 EFFECT OF NOBLE METAL PROMOTERS

Noble metal promoters can modify LTFT catalytic performance in FT synthesis, cobalt reducibility, cobalt dispersion, and hydrogen activation. Note however that promotion with small amounts of noble metal does not usually affect the mechanical properties of cobalt-supported catalysts [65]. The effects of promotion with noble metals on catalytic performance and structure of cobalt LTFT catalysts are summarized in the following sections.

5.2.4.1 Effect of Pt Promotion

Guczi and coworkers [66–68] reported on the effect that Pt doping has 10 wt% Co/Al_2O_3 catalysts with high Pt/Co ratios. The addition of Pt increased Co reducibility and a stabilization of Co ions on the Al_2O_3 surface. They postulated the presence of Co–Pt bimetallic particles, which were inferred from a lower catalytic activity (compared with the monometallic Co catalyst) in the synthesis of hydrocarbons from CO and H_2 [67]. The Pt–Co systems showed high activity for methanol formation. Zyade et al. [69] characterized similar catalysts with EXAFS (extended X-ray absorption fine structure) and observed the formation of bimetallic particles with a Pt–Co interatomic distance of 0.271 nm. Dees and Ponec [70] studied Co–Pt catalysts supported on silica and alumina, with 5 wt% metal loading and varying Pt/Co ratios. They reported the formation of metal alloys on silica based on XRD (X-ray diffraction) measurements.

Schanke et al. [71] tried to rationalize the role of Pt promoter in hydrocarbon synthesis on cobalt catalysts supported on alumina and silica. They prepared catalysts via impregnation with 9 wt% Co and variable Pt loading (0 or 0.4 wt%). TPR (temperature programmed reduction) studies revealed that Pt shifts the reduction peaks to lower temperatures for all catalysts (Figure 5.3). H_2 chemisorption showed that Pt also promoted metallic cobalt dispersion in comparison to the unpromoted catalyst. The largest effect was found with alumina-supported catalysts due to the reduction of highly dispersed surface cobalt oxide. Finally, CO hydrogenation rates (based on weight of cobalt) for Pt-promoted catalysts were 3–5 times higher than those on unpromoted catalysts. By means of the SSITKA (steady-state isotopic-transient kinetic analysis), the authors found that the true turns over numbers (TONs) were constant for all catalysts and that Pt-promoted catalysts presented a high coverage of reactive intermediates resulting in an increase in apparent turnover numbers. The selectivity was not influenced by the presence of Pt. The influence of Pt was explained by considering a

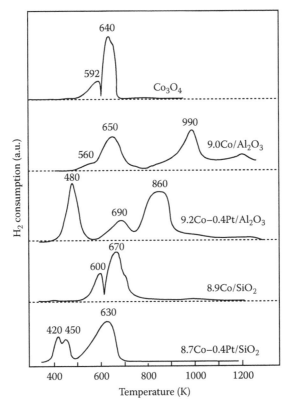

FIGURE 5.3 TPR of supported cobalt catalysts and bulk Co_3O_4. (Reproduced from Schanke, D. et al., *J. Catal.*, 156, 85, 1995. With permission.)

possible hydrogen spillover effect from Pt to Co or a more direct interaction between Pt and Co in the form of a Co–Pt interface or bimetallic Co–Pt particles.

A broad high temperature reduction feature around 600°C in unpromoted catalyst, associated with Co species, which were extremely difficult to reduce was shifted to approximately 450°C in the Ru-promoted catalyst. The extent of reduction after standard reduction treatment in 1 atm of flowing H_2 (50 cm³/min) at 350°C for 10 hours was approximately 60% for unpromoted catalysts compared to 85%–100% for ruthenium-promoted catalysts. H_2 chemisorption measurement showed that addition of ruthenium not only increased the extent of reduction but also increased threefold the number of exposed surface cobalt atoms catalysts compared to unpromoted catalysts. The average Co metal particle size for Ru-promoted catalysts decreased to roughly half that of particles formed for the unpromoted Co/Al₂O₃ catalysts. This was probably due to the presence of additional small particles due to reduction enhancement by Ru.

5.2.4.2 Effect of Ru Promotion

Kogelbauer et al. [72] conducted a study to determine the manner in which Ru promotes LTFT Co/Al₂O₃ catalysts. The catalysts contained 0.5 wt% Ru and 20 wt% Co.

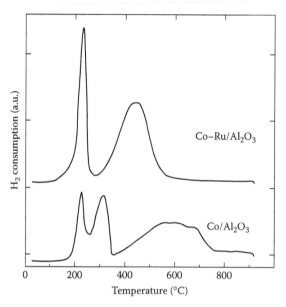

FIGURE 5.4 TPR of Co/Al$_2$O$_3$ and Co–Ru/Al$_2$O$_3$ after calcination at 300°C for 2 h. (Reproduced from Kogelbauer, A. et al., *J. Catal.*, 160, 12, 1996. With permission.)

TPR analysis, after complete decomposition of Co nitrate precursor, showed that both promoted and unpromoted Co/Al$_2$O$_3$ catalysts reduced in two steps (Figure 5.4). A shift of approximately 100°C to lower reduction temperature was observed for the Ru-promoted catalysts.

The rate of FTS at 220°C increased by a factor of 3 in the Co–Ru/Al$_2$O$_3$ catalyst. Both promoted and unpromoted catalysts had similar chain growth probability and methane selectivity. The authors proposed that noble metals activate hydrogen spillover to Co$_3$O$_4$, thus promoting its reduction at lower temperature. They also postulated that Ru could prevent the formation of highly irreducible Co compounds or promote their reduction, that is, Ru acts only as reduction promoter for Co leading to an increase in active metal reducibility and dispersion.

Similar results were observed on other Ru–Co systems supported on titania [36d] and alumina [73,74]. Hosseini et al. [74] varied the amount of Ru (0.5, 1.0, 1.5, 2.0 wt%) added to alumina-supported cobalt catalysts (20 wt%). The effect of Ru on the catalytic behavior of Co/γ-Al$_2$O$_3$ for CO hydrogenation was investigated in a continuous stirred tank reactor (CSTR). Characterization studies [XRD, TGA (thermo gravimetric analysis), TPR, H2 chemisorption, and BET surface area] showed that catalysts promoted with 0.5% and 1.0% Ru presented a higher extent of reduction, higher Co dispersion, and smaller catalyst pore volume. These catalysts showed higher CO conversions. Further increase in Ru content (1.5, 2 wt%) reverted the results, that is, efficient promotion is achieved only at low promoter loadings. C$_{5+}$ selectivity of the products was not affected.

At higher Ru content, they observed a shift in complete reduction temperature to lower values in comparison to unpromoted catalyst. The authors suggested that interactions between cobalt and ruthenium induced by the higher mobility of Ru as well as the formation of a Co–Ru oxide to be the corporate of the findings. The proposed explanation was supported by the assumption that Co_2RuO_4 formed a spinel isostructural with Co_3O_4, and that this spinel could be reduced at lower temperatures than cobalt aluminates [75]. It was also assumed that Ru accelerates Co reduction due to H_2 spillover process.

Nagaoka et al. [76] studied the influence of the addition of trace amounts of Pt (Pt/Co = 0.005–0.05 in atomic ratio) or Ru (Ru/Co = 0.01–0.05) to Co/TiO_2. They indicated that the number of Co oxide particles interacting strongly with TiO_2 was decreased by the addition of noble metals. Also, the addition of both noble metals resulted in a decrease of the reduction temperature of Co oxides and titania, presumably due to hydrogen spillover from the noble metal surface. It was noted that the peak for Co oxides interacting strongly with TiO_2 was shifted only by the addition of Pt and concluded that Pt promoted the reduction of Co oxides more effectively than Ru.

5.2.4.3 Effect of Pd Promotion

Sarkany et al. [77] investigated the effect of Pd on Co–Pd/alumina catalysts. The catalysts contained 5 wt% Co and varying amounts of Pd (0.1 to 1.0 wt%). Based on XRD, XPS (X-ray photoelectron spectroscopy), TPR, and CO and H_2 chemisorption studies, they concluded that the presence of Pd increased both the reducibility and dispersion of Co. Pd was believed to act as a hydrogen source and did not affect significantly the reduction behavior of Co_3O_4 in catalysts with poor contact between PdO and Co_3O_4, that is, Pd is only effective as a promoter when it is in close contact with Co structure.

Guczi et al. [78] studied the effect of Pd on silica-supported Pd/Co catalysts prepared by the sol–gel method. Using TPR and XPS characterization, they established that Pd facilitates Co reduction and segregation to the catalyst surface. A CO hydrogenation reaction study (1 bar and between 200°C and 300°C, $H_2/CO = 2$), in a plug flow reactor using differential conditions, showed a synergistic effect for bimetallic Co/Pd catalysts (ratio = 2) when compared to the use of monometallic Co or Pd catalysts. Also, the presence of Pd enhanced the amount of alkanes and the chain length of the products increased up to C_8–C_9. They established that palladium acted in the bimetallic system not only as a component, which helped cobalt reduction, but also as sites activating hydrogen participating in the reaction. Similar results were obtained from the same research group using XPS, XRD, XANES (X-ray absorption near edge structure), CO hydrogenation, and low temperature methane activation under nonoxidative conditions [79].

Tsubaki et al. [80] also investigated the role of noble metals when added to FTS catalysts. Silica-supported catalysts were prepared by incipient wetness impregnation of the support with mixed cobalt salts, $Co(NO_3)_2.6H_2O$ and $Co(CH_3COOH)_2.4H_2O$ as well as $Pd(NH_3)_2(NO_2)_2$, $Pt(NH_3)_2(NO_2)_2$, and $Ru(NO_3)_3$ to obtain catalysts containing 10 wt% Co (5 wt% from cobalt nitrate and 5 wt% from cobalt acetate) and 0.2 wt% of Ru, Pt, or Pd. The addition of small amounts of Ru to Co/SiO_2 catalyst increased both the catalytic activity and the reduction degree remarkably. The TOF (turn over frequency) increased but the CH_4 selectivity was unchanged. The CO hydrogenation rate followed the order RuCo > PdCo > PtCo > Co. The Pt or Pd catalyst exhibited higher CH_4 selectivity. Pt and Pd hardly exerted any effect on the degree of cobalt reduction; the metals promoted cobalt dispersion and decreased the TOF. Characterization studies [TPR, XRD, EDS (energy dispersive spectroscopy), FT/IR, XPS] suggested the existence of different contacts between Co and Ru, Pt, or Pd. Ru was enriched on cobalt while Pt or Pd dispersed well to form Pt–Co or Pd–Co alloys.

The promotional effect of Pd on both methane activation and CO hydrogenation prompted Carlsson et al. [81] to investigate the fundamental properties of Co–Pd bimetallic catalyst supported on a thin alumina film using STM (scanning tunneling microscopy), CO–TPD, and XPS characterization techniques. They found that the binding energy of CO to both Pd and Co sites is lowered by the presence of the other metal. CO binds preferentially to Co-atop sites and Pd-3-fold hollow sites (Figure 5.5). They also suggested a net polarization of charge or redistribution of d-band states in the bimetallic particles as they observed a shift in the Pd 3d level to higher binding energy concurrent with a shift in the Co 2p level to lower binding energy.

5.2.4.4 Effect of Re Promotion

Das et al. [82] investigated the effect of rhenium on alumina-supported cobalt catalysts (15% Co/Al_2O_3) prepared by a three-step incipient wetness impregnation (IWI) of cobalt nitrate followed by IWI of an aqueous solution of rhenium oxide. A Re loading of 0.2, 0.5, and 1.0 wt% was used in their study. Catalyst characterization [XRD, TPR, XAS (X-ray absorption), BET, H_2 chemisorption] showed that addition of small amounts of rhenium decreased the reduction temperature of Co oxides compared to the unpromoted catalyst but did not alter Co dispersion and cluster size. They established from TPR studies that rhenium had no effect on the low temperature reduction peak responsible for the reduction of cobalt oxide. They also showed that Co_3O_4 crystallites are essentially reduced (227°C–377°C) during rhenium oxidation (350°C) so that no spillover effect can operate to aid in reducing those species. However, H_2 spillover from the reduced rhenium metal could occur to facilitate reduction of cobalt species interacting with the support, as this phenomenon occurs after reduction of rhenium oxide to rhenium metal is achieved. Catalytic activity tests for the FTS reaction using a CSTR showed that addition of Re increased the synthesis gas conversion, based on catalyst weight, but TOF results were similar to that obtained in the absence of Re. *In situ* EXAFS data of Co-based catalysts promoted with Re confirmed the formation of an alloy (or direct contact between promoter and cobalt), which is necessary for rhenium to affect Co catalytic behavior [83] (Table 5.2).

Jacobs et al. [84] studied the reduction of a series of Re-promoted and unpromoted Co/Al_2O_3 catalysts using a combination of TPR, H_2 chemisorption, *in situ* XPS, and EXAFS/XANES. *In situ* EXAFS study at the L_{III} edge of Re revealed that there was a direct contact of Re with cobalt atoms but no evidence of Re–Re bonds. Even though direct atom-to-atom contact was found, their TPR data suggested that hydrogen spillover from the promoter to cobalt oxide clusters is important for cobalt oxide reduction.

Reduction of the promoter is required before the reduction of active metal by hydrogen spillover can occur. They established that this is presumably the reason

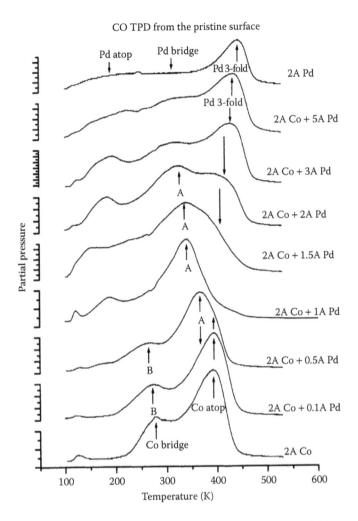

FIGURE 5.5 TPD spectra of CO from various pristine Co + Pd particles supported on Al$_2$O$_3$/NiAl(110). Metals were deposited at 300 K and 20 L CO were dosed at 100 K prior to TPD. The heating rate was 1.5 K s^{-1}. (Reproduced from Carlsson, A. F. et al., *J. Phys. Chem. B*, 107, 778, 2003. With permission.)

why Pt and Ru promoters shift the profiles for the reduction of cobalt oxide species to lower temperature for both peaks, while Re affects only the second broad peak. Re reduction occurs at the same temperature as the first stage of cobalt oxide reduction. The results were corroborated by Storsæter et al. [86]. However, they found that Re increased the degree of reduction of Co supported on Al$_2$O$_3$ but not on catalysts supported on SiO$_2$ or TiO$_2$.

5.2.4.5 Effect of Au Promotion

There are scarce studies on Au promotion of FTS cobalt catalysts in the literature to date. Most of the studies involving Au/Co system used cobalt oxide as a support or

TABLE 5.2

Results from the EXAFS Analysis of Data from the Re L_{III} Edge of the Sample Containing 4.6% Co–2% Re/Al$_2$O$_3$*

Reduction at 450°C	Coordination shell	E_0 (eV)	N	$2\sigma^2$ (Å^2)	R (Å)
No reduction	Re–O	15.1(2)	4.0(1)	0.006(1)	1.74(1)
1 h	Re–O	15.1(3)	3.4(1)	0.006(1)	1.74(1)
6 h	Re–O	13.0(3)	0.7(1)	0.003(1)	1.74(1)
	Re–O$_{support}$		1.7(2)	0.020(2)	2.00(1)
	Re–Co		2.1(2)	0.022(2)	2.53(2)
	Re–Re		2.4(3)	0.016(2)	2.71(2)
12 h	Re–O	13.3(3)	0.2(1)	0.002(0)	1.76(1)
	Re–O$_{support}$		1.4(1)	0.018(2)	1.99(1)
	Re–Co		3.2(1)	0.024(2)	2.54(1)
	Re–Re		4.0(3)	0.024(2)	2.69(1)

Source: Reproduced from Rønning, M. et al., *Catal. Lett.*, 72, 141, 2001. With permission.

* Each bonding distance (R) is associated with a coordination number (N) and thermal vibration and static disorder (Debye–Waller-like factor, $2\sigma^2$). E_0 is the refined correction to the threshold energy of the absorption edge. The standard deviation in the least significant digit as calculated by EXCURV90 is given in parentheses. However, note that such estimates of precision (which reflect statistical errors in the fitting) overestimate the accuracy, particularly in cases of high correlation between parameters. The estimated standard deviations for the distances are 0.01–0.02 Å, with ±20% accuracy for N and $2\sigma^2$. Although the accuracy for these is increased by refinements using k^0 vs. k^3 weighting.

promoter for highly dispersed Au, which is active in a plethora of reactions such as low temperature CO oxidation, oxidative destruction of dichloromethane, selective catalytic oxidation (SCO) of NO in flue gases at a low temperature, automotive pollution abatement.

The promoting effect of gold on the structure and activity of a Co/kaolin catalyst was reported by Leite et al. [87]. Catalysts were prepared by precipitation of Co(NO$_3$)$_2$.6H$_2$O or coprecipitation of HAuCl$_4$.3H$_2$O/Co(NO$_3$)$_2$.6H$_2$O to give Au–Co containing catalysts. Sodium carbonate was used as precipitant. These catalysts were characterized using XRD, TGA, TPR, and tested in the synthesis of 2,3-dihydrofuran. The authors concluded that modification by gold leads to the formation of new cobalt species, reducible at significantly lower temperatures in comparison to those of the unpromoted catalysts.

Jalama et al. [88] performed a study on the effect of Au addition to LTFT 10% Co/TiO$_2$ catalyst. Gold improved the cobalt dispersion and reducibility. The catalysts

TABLE 5.3
Summary of LTFT Reactions Depicting the Effect of Au Addition
to Co/TiO$_2$ Catalyst[a]

Catalyst	CO Conv[b]	Selectivity[c]			Olefin to Paraffin Ratio			
		CH$_4$	C$_2$-C$_4$	C$_{5+}$	C$_2$	C$_3$	C$_4$	C$_5$
10%Co/TiO$_2$	13	12	8	80	0.15	1.28	1.23	0.64
0.2%Au/10%Co/TiO$_2$	16	14	7	79	0.09	1.12	1.03	0.35
0.7%Au/10%Co/TiO$_2$	16	15	7	78	0.09	1.31	0.98	0.43
1%Au/10%Co/TiO$_2$	22	18	16	66	0.10	1.49	1.59	1.10
2%Au/10%Co/TiO$_2$	18	24	20	56	0.16	1.56	1.56	0.97
5%Au/10%Co/TiO$_2$	15	28	23	49	0.10	1.45	1.60	1.00

Source: Reproduced from Jalama, K. et al., *Top. Catal.*, 44, 129, 2007. With permission.
[a] Reaction conditions: H$_2$:CO mol ratio = 2 with 10% N$_2$ as internal standard, total pressure 20 bar, space velocity 3 NL/(g$_{cat}$ h), 220°C, data collected after 70 h time oil stream.
[b] Percentage CO conversion.
[c] Selectivity based on carbon moles.

activity for LTFT reaction increased with an increase in Au loading and passed through a maximum in activity at 1 wt% Au, while the methane and light product selectivity monotonically increased with Au loading (Table 5.3).

5.2.5 Catalyst Pretreatment

The final catalyst after preparation usually contains cobalt in an oxidic form that is not active for FTS. It must be activated, that is, reduced to metallic cobalt before the FT reaction is commenced. Pure or diluted hydrogen in an inert gas such as nitrogen is usually used for cobalt catalyst reduction. The process is usually performed between 250°C and 400°C [89] at low pressures and involves two steps:

$$\text{Step 1:} \quad Co_3O_4 + H_2 \rightarrow 3CoO + H_2O \tag{5.1}$$

$$\text{Step 2:} \quad CoO + H_2 \rightarrow Co + H_2O \tag{5.2}$$

Cobalt catalyst pretreatment with CO containing gases has also been reported. Increased activity [37b,90,91] and greater selectivity toward producing C$_{5+}$ hydrocarbons [90,91] were obtained when a Co catalyst was activated with a gas containing CO. On their side, Dyer et al. [92] have reported some negative effects when they used a synthesis gas with a H$_2$/CO ratio of 1/1 to reduce their Co catalyst. Pretreatment in H$_2$-free CO gas has been reported to lead to lower activity, higher methane selectivity [93,94], and good stability [93] compared to H$_2$-reduced catalyst.

5.3 FUTURE PERSPECTIVES

Shale gas has fundamentally changed the US energy picture, providing a blessing in an otherwise waning economy. A decade ago, shale gas and liquids production were unimportant. As the gas supply "bubble" of the 1990s ended and crude oil prices accelerated, so did wellhead natural gas prices. In 2005, the damage caused by Hurricanes Katrina and Rita to the US Gulf natural gas supply infrastructure caused a further spike in wellhead prices, and concerns grew that natural gas prices would continue to soar. As shale gas production has accelerated, US natural gas prices have plummeted. Although the severe economic recession that began in late 2008 and the resulting decrease in the demand for natural gas have contributed to lower wellhead natural gas prices, much of that price decrease stems from the rapid increase in domestic shale gas supplies, which increased almost 10-fold between 2005 and 2010. While conventional natural gas production in the United States has diminished over time, shale gas has become a rapidly increasing source of US gas supplies, accounting for about 20% of total US onshore domestic natural gas production in 2010. The US Energy Information Administration (EIA) forecasts that, by 2035, shale gas could account for over 50% of onshore natural gas production [95]. Furthermore, several efforts have been done to produce more biosyngas from biomass gasification to meet energy targets. For example, in the United States, the Energy Independence and Security Act (EISA) of 2007 has increased the volume of renewable fuel required to be blended into transportation fuel from 9 billion gallons in 2008 to 36 billion gallons by 2022 [96], some of which will come from FTS with biosyngas. Thus, methane upgrade to liquid fuels became once more a central research topic for academy and industry. The research topics are as follows:

1. Enhance catalyst activity
2. Increase selectivity
3. Increase carbon utilization (CO_2 hydrogenation)
4 Minimize catalyst deactivation

While the first two research topics are common to any catalytic process improvement, where the most promising strategy for FTS catalysts activity and selectivity improvement seems to be addition of dopants, the remaining two are particularly important to the development of a new generation of FTS catalysts, and will be discussed in the following.

Any process that is used to produce syngas invariably produces significant amounts of CO_2, which is separated before the syngas is fed into FTS reactors leading to a substantial carbon loss. This is undesired both from an economical and environmental point of view. In order to increase carbon utilization, hydrogenation of CO_2 in the syngas into liquid hydrocarbons needs to be investigated [97], which reduces CO_2 emission into the environment and also helps with bringing down the capital investment and operation cost of the FTS process. The mechanism of FTS reaction using CO_2-rich syngas is proposed as [98]:

$$CO_2 + 3H_2 \rightarrow -CH_2- + 2H_2O + 125 \text{ kJ/mol} \tag{5.3}$$

According to Riedel et al. [99], Fe and Co FTS catalysts behaved differently in CO_2 hydrogenation. In the case of cobalt catalysts, CO_2 acts only as a diluent, and CO_2 richer feeds led to the increased methane formation. However, with iron catalysts, the composition of the hydrocarbon products of H_2/CO_2 feed gas is the same as obtained from H_2/CO feed gas, with no excessive methane formed, suggesting that Fe catalysts can perform FTS CO_2 hydrogenation rather than cobalt catalyst. The result was later corroborated by Zhang et al. [100], who observed that with cobalt catalysts, the products of CO_2 hydrogenation are 70% or more methane. However, Dorner et al. [101] demonstrate that by doping Co catalysts with Pt, it was possible to shift the product distribution from methane to higher hydrocarbons, revitalizing the possibility of using Co catalysts to perform CO_2 hydrogenation.

The catalyst lifetime is a major concern in the large-scale catalytic process, since they can greatly affect the productivity and the economic aspect of the whole process. Thus, it is essential to study how to avoid catalyst deactivation during the FTS process. There are several possible deactivation mechanisms; however, in the case of FTS there are three prime deactivation pathways:

1. *Carbon deposition* leading to catalysts fouling. Fouling is the mechanical deposition of impurities from the feed gas, which blocks the active sites or catalyst channels resulting in a decrease of catalytic performance. The organic impurities, such as tar, when condensed could be a source of catalyst fouling.
2. *Sintering (aging).* Sintering, or aging, relates to the loss of catalytic activity, due to the reduction of active metal surface area caused by crystallite growth and/or loss of support surface area triggered by support or pore collapse.
3. *Poisoning.* It is the strong chemisorption of impurities on active sites, which retards catalytic activity. The most common poison is sulfur that adsorbs selectively on many metal catalysts to form either reversible or irreversible sulfides. Other contaminants that may also cause poisoning during the catalytic conversion are Cl, Mg, Na, K, P, Si, Al, Ti, and Si [102].

In conclusion, FTS catalysts with high activity and selectivity to desired product selection remain topics of intense research and development. Minimization of catalyst decay should be the priority of the FTS catalyst design in future research and more attention should be paid to increase the carbon utilization, in order to reduce greenhouse emissions and to promote the overall rate of carbon conversion into liquid fuels.

REFERENCES

1. K. Kaygusuz, *Renew. Sustain. Energy Rev.* 16 (2012) 1116.
2. A. R. Brandt, A. Millard-Ball, M. Ganser, S. M. Gorelick, *Environ. Sci. Technol.* 47 (2013) 8031.
3. T. N. Veziroğlu, S. Şahi'n, *Energy Convers. Manag.* 49 (2008) 1820.
4. a) J. Street, F. Yu, *Biofuels* 2 (2011) 677; b) C. Le Quere, M. R. Raupach, J. G. Canadell, G. Marland, *Nat. Geosci.* 2 (2009) 831.

5. http://en.wikipedia.org/wiki/Synthetic_fuel

6. a) http://www.exxonmobil.com/Apps/RefiningTechnologies/presentations.aspx; b) I. Mochida, O. Okuma, S.-H. Yoon, *Chem. Rev.* doi: 10.1021/cr4002885.

7. G. Olive-Henrichi, S. Olive, in *The Chemistry of the Metal-Carbon Bond,* Vol. 3, Hertley and Patai (Eds.), John Wiley and Sons, New York, 1985.

8. P. Sabatier, J. B. Senderens, *Hebd. Seances Acad. Sci.* 134 (1902) 514, 680.

9. a) BASF, German Patent, (1913) 293, 787; b) BASF, German Patent, (1914) 295, 202; c) BASF, German Patent, (1914) 295, 203.

10. F. Fischer, H. Tröpsch, *Brennstoff. Chem.* 4 (1923) 276.

11. F. Fischer, H. Tröpsch, German Patent, (1925) 484, 337.

12. F. Fischer, H. Tröpsch, *Brennstoff. Chem.* 7 (1926) 97.

13. R. B. Anderson, *The Fischer-Tröpsch Synthesis*, Academic Press, Inc., 1984.

14. H. Pichler, *Brennstoff. Chem.* 19 (1938) 226.

15. R. B. Anderson, in *Catalysis*, Vol. 4, P.H. Emmet (Ed.), Von Nostrand-Reinhold, New Jersey, 1956.

16. M. E. Dry, *Catal. Today* 71 (2002) 227.

17. M. E. Dry, in *Applied Industrial Catalysis*, Vol. 2, B. E. Leach (Eds.), Academic Press, pp. 167–213.

18. M. F. M. Post, Shell International Research, Eur Pat, EP 0,174,696 (1985).

19. J. Haggin, *Chem. Eng. News* 27 (1990) 35.

20. www.shell.com/global/aboutshell/major-projects-2/pearl/overview.html.

21. www.sasol.co.za/innovation/gas-liquids/projects.

22. www.gastechnews.com/unconventional-gas/gas-to-liquids-on-the-threshold-of-a-new-era/

23. J. H. Yang, H. J. Kim, D. H. Chun, H. T. Lee, J. C. Hong, H. Jung, J. I. Yang, *Fuel Process. Technol.* 91 (2010) 285.

24. L. Wei, J. A. Thomasson, R. M. Bricka, R. Sui, J. R. Wooten, E. P. Columbus, *Trans. ASABE* 52 (2009) 21.

25. A. Demirbas, *Energy Educ. Sci. Technol.* 17 (2006) 27.

26. S. Gamba, L. A. Pellegrini, V. Calemma, C. Gambaro, *Catal. Today* 156 (2010) 58 (and references within).

27. B. H. Davis, *Top. Catal.* 32 (2005) 143.

28. T. Takeshita, K. Yamaji, *Energy Policy* 36 (2008) 2773.

29. H. Schulz, *Appl. Catal. A* 186 (1999) 3.

30. D. Xu, W. Li, H. Duan, Q. Ge, H. Xu, *Catal. Lett.* 102 (2005) 229.

31. G. C. Bond, *Catalysis by Metals*, Academic Press, Inc., New York, NY, 192, pp. 38.

32. M. A. Vannice, *J. Catal.* 40 (1975) 129.

33. R. C. Reuel, C. H. Bartholomew, *J. Catal.* 85 (1984) 78.

34. M. A. Vannice, R. L. Garten, *J. Catal.* 56 (1979) 236.

35. C. H. Bartholomew, R. B. Pannell, J. L. Butler, *J. Catal.* 65 (1980) 335.

36. a) D. J. Duvenhage, N. J. Coville, *Appl. Catal. A* 233 (2002) 63; b) J. Li, N. J. Coville, *Appl. Catal. A* 181 (1999) 201; c) J. Li, N. J. Coville, *Appl. Catal. A* 208 (2001) 177; d) J. Li, G. Jacobs, Y. Quig, T. Das, B. H. Davis, *Appl. Catal. A* 223 (2002) 195; e) J. Li, Y. Jacobs, T. Das, B. H. Davis, *Appl. Catal. A* 223 (2002) 255; f) K. Sato, Y. Inoue, I. Kojima, E. Miyazaki, I. Yasumori, *J. Chem. Soc. Faraday Trans.* 180 (1984) 841; g) B. Jongsomjit, C. Sakdamnuson, J. G. Goodwin Jr, P. Praserthdam, *Cat. Lett.* 94 (2004) 209.

37. a) B. Jongsomjit, J. Panpranot, J. G. Goodwin Jr, *J. Catal.* 204 (2001) 98; b) B. Jongsomjit, J. G. Goodwin Jr, *Catal. Today* 77 (2002) 191; c) B. Jongsomjit, J. Panpranot, J. G. Goodwin Jr, *J. Catal.* 205 (2003) 66; d) Y. Zhang, D. Wei, S. Hammache, J. G. Goodwin Jr, *J. Catal.* 188 (1999) 281.

38. A. Kogelbauer, J. C. Weber, J. G. Goodwin Jr, *Catal. Lett.* 34 (1995) 269.

39. L. Fu, C. H. Bartholomew, *J. Catal.* 92 (1985) 376.

40. M. I. Fernandez, R. A. Guerrero, G. F. J. Lopez, R. I. Rodriguez, C. C. Moreno, *Appl. Catal.* 14 (1985) 159.
41. S. W. Ho, M. Houalla, D. M. Hercules, *J. Phys. Chem.* 94 (1990) 6396.
42. B. G. Johnson, C. H. Bartholomew, D. W. Goodman, *J. Catal.* 128 (1991) 231.
43. E. Iglesia, S. L. Soled, R. A. Fiato, G. H. Via, *Stud. Surf. Sci. Catal.* 81 (1994) 433.
44. E. Iglesia, S. L. Soled, R. A. Fiato, *J. Catal.* 137 (1992) 212.
45. E. Iglesia, *Appl. Catal. A* 161 (1997) 59.
46. B. Ernst, C. Hilaire, A. Kiennemann, *Catal. Today* 50 (1999) 413.
47. A. Y. Khodakov, A. Griboval-Constant, R. Bechara, V. L. Zholobenko, *J. Catal.* 206 (2002) 230.
48. A. Martinez, C. Lopez, F. Marquez, I. J. Diaz, *J. Catal.* 220 (2003) 486.
49. W. P. Ma, Y. J. Ding, L. W. Lin, *Ind. Eng. Chem. Res.* 43 (2004) 2391.
50. S. Storsaeter, B. Tøtdal, J. C. Walmsley, B. S. Tanem, A. Holmen, *J. Catal.* 236 (2005) 139.
51. E. Kikuchi, R. Sorita, H. Takahashi, T. Matsuda, *Appl. Catal. A* 186 (1999) 121.
52. D. Song, J. Li, *J. Mol. Catal. A* 247 (2006) 206.
53. A. Barbier, A. Tuel, I. Arcon, A. Kodre, G. A. Martin, *J. Catal.* 200 (2001) 106.
54. G. L. Bezemer, J. H. Bitter, H. P. C. E. Kuipers, H. Oosterbeek, J. E. Holewijn, X. Xu, F. Kapteijn, A. J. van Dillen, K. P. de Jong, *J. Am. Chem. Soc.* 128 (2006) 3956.
55. F. Diehl, A. Y. Khodakov, *Oil Gas Sci. Technol.—Rev. IFP* 64 (2009) 11.
56. a) B. G. Johnson, C. H. Bartholomew, D. W. Goodman, *J. Catal.* 128 (1991) 231; b) E. Iglesia, S. L. Soled, R. A. Fiato, G. H. Via, *Stud. Surf. Sci. Catal.* 81 (1994) 433; c) K. Takeuchi, T. Matsuzaki, H. Arakawa, T. Hanaoka, Y. Sugi, *Appl. Catal.* 48 (1989) 149; d) K. Takeuchi, T. Matsuzaki, T. Hanaoka, H. Arakawa, Y. Sugi, *J. Mol. Catal.* 55 (1989) 361; e) C. H. Bartholomew, *Stud. Surf. Sci. Catal.* 64 (1991) 158; f) T. Matsuzaki, K. Takeuchi, T. Hanaoka, H. Arakawa, Y. Sugi, *Appl. Catal.* 105 (1993) 159; g) M. K. Niemela, A. O. I. Krause, T. Vaara, J. Lathinen, *Top. Catal.* 2 (1995) 45.
57. a) K. Takeuchi, T. Matsuzaki, T. Hanaoka, H. Arakawa, Y. Sugi, *J. Mol. Catal.* 55 (1989) 361; b) T. Matsuzaki, K. Takeuchi, T. Hanaoka, H. Arakawa, Y. Sugi, *Appl. Catal.* 105 (1993) 159.
58. M. K. Niemela, A. O. I. Krause, T. Vaara, J. J. Kiviaho, M. K. O. Reinikainen, *Appl. Catal. A: General* 147 (1999) 32.
59. a) M. P. Rosynek, C. A. Polansky, *Appl. Catal.* 73 (1991) 97; b) T. Matsuzaki, K. Takeuchi, T. Hanaoka, H. Arakawa, Y. Sugi, *Catal. Today* 28 (1996) 251; c) S. Sun, N. Tsubaki, K. Fujimoto, *Appl. Catal. A: General* 202 (2000) 121; d) Y. Wang, M. Noguchi, Y. Takahashi, Y. Ohtsuka, *Catal. Today* 68 (2001) 3; e) J. Panpranot, S. Kaewkun, P. Praserthdam, J. G. Goodwin Jr, *Cat. Lett.* 91 (2003) 95; f) J. S. Girardon, A. S. Lermontov, L. Gengembre, P. A. Chernavskii, A. Griboval-Constant, A. Y. Khodakov, *J. Catal.* 230 (2005) 339.
60. E. van Steen, G. S. Sewel, R. A. Makhote, C. Micklethwaite, H. Manstein, M. de Lange, C. T. O'Connor, *J. Catal.* 162 (1995) 220.
61. H. Ming, B. G. Baker, *Appl. Catal.* 123 (1995) 23.
62. J. van der Loosdrecht, M. van der Haar, A. M. van der Kraan, A. J. van Dillen, J. W. Geus, *Appl. Catal. A: General* 150 (1997) 365.
63. M. Kraun, M. Baerns, *Appl. Catal.* 186 (1999) 189.
64. A. Y. Khodakov, W. Chu, P. Fongarland, *Chem. Rev.* 107 (2007) 1692.
65. D. Wei, J. G. Goodwin Jr, R. Oukaci, A. H. Singleton, *Appl. Catal. A* 210 (2001) 137.
66. L. Guczi, T. Hoffer, Z. Zsoldos, S. Zyade, G. Maire, F. Garin, *J. Phys. Chem.* 95 (1991) 802.
67. Z. Zsoldos, T. Hoffer, L. Guczi, *J. Phys. Chem.* 95 (1991) 795.
68. Z. Zsoldos, L. Guczi, *J. Phys. Chem.* 96 (1992) 9393.
69. S. Zyade, F. Garin, G. Maire, *New J. Chem.* 11 (1987) 429.
70. M. J. Dees, V. Ponec, *J. Catal.* 119 (1989) 376.
71. D. Schanke, S. Vada, E. A. Blekkan, A. M. Hilmen, A. Hoff, A. Holmen, *J. Catal.* 156 (1995) 85.
72. A. Kogelbauer, J. G. Goodwin Jr, R. Oukaci, *J. Catal.* 160 (1996) 125.

73. S. A. Hosseini, A. Taeb, F. Feyzi, *Catal. Comm.* 6 (2005) 233.
74. S. A. Hosseini, A. Taeb, F. Feyzi, F. Yaripour, *Catal. Comm.* 5 (2004) 137.
75. J. Dullac, *Bull. Soc. Fr. Mineral. Crystallogr.* 92 (1969) 487.
76. K. Nagaoka, K. Takanabo, K. Aika, *Appl. Catal.* 268 (2004) 151.
77. A. Sarkany, Z. Zsoldos, G. Stefler, J. W. Hightower, L. Guczi, *J. Catal.* 157 (1995) 179.
78. L. Guczi, Z. Schay, G. Stefler, F. Mizukami, *J. Mol. Catal. A* 141 (1999) 177.
79. L. Guczi, L. Borko, Z. Schay, D. Bazin, F. Mizukami, *Catal. Today* 65 (2001) 51.
80. N. Tsubaki, S. Sun, K. Fujimoto, *J. Catal.* 199 (2001) 236.
81. A. F. Carlsson, M. Naschitzki, M. Bäumer, H. J. Freund, *J. Phys. Chem. B* 107 (2003) 778.
82. T. K. Das, G. Jacobs, P. M. Patterson, W. A. Conner, J. Li, B. H. Davis, *Fuel* 82 (2003) 805.
83. M. Rønning, D. G. Nicholson, A. Holmen, *Catal. Lett.* 72 (2001) 141.
84. G. Jacobs, J. A. Chaney, P. M. Patterson, T. K. Das, B. H. Davis, *Appl. Catal. A* 264 (2004) 203.
85. D.C. Koningsberger and R. Prins (Eds.), *X-ray Absorption: Principles, Applications, Techniques of EXAFS, SEXAFS and XANES,* Wiley, New York, 1988.
86. S. Storsæter, Ø. Borg, E. A. Blekkan, A. Holmen, *J. Catal.* 231 (2005) 405.
87. L. Leite, V. Stonkus, L. Llieva, L. Plyasova, T. Tabakova, D. Andreeva, E. Lukevics, *Catal. Comm.* 3 (2002) 341.
88. K. Jalama, N. J. Coville, D. Hildebrandt, D. Glasser, L. L. Jewell, J. A. Anderson, S. Taylor, D. Enache, G. J. Hutchings, *Top. Catal.* 44 (2007) 129.
89. A. Steynberg, M. Dry, *Stud. Surf. Sci. Catal.* 152 (2004) 533.
90. B. Nay, M. R. Smith, C. D. Telford, US Patent 5585316 (1996), to British Petroleum.
91. K. Jalama, J. Kabuba, H. Xiong, L. L. Jewell, *Catal. Comm.* 17 (2012) 154.
92. a) P. N. Dyer, R. Pietantozzi, US Patent 4619910 (1986), to Air Products & Chemicals; b) P. N. Dyer, R. Pietantozzi, H. P. Withers, US Patent 4670472 (1987), to Air Products & Chemicals; c) P. N. Dyer, R. Pietantozzi, H. P. Withers, US Patent 4681867 (1987), to Air Products & Chemicals.
93. J. Li, L. Xu, R. Keogh, B. Davis, *Catal. Lett.* 70 (2000) 127.
94. Z. Pan, D. B. Bukur, *Appl. Catal. A: General* 404 (2011) 74.
95. http://energyindepth.org/wp-content/uploads/ohio/2012/02/Economic-Impacts-of-Shale-Gas-Production_Final_23-Jan-2012.pdf.
96. D. Tilman, R. Socolow, J. A. Foley, J. Hill, E. Larson, L. Lynd, S. Pacala et al., *Science* 325 (2009) 270.
97. a) D. Unruh, M. Rohde, G. Schaub, *Stud. Surf. Sci. Catal.* 153 (2004) 91; b) O. O. James, A. M. Mesubi, T. C. Ako, S. Maity, *Fuel Process. Technol.* 91 (2010) 136; c) Y. Yao, D. Hildebrandt, D. Glasser, *Ind. Eng. Chem. Res.* 49 (2010) 11061; d) D. W. Robert, D. R. Hardy, F. W. Williams, H. D. Willaue, Catalytic CO_2 hydrogenation to feedstock chemicals for jet fuel synthesis using multi-walled carbon nanotubes as support. In *Advances in CO_2 Conversion and Utilization,* Y. H. Hu Ed., ACS Symposium Series 1056, American Chemical Society: Washington, DC, 2010; pp. 125–139; e) C. G. Visconti, L. Lietti, E. Tronconi, P. Forzatti, R. Zennaro, E. Finocchio, *Appl. Catal. A* 255 (2009) 61.
98. K. W. Jun, H. S. Roh, K. S. Kim, J. S. Ryu, K. W. Lee, *Appl. Catal. A* 259 (2004) 221.
99. T. Riedel, M. Claeys, H. Schulz, G. Schaub, S. S. Nam, K. W. Jun, M. J. Choi, G. Kishan, K. W. Lee, *Appl. Catal. A* 186 (1999) 201.
100. Y. Zhang, G. Jacobs, D. E. Sparks, M. E. Dry, B. H. Davis, *Catal. Today* 71 (2002) 411.
101. R. W. Dorner, D. R. Hardy, F. W. Williams, B. H. Davis, H. D. Willauer, *Energy Fuels* 23 (2009) 4190.
102. R. L. Bain, D. C. Dayton, D. L. Carpenter, S. R. Czernik, C. J. Feik, R. J. French, K. A. Magrini-Bair, S. D. Phillips, *Ind. Eng. Chem. Res.* 44 (2005) 7945.

6 Syngas to Methanol and Ethanol

Martin Muhler and Stefan Kaluza

CONTENTS

6.1 Methanol 169
    6.1.1 History and Current Situation 169
    6.1.2 Catalysts 170
    6.1.3 Thermodynamics and Mechanistic Considerations 173
    6.1.4 Industrial Methanol Synthesis 175
6.2 Ethanol 177
    6.2.1 History and Current Situation 177
    6.2.2 Catalysts 178
        6.2.2.1 Noble Metal-Based Catalysts 179
        6.2.2.2 Mo-Based Catalysts 180
        6.2.2.3 Modified Methanol Synthesis Catalysts 180
        6.2.2.4 Modified Fischer–Tropsch Synthesis Catalysts 181
        6.2.2.5 Bimetallic Cu–Co-Based Catalysts 182
        6.2.2.6 Comparison 183
    6.2.3 Thermodynamics and Mechanistic Considerations 183
    6.2.4 Ethanol Production from Synthesis Gas 186
Acknowledgments 188
References 188

This chapter gives a brief review and compiles recent developments in the heterogeneously catalyzed conversion of synthesis gas to the important short-chain alcohols methanol and ethanol. While methanol synthesis has been known for decades and is applied as large-scale industrial process, the research on the formation of ethanol from synthesis gas is still explorative and just a few developments have made it so far to the pilot plant.

6.1 METHANOL

6.1.1 HISTORY AND CURRENT SITUATION

In 1913, the technical methanol synthesis was mentioned for the first time. Mittasch and coworkers described the formation of methanol among several oxygen-containing compounds from carbon monoxide and hydrogen in the presence of iron oxide catalysts originally developed for the improvement of the ammonia synthesis. However, Pier et al. made the most important step in terms of a large-scale methanol production in the early 1920s [1]. They developed a ZnO–Cr$_2$O$_3$ catalyst that was resistant against sulfuric compounds present in the unpurified synthesis gas. This achievement resulted in the first industrial application of the process that was carried out at high-pressure conditions (25–35 MPa, 590–720 K) and dominated the field of methanol production for the following 40 years. Already in the 1920s, the high activity of copper-containing catalysts in methanol synthesis was mentioned, especially copper combined with zinc oxide and stabilized by alumina [2,3]. But due to its high sensitivity against poisoning by sulfur and chlorine impurities in the synthesis gas, an industrial application was not feasible. Finally, in the early 1960s, careful purification of the feedstock originating from steam reforming of naphtha and natural gas led to the breakthrough of copper-based catalysts for methanol synthesis. It was first implemented by Imperial Chemical Industries (ICI) in 1966, who developed the low-pressure, low-temperature process (5–10 MPa, 470–570 K) based on Cu [4]. It was the beginning of a series of new patents concerning copper-based catalysts for methanol synthesis submitted by several catalyst manufacturers such as Catalysts & Chemicals, Inc. (1969) [5], DuPont (1973) [6], Shell (1973) [7], BASF (1978) [8], Süd-Chemie AG (1984) [9], or IFP (1986) [10]. Today, all industrially applied catalysts contain Cu/ZnO promoted by several other oxides serving as structural stabilizers.

Methanol is one of the most important basic chemicals. In 2013, the worldwide demand was expected to exceed 65 million tons [11]. More than 90 methanol plants located all over the world have a combined production capacity of about 100 million tons per year. Methanol is not only used as a solvent and a gasoline extender, but provides a feedstock for several important chemicals such as formaldehyde, methyl *tert*-butyl ether (MTBE) and *tert*-amyl methyl ether (TAME), acetic acid, methyl methacrylate (MMA), methylamines, chloromethanes, dimethyl terephthalate (DMT), and dimethyl ether (DME) [12].

In the recent years, the interest in methanol as an alternative energy storage has increased significantly. It has already been used for fuel-cell applications, particularly, in transportation and mobile devices and as a clean fuel in combustion engines for several years [13]. The demand of methanol will further increase, which will require further improvement of the process and the catalyst design in the near future.

6.1.2 CATALYSTS

The most frequently applied industrial catalyst for methanol synthesis is a ternary system containing 50–70 mol% CuO, 20–50 mol% ZnO, and 5–20 mol% Al$_2$O$_3$ in an unreduced state [12,14]. In general, metallic copper is accepted to be the active component for the formation of methanol from synthesis gas, while alumina was identified as a structural promoter in order to achieve a stable activity and a slower deactivation under methanol synthesis conditions [15,16]. However, the special role of ZnO is still controversial. Undoubtedly, the interaction of Cu and ZnO plays an important role for achieving a highly active catalyst, but different models exist including a Schottky junction between Cu and the semiconducting ZnO as proposed by Frost [17]. Charged oxygen vacancies in ZnO were suggested as active sites in methanol synthesis, whose number is enhanced by a perturbation of the oxide defect equilibrium [17]. Another explanation for the special role of ZnO in the ternary catalysts was given by a reverse spillover of hydrogen, which first is dissociated on Cu, transported to and stored at ZnO, followed by a reverse spillover to Cu [18,19]. Nakamura and Fujitani supported a model, where a Cu–Zn surface alloy is the active site for methanol synthesis [20–22].

Muhler and coworkers confirmed the special role of ZnO by the investigation of a series of ternary (Cu/ZnO/Al$_2$O$_3$) and binary (Cu/ZnO, Cu/Al$_2$O$_3$) catalysts [23,24]. Ternary catalysts are more active in methanol synthesis compared to binary catalysts with the same Cu surface area as illustrated in Figure 6.1 [23].

Novel preparation routes were applied to enhance the interface between Cu and ZnO resulting in deviations in the Cu surface area–activity relationship. Adding ZnO via chemical vapor deposition (CVD) to a binary Cu/Al$_2$O$_3$ catalyst enhanced the activity significantly, although the Cu surface area was decreased by ZnO$_x$ species [see Figure 6.1 (★)] [23]. At the highly reducing atmosphere during methanol synthesis, the formation of Cu–ZnO$_x$ surface structures is proposed in a strong metal–support interaction (SMSI) model leading to highly active catalysts. The superiority of the novel CVD-derived catalysts lies in the optimized Cu–ZnO interface. Those results were underlined by microcalorimetric measurements aiming at the heat of adsorption of CO on binary and ternary Cu-based catalysts [24]. ZnO-containing catalysts show a decreased fractional coverage and a decreased initial heat of adsorption of about 10 kJ mol^{-1} compared to Cu/Al$_2$O$_3$ as shown in Figure 6.2.

The proposed SMSI effect is well in line with the results of the Topsøe group [25,26]. They observed a change of the morphology of the Cu particles depending on the reduction potential of the atmosphere by means of *in situ* extended X-ray

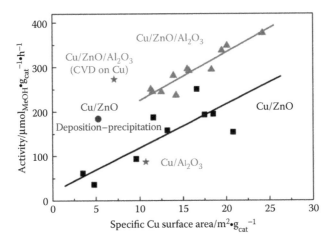

FIGURE 6.1 Area–activity relationship for various Cu-based catalysts. Binary (■) and ternary (▲) catalysts synthesized by coprecipitation; Cu/ZnO (●) by deposition–precipitation; Cu/Al$_2$O$_3$ (★) by coprecipitation, Cu/ZnO/Al$_2$O$_3$ (★) by coprecipitation + CVD. (With kind permission from Springer Science+Business Media: *Catal. Lett.*, Deactivation of Supported Copper Catalysts for Methanol Synthesis, 92, 2004, 49, Kurtz, M. et al.)

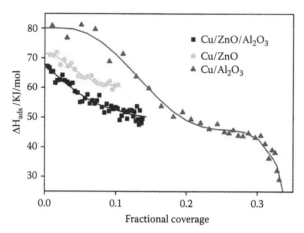

FIGURE 6.2 Differential heats of adsorption of CO on Cu catalysts at 300 K. (Reprinted from *J. Catal.*, 220, d'Alnoncourt, R. N. et al., The influence of ZnO on the differential heat of adsorption of CO on Cu catalysts: A microcalorimetric study, 249, Copyright 2003, with permission from Elsevier.)

absorption fine structure spectroscopy (EXAFS) and X-ray diffraction (XRD) [25] and *in situ* transmission electron microscopy (TEM) [26] for Cu/ZnO catalysts. Under severe reducing conditions, the Cu particles spread out on ZnO and change their activity [26]. If the atmosphere is changed to a more oxidizing gas, the reversible effect is observed and the Cu particles agglomerate resulting in a more rounded shape [26]. Additionally, under more severe reducing conditions [26], Cu–Zn surface structures are formed as observed via infrared spectroscopy. If the reduction potential exceeds a critical grade of severity, this would lead to a brass alloy formation and, therefore, to deactivation of the catalyst. Recently, Behrens et al. underlined the role of Cu–ZnO synergy for methanol synthesis [27]. They concluded that a defective form of nanoparticulate Cu rich in planar defects like stacking faults explains the activity of ZnO-free Cu catalysts. For Cu/ZnO systems they observed an additional drop of the Cu:Zn surface ratio during reductive activation of the catalyst supporting the idea of the SMSI effect. They concluded that besides the presence of steps at the Cu surface, the presence of $Zn^{\delta+}$ at the defective Cu surface is required.

All results indicate that in order to prepare an active catalyst a synthesis route is required, which provides intimate contact between all three components as well as a high dispersion of the active component. Precipitation and coprecipitation processes provide a good dispersion and a high homogeneity of the components in the catalyst precursor. Therefore, precipitation is one of the most frequently applied methods in large-scale catalyst preparation [28]. The ternary Cu/ZnO/Al$_2$O$_3$ catalyst for methanol synthesis is usually prepared by coprecipitation of a metal nitrate solution with sodium or ammonium carbonate generating mixed hydroxy carbonates [29]. The quality and properties of the precipitate strongly depend on various parameters, such as concentrations, pH, and temperature, to mention just a few. Large-scale precipitations are usually operated in batch reactors. However, a continuously operating coprecipitation synthesis was recently developed, providing an improved process control and a high degree of reproducibility [30,31]. Precipitation and aging of the hydroxy carbonate precursor are followed by washing, drying, calcinations, and reduction, with every single process step influencing the properties of the final catalyst [29,32,33].

Although the ternary $Cu/ZnO/Al_2O_3$ system has been applied in methanol synthesis for about 40 years and in spite of the importance of the process, the nature of the active site, the role of the different components, and the mechanism of the methanol synthesis are still the subject of ongoing research. Technical challenges are the improvement of the catalyst's long-term stability and a higher resistance against impurities that might occur by switching to alternative sources for synthesis gas.

6.1.3 THERMODYNAMICS AND MECHANISTIC CONSIDERATIONS

The formation of methanol from synthesis gas can occur via the hydrogenation of both carbon monoxide and carbon dioxide [12]:

$$CO + 2\,H_2 \rightleftharpoons CH_3OH \qquad \Delta H_{298\,K,5\,MPa} = -90.7\,kJ\,mol^{-1} \qquad (6.1)$$

$$CO_2 + 3\,H_2 \rightleftharpoons CH_3OH + H_2O \qquad \Delta H_{298\,K,5\,MPa} = -40.9\,kJ\,mol^{-1} \qquad (6.2)$$

Additionally, CO and CO_2 are linked by the reverse water–gas shift (RWGS) reaction:

$$CO_2 + H_2 \rightleftharpoons CO + H_2O \qquad \Delta H_{298\,K,5\,MPa} = +49.8\,kJ\,mol^{-1} \qquad (6.3)$$

These three equilibrium reactions form the reaction network for methanol synthesis. Quite a number of publications focus on the determination of the equilibrium constants, with the work by Graaf et al. [34] assuming ideal gases being the most cited and applied one. From the thermodynamic data it can be seen that high temperatures limit the formation of methanol, while the endothermic RWGS is favored. Both hydrogenation reactions are accompanied by a decreasing molar volume leading to preferred methanol formation with increasing pressure (Le Châtelier's principle). Therefore, low-temperature and high-pressure conditions are preferred for high product yields in methanol synthesis. Besides temperature and pressure, the equilibrium is influenced by the composition of the synthesis gas. The addition of inert gas components as well as higher contents of CO_2 led to a decrease in the maximum equilibrium conversion (Figure 6.3). Therefore, synthesis gases with higher CO content are usually applied in industrial methanol synthesis.

During the past few decades, there was an ongoing controversial discussion whether methanol is preferentially formed from CO or CO_2. Based on investigations by Natta [35] applying ZnO/Cr_2O_3 catalyst, it was assumed that methanol was mainly formed by hydrogenation of CO. In the 1970s, Russian scientists revealed by the use of kinetic measurements and isotopic tracers that CO_2 is the carbon source in methanol synthesis [36]. However, the mechanism based on CO dominated the scientific community for many years [37,38]. This was mainly attributed to the work of Klier et al. [39]. By applying different syngas compositions they

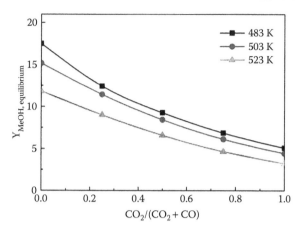

FIGURE 6.3 Equilibrium yields of methanol with increasing CO_2/CO_x ratio (calculated using AspenPlus® software).

revealed an optimal CO_2 content of about 2% for the highest methanol yield [40]. They assumed CO to be the carbon source for methanol formation and CO_2 just stabilizing the optimal oxidation state of the catalyst. Lower concentrations of CO_2 would lead to an over-reduction of the catalyst, while high CO_2 contents cause competitive adsorption at the active sites. Only when novel isotopic measurements together with kinetic studies were performed, it was possible to show that methanol is preferentially formed from carbon dioxide and that its hydrogenation occurs much faster than the hydrogenation of CO [41–44]. Moreover, measurements with varied CO_2 partial pressure at differential conversions revealed an increase in the methanol formation rate with increasing CO_2 content [41,45,46]. By comprehensive kinetic studies, Graaf et al. [47] developed a model, which in combination with experimental data proved the formation of methanol almost exclusively from CO_2. Based on these results, following models consider only CO_2 as the carbon source in methanol synthesis [48–51].

The contribution of the water–gas shift (WGS) reaction to the reaction network was also discussed controversially during the years, but recent results indicate that the rate of the WGS reaction is one order of magnitude faster compared to the hydrogenation of CO_2, and therefore, is in equilibrium as long as the syngas comprises at least 5% CO [45,52].

A huge number of publication deals with the investigation of the intermediate species during the reaction in order to unravel the elementary steps and to determine the overall mechanism. Formate (HCOO*) was found as stable intermediate on Cu- [53–57] or ZnO- [58–62] containing systems as well as on binary Cu/ZnO-based catalysts [63–67] and can be formed from both CO [63] and CO_2 [56,57]. Methanol is assumed to be formed by successive hydrogenation of adsorbed formate species, which is regarded as the rate-determining step (RDS) in methanol synthesis [53,68]. Methoxy (H_3CO*) was found to be another important intermediate and was also detected on Cu- [69,70], ZnO- [60–62], and Cu/ZnO-based systems [58,60,63–65,67]. Further surface species are formyl (HOC*) [71–73], dioxomethylene (H_2COO*) [67], and hydroxycarbene (HOHC*) [74]. Based on

TABLE 6.1

Elementary Steps of the Assumed Reaction Mechanism for Methanol Synthesis

Surface Reaction[a]			Step	Surface Reaction[b]		
$H_2O(g) + *$	$\rightleftharpoons$	H_2O*	(1)	$H_2(g) + 2*$	$\rightleftharpoons$	$2H*$
$H_2O* + *$	$\rightleftharpoons$	$OH* + H*$	(2)	$\mathbf{CO_2(g) + *}$	$\rightleftharpoons$	$\mathbf{O* + CO(g)}$
$2OH*$	$\rightleftharpoons$	$H_2O* + H*$	(3)	$CO_2(g) + O* +$	$\rightleftharpoons$	CO_3**
$OH* + *$	$\rightleftharpoons$	$O* + H*$	(4)	$CO_3** + H*$	$\rightleftharpoons$	$HCO_3** + *$
$2H*$	$\rightleftharpoons$	$H_2 + 2*$	(5)	$HCO_3** + *$	$\rightleftharpoons$	$HCOO** + O*$
$CO(g) + *$	$\rightleftharpoons$	$CO*$	(6)	$\mathbf{HCOO** + H*}$	$\rightleftharpoons$	$\mathbf{H_2COO** + *}$
$CO* + O*$	$\rightleftharpoons$	$CO_2* + *$	(7)	H_2COO**	$\rightleftharpoons$	$H_2CO* + O*$
CO_2*	$\rightleftharpoons$	$CO_2(g) + *$	(8)	$H_2CO* + H*$	$\rightleftharpoons$	$H_3CO* + *$
$CO_2* + H*$	$\rightleftharpoons$	$HCOO* + *$	(9)	$H_3CO* + H*$	$\rightleftharpoons$	$CH_3OH(g) + 2*$
$HCOO* + H*$	$\rightleftharpoons$	$H_2COO* + *$	(10)	$O* + H*$	$\rightleftharpoons$	$OH* + *$
$\mathbf{H_2COO* + H*}$	$\rightleftharpoons$	$\mathbf{H_3CO* + O*}$	(11)	$OH* + H*$	$\rightleftharpoons$	$H_2O* + *$
$H_3CO* + H*$	$\rightleftharpoons$	$H_3COH* + *$	(12)	H_2O*	$\rightleftharpoons$	$H_2O(g) + *$
H_3COH*	$\rightleftharpoons$	$H_3COH(g)$	(13)			

* Represents one adsorption site on the catalyst.

Source: [a] Askgaard, T. S. et al., *J. Catal.*, 156, 229, 1995.

[b] Vanden Busche, K. M. and Froment, G. F., *J. Catal.*, 161, 1, 1996.

Note: The respective rate-determining steps are given in bold.

these intermediate species and in combination with investigations on Cu single crystals, Askgaard et al. [48] developed a mechanistic model (Table 6.1, left column). The first steps describe the WGS reaction, while the latter concern the formation of methanol from CO_2 with the hydrogenation of dioxomethylene to methoxy as the RDS. A similar model based on the intermediate species was developed by Vanden Bussche and Froment [49]. The authors assumed the formation of a carbonate species (CO_3*) by the adsorption of CO_2 at surface oxygen, which is then successively hydrogenated to methanol (Table 6.1, right column). Due to the high stability of the formate species, its hydrogenation is regarded as the RDS. Both models are generally accepted throughout the literature and were successfully applied in describing experimental data obtained under industrially relevant reaction conditions. However, recent calculations by Norskov and coworkers [27] found the hydrogenation of CO_2 to occur via formation of HCOO, HCOOH, and H_2COOH. The latter is then split into adsorbed OH and H_2CO, which is finally hydrogenated to methanol via methoxy.

6.1.4 INDUSTRIAL METHANOL SYNTHESIS

Nowadays, industrially applied catalysts are exclusively based on Cu. Typical catalysts as supplied by Clariant, Haldor Topsøe, or Johnson Matthey Catalysts are based on the ternary system $Cu/ZnO/Al_2O_3$ [14].

The industrial application of methanol synthesis is performed either in an adiabatic or isothermal process. For both processes it is important that the heat arising from the exothermic hydrogenation of the carbon oxides to methanol has to be removed to achieve high methanol yields [12]. An adiabatic process, as it is distributed for example by Topsøe and Kellogg, was designed for plants with large capacities and consists of a series of adiabatic fixed-bed flow reactors with cooling of the process gases in-between the reactors. Lurgi developed the energy-efficient and quasi-isothermal MegaMethanol® process, which is schematically shown in Figure 6.4.

The first reactor is a boiling-water reactor, while the outlet gas is cooled in the second reactor by the synthesis gas fed to the first reactor. This concept allows a quasi-isothermal process control, which reduces the thermal strain and, therefore, extends the catalyst lifetime [75,76].

The application of CO_2-rich synthesis gases in methanol synthesis is restricted because of the increasing equilibrium limitation. The higher the CO_2 content the lower is the thermodynamically allowed carbon oxide conversion (Figure 6.5).

Thus, to increase the methanol yield high recycle streams and additional condensation steps to remove the liquid products are necessary. To avoid increasing investment and operation costs, the conversion of CO_2-rich syngas mixtures to DME is a possible alternative as demonstrated by the Haldor Topsøe TIGAS® process. By applying a multifunctional catalyst, methanol formed in the first step can be converted in the same reactor to DME reducing the equilibrium limitation and, therefore, allowing a higher carbon oxide conversion per pass [77,78] (Figure 6.5). A downstream reactor finally converts DME to gasoline. The TIGAS process requires a lower pressure compared to the conventional methanol synthesis leading to further improvement in the cost efficiency.

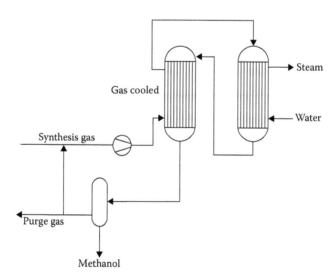

FIGURE 6.4 Scheme of the LURGI MegaMethanol® process. (Adapted from Wurzel, T., DGMK conference "Synthesis Gas Chemistry," Dresden, Germany, 2006.)

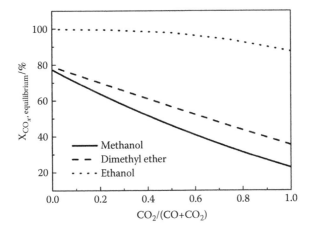

FIGURE 6.5 Equilibrium degrees of conversion of carbon oxides to different products as a function of the $CO_2/(CO + CO_2)$ ratio (523 K, 60 bar, $H_2:CO_x = 3:1$; calculated using AspenPlus® software).

Similarly, the direct conversion of CO_2-rich syngas mixtures to higher alcohols like, for example, ethanol provides a good opportunity to circumvent the equilibrium limitation of conventional methanol synthesis.

6.2 ETHANOL

6.2.1 History and Current Situation

The formation of ethanol by fermentation of sugars is one of the oldest processes of mankind and has been performed for ages. Ethanol has been mainly used for alcoholic beverages or as antiseptic agent. The use of ethanol as transportation fuel has been known since 1908, but was almost completely replaced when fuels based on crude oil were developed [79]. However, due to the depletion of the fossil resources and the increasing demand of sustainable and renewable fuels, the importance of ethanol increased hugely in recent years. In 2012, the worldwide production rate exceeded 84,000 million tons of ethanol, with North and Central America covering by far the highest values followed by South America and Brazil [80]. The use of ethanol as gasoline additive was found to decrease the emission of greenhouse gases [81], but at the expense of a lower overall efficiency [82].

Industrial-grade pure ethanol can be produced by the hydration of ethylene. However, this process is quite costly and depends on fossil resources for the formation of the ethylene feedstock [83]. Therefore, ethanol is mainly produced by the fermentation of sugars derived from corn or starch (first-generation bioethanol) [84]. Although this is the most frequently applied process, it comprises several disadvantages. On the one hand, the fermentation leads to an aqueous mixture, which requires expensive and energy-inefficient distillation steps. On the other hand, the process is not suitable for sugars derived from lignocellulose or woody biomass. This drawback restricts the fermentation process to food-based biomass and, therefore, leads to a direct competition with the food and nutrition industry [85,86].

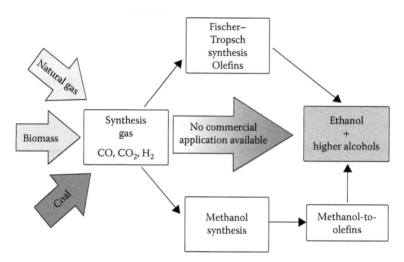

FIGURE 6.6 Process scheme combining biomass gasification and downstream synthesis gas conversion for the production of ethanol and higher alcohols.

Due to the increasing demand during the past years, intense research and development activities were begun in the field of nonfood biomass-derived ethanol production (second-generation bioethanol). One of the research strategies deals with the fermentation of nonfood biomass, which requires special enzymes and novel methods for the digestion of lignocellulosic biomass [87,88]. Gasification of nonfood or waste biomass to synthesis gas followed by downstream heterogeneously catalyzed conversion to higher alcohols represent another promising alternative for the production of second-generation bioethanol [86,89] (Figure 6.6).

The main advantages of this process are the almost complete conversion of biomass including lignocellulose into synthesis gas, as well as the already existing expertise and infrastructure for syngas-based production chains. However, a process for the direct and selective conversion of synthesis gas to ethanol has not yet been established.

6.2.2 CATALYSTS

A huge variety of elements and different combinations of them can be found in literature that were investigated for their application as catalyst in the direct conversion of synthesis gas to ethanol and higher alcohols. In general, the following four different systems can be distinguished:

- Rh-based catalysts
- Modified Mo- and MoS_2-based catalysts
- Modified methanol synthesis catalysts
- Modified Fischer–Tropsch (FT) synthesis catalysts

This chapter provides a brief overview and summarizes general aspects concerning syngas conversion to ethanol as the main product without claiming to cover the entire research in this field. In 2007 and 2008, two reviews were independently published, which are highly recommended for further reading on this topic [90,91].

6.2.2.1 Noble Metal-Based Catalysts

Rhodium is regarded as the most suitable noble metal for the direct synthesis of ethanol and higher alcohols from syngas due to its unique property to facilitate the dissociative as well as the nondissociative adsorption of CO [92]. Both steps are assumed to be essential for the synthesis of alcohols and other oxygenates from syngas. The general reaction mechanism assumes the dissociation of CO followed by hydrogenation to form a CH_x species. Insertion of nondissociated CO in the $Rh–CH_x$ bond and reaction with adsorbed atomic hydrogen finally leads to the formation of ethanol (see Section 6.2.3).

However, pure Rh supported on different oxides shows only poor performance toward ethanol formation. In order to increase the catalytic activity and ethanol selectivity a variety of different promoters have been investigated [92]. The promoter is assumed to interact with the oxygen atom of the adsorbed CO molecule, enhancing the tendency for dissociative adsorption. The generation of oxygen vacancies in the promoter phase under reducing conditions is favorable as they provide adsorption sites for CO through the oxygen atom [93,94]. Moreover, the promoter phase should partly cover the Rh clusters during reduction leading to a high number of $Rh–MO_x$ interface sites and, therefore, to enhanced CO dissociation [95,96]. Fe was found to be an appropriate promoter for Rh-based catalysts that meets the described requirements [97,98]. Moreover, the addition of Fe_2O_3 facilitates the insertion of nondissociated CO into the $Rh–CH_x$ bond, which leads to a higher ethanol selectivity and simultaneously suppresses the undesired formation of methane [99]. Also, other promoters such as Mn, La, Ce, and Y were found to increase the selectivity toward ethanol, whereas Zr, Ti, and V enhance the catalytic activity [91,100–102]. Moreover, the addition of a second promoter such as Li, Na, or K leads to an increase in the selectivity to C_2-oxygenates by suppressing the formation of hydrocarbons [103].

The role of the support has also been intensively studied, with two different effects being distinguished: the direct effect describes support materials that influence the product formation due to their chemical properties. High ethanol yields were found for oxidic supports exhibiting weak basicity, such as La_2O_3, CeO_2, or TiO_2 [100,104]. More basic supports such as MgO or ZnO lead to the formation of methanol as major product, while acidic supports favor the formation of methane and other hydrocarbons [91,104]. The influence of the support on the Rh dispersion, which in turn influences the type of CO adsorption, is described as an indirect support effect. For example, Rh supported on SiO_2 forms particles that favor the associative CO adsorption, while highly dispersed Rh clusters are formed on Al_2O_3 or TiO_2 leading to the dissociation of CO [105].

Although Rh-based catalysts show quite high selectivities for ethanol (up to 50%), the achieved degrees of conversion are very low. Moreover, the application on the industrial scale is unfavorable. The annual availability of rhodium is about 22 tons, out of which 80% is needed for the production of automotive catalytic converters [106], resulting in very high costs. As a consequence, large-scale production of

Rh-based catalysts for the conversion of synthesis gas to ethanol is uneconomic due to the high Rh content required at the present stage of catalyst development.

6.2.2.2 Mo-Based Catalysts

Bulk and supported Mo-based catalysts have also been extensively studied and applied in higher alcohol synthesis. It is generally accepted that the reaction mechanism for the formation of ethanol and higher alcohols from synthesis gas on modified Mo-based catalysts involves the same steps as already discussed for noble metal-based systems, that is, insertion of nondissociated CO into a surface CH_x species formed by hydrogenation of dissociatively adsorbed CO [107].

By doping of Mo supported on carbon or clays with alkali or transition metals the product selectivity can be shifted from hydrocarbons to alcohols. For example, potassium suppresses the reduction of Mo to the metal by stabilizing a MoO_2 phase, which is supposed to be responsible for the formation of alcohols [108]. The addition of Ni to a K-doped β-Mo_2C leads to enhanced conversion and increased selectivity, especially toward C_{2+} alcohols [109]. The addition of Co to the K-promoted Mo catalyst results in an additional increase in hydrocarbon selectivity, which was found to be caused by a Co_2C phase formed under the reaction conditions [110].

Systems based on MoS_2, which are industrially applied as hydrodesulfurization catalyst to remove sulfur from natural gas or refined petroleum products, have also been investigated for higher alcohol synthesis. Alkali doping again leads to a shift in the product distribution with higher selectivity toward alcohols. The addition of Co to a K-doped MoS_2 catalyst increases the total alcohol yield as well as the ethanol selectivity, reaching a maximum at a Mo/Co molar ratio of 2 [111]. However, during the reaction CoS_x phases are formed, which were found to be responsible for the observed catalytic deactivation [112]. The formation of these CoS_x phases was inhibited by the addition of La leading to enhanced stability of the catalyst. Moreover, a further increase in the conversion and alcohol selectivity was observed in contrast to the La-free system [113]. Ni also leads to higher conversion and alcohol formation, but simultaneously favors the formation of methane. The addition of Mn promotes the dispersion of Ni on the surface avoiding the formation of larger Ni clusters, which were found to be the active sites for methanation [114].

MoS_2-based systems are considered promising catalysts for the synthesis of ethanol and higher alcohols from synthesis gas. The catalysts were found to show high resistance against sulfur poisoning [91], which is a severe topic especially in the case of syngas derived from biomass gasification. However, significant amounts of sulfur are found in the products and require extensive and costly posttreatment in order to fulfill standard fuel regulations [115].

6.2.2.3 Modified Methanol Synthesis Catalysts

Although the synthesis of methanol over ZnO/Cr_2O_3 or Cu/ZnO-based catalysts is highly selective, significant amounts of ethanol and other oxygenates are detectable in some cases. The catalysts are usually prepared by precipitation and the formation of by-products during methanol synthesis is attributed to a certain content of alkali ions remaining in the catalyst [116]. This observation led to an increase in the research on alkali-promoted Cu-based catalysts for the synthesis of ethanol and higher alcohols from synthesis gas. Different alkali metals and loadings were investigated and it was found that the tendency for the formation of higher alcohols increased with increasing alkali atomic radius Li < Na < K <Rb < Cs [117]. Two main effects were attributed to alkali doping: (1) neutralization of acid sites on the catalyst surface leading to inhibited ether formation [118] and (2) provision of basic sites for the formation of C–C and C–O bonds [119]. The optimum loading depends on several factors, such as preparation, support, and promoter concentration. In most cases, a maximum in the formation of alcohols is observed with increasing alkali loading. It is assumed that increasing the alkali loading increases the alcohol productivity, but at a certain loading further addition leads to a blocking of the Cu/ZnO sites, which results in a decrease of the overall catalytic activity [120].

Alkali-doped Cu-based catalysts supported on ZnO/Al_2O_3, ZnO/Cr_2O_3, or MgO/ CeO_2 were successfully applied in the synthesis of higher alcohols [118,121–124]. All systems produced mixtures of linear and branched C_1–C_6 alcohols with only small amounts of other oxygenates or hydrocarbons.

However, methanol was still found to be the major product, while the selectivity to ethanol was quite low. This observation is rationalized by the proposed reaction mechanism on Cu-based catalysts, which describes the formation of higher oxygenates by the condensation of two lower alcohol molecules. In this mechanism, the coupling of two C_1 species to form a C_2 intermediate is regarded as the RDS, while the reaction of the C_2 intermediate to form higher alcohols proceeds fast, leading to the high amounts of methanol and C_{2+} alcohols at the expense of ethanol [122].

6.2.2.4 Modified Fischer–Tropsch Synthesis Catalysts

The conventional Fischer–Tropsch (FT) synthesis based on Co, Fe, Ni, or Ru catalysts is a well-known process that selectively converts syngas to long-chain hydrocarbons. However, minor amounts of alcohols and other oxygenates are simultaneously formed during the reaction [125,126]. For the application in the synthesis of ethanol and higher alcohols, Co, Fe, and Ru catalysts supported on Al_2O_3, SiO_2, or carbon nanotubes were modified by a variety of different promoters including transition metals, such as Cu, Mo, Mn, Pd, and La as well as alkali metals, such as Li, Na, K, and Cs leading to enhanced formation of alcohols and other oxygenates [127–133].

In general, the activity of modified FT-catalysts for the formation of alcohols and oxygenates is attributed to a synergistic effect by the combination of an FT metal that adsorbs CO dissociatively and a promoter metal that facilitates the nondissociative adsorption of CO [134,135]. The synergistic effect requires a close interaction between the catalyst and the promoter phase and, therefore, strongly depends on the preparation method, the loading and dispersion of the promoter phase as well as its precursor and the pretreatment of the catalyst [136,137]. Additional alkali doping of the modified FT-catalysts leads to a further increase in the selectivity to C_{2+} oxygenates by suppressing the formation of hydrocarbons [136].

Modified FT-catalysts mainly form linear primary alcohols. The proposed reaction mechanism is the same as described for Rh-based catalysts, assuming the insertion of nondissociated CO into a surface CH_x species as the essential step [130]. Nevertheless, the amount of hydrocarbons is normally higher than the amount of the respective oxygenates with methane being the major product. Some systems showed a shift in this tendency only for the C_2 compounds and possessed a higher selectivity to oxygenates compared to the C_2 hydrocarbons. However, the overall product formation generally follows an Anderson–Schulz–Flory distribution [91,138].

6.2.2.5 Bimetallic Cu–Co-Based Catalysts

Catalysts based on Cu and Co for the conversion of syngas to ethanol and higher alcohols attracted much attention during the last decades and are still extensively studied. The catalysts comprise Cu as the active site in methanol synthesis as well as the typical FT metal Co and, therefore, provide the opportunities for the synergistic effect described earlier.

Between 1975 and 1990, a large number of patents were filed by the Institut Français du Pétrole (IFP) [139–144] dealing with the development of Cu–Co-based catalysts for the synthesis of higher alcohols from synthesis gas. In addition to Cu and Co, the described catalysts comprised a huge number of different transition metals as well as alkali and alkaline earth metals, which are assumed to act as structural or electronic promoters and lead to increased catalytic activity and alcohol selectivity. In contrast to other systems, the Cu–Co-based catalysts were found to facilitate especially the formation of C_{2+} oxygenates leading to ethanol as the major product of all alcohols including methanol [139,140]. The catalysts are prepared by coprecipitation in order to achieve a preferably homogeneous distribution of all components in the material. After calcination, the catalyst is impregnated with the respective alkali or alkaline earth metals [145].

Based on the early work of IFP, the system was extensively investigated during the past years focusing especially on the interaction between Cu and Co. The synergistic effect was found to play a major role for the performance of Cu–Co-catalysts with respect to higher alcohol formation [146]. The precipitation of mixed crystalline phases, such as layered double hydroxides or perovskites, can improve the interaction between Cu and Co by creating close proximity of the active sites already during the preparation step [147–150]. TPR studies revealed a synergistic effect during reduction of the bimetallic catalyst by metallic Cu enhancing the reducibility of cobalt oxide [137,151]. Spivey and coworkers [152] reported that mixed Cu–Co nanoparticles exhibit higher selectivity to ethanol and other oxygenates as the respective core–shell nanoparticles, which might be due to a higher amount of bimetallic interfaces in the former case. Kruse and coworkers applied an oxalate coprecipitation route for the synthesis of unsupported Cu–Co catalysts. They observed a high selectivity to long-chain alcohols, especially $C_8–C_{14}$, in the CO hydrogenation reaction. Compared to the bimetallic system, the addition of a third metal, for example, Mn or Mo, strongly increased the catalytic activity and selectivity [153–155]. Similar observations were made by Goodwin Jr. and coworkers [156], who investigated combinations of Cu, Co, and ZnO as catalysts for CO hydrogenation. Only the ternary composition, that is, Co/Cu–ZnO, showed significant selectivity to ethanol and higher oxygenates [156]. In addition to several reports about the addition of transition, alkali, or alkaline earth metal for improving Cu-Co-based catalysts, the use of carbon nanotubes either as promoter or as support was recently published [157,158].

6.2.2.6 Comparison

The comparison of different catalysts for the synthesis of ethanol and higher alcohols from syngas is difficult for several reasons. The data presented in the literature are usually obtained under different reaction conditions such as temperature, pressure, H_2/CO ratio, and flow rates, which strongly influence the degrees of conversion, yields, and selectivities. Some authors omit CO_2 formation or other important values, such as the degree of conversion, or the selectivities of by-products are missing. Moreover, the setups used for the test measurements range from differential reactors up to pilot-plant scale facilities. Table 6.2 summarizes the performance of some selected catalysts.

6.2.3 THERMODYNAMICS AND MECHANISTIC CONSIDERATIONS

The conversion of synthesis gas to ethanol and higher alcohols can thermodynamically occur by the hydrogenation of CO and CO_2 described by Equations 6.4 and 6.5, respectively.

$$n\,CO + 2n\,H_2 \rightleftharpoons C_nH_{2n+1}OH + (n-1)H_2O \tag{6.4}$$

$$n\,CO_2 + 3n\,H_2 \rightleftharpoons C_nH_{2n+1}OH + (2n-1)H_2O \tag{6.5}$$

The reactions are both exothermic and accompanied by decreasing molar volume. Therefore, low-temperature and high-pressure conditions favor higher alcohol yields. In contrast to the highly selective methanol synthesis, the formation of side-products, such as methanol, methane, and higher hydrocarbons, has to be considered [90,91]. Especially, methanation is a severe competitive reaction as it is thermodynamically favored under the applied reaction conditions [159]. Figure 6.7 shows the equilibrium yields of ethanol for different syngas compositions. However, when the methanation reaction is allowed to occur, it completely suppresses the formation of ethanol. In order to achieve high ethanol selectivity, it is essential to kinetically limit the formation of methane.

As already mentioned most catalysts applied in the conversion of syngas to ethanol and higher alcohols provide active sites in close proximity that adsorb CO dissociatively as well as nondissociatively. Both are important for the mechanism assumed for Rh-, FT-, and Mo-based catalysts [107,130,160,161]. Despite some minor differences, a general reaction mechanism can be described as schematically depicted in Figure 6.8.

The first reaction steps comprise the dissociative adsorption of CO or CO_2, respectively, followed by hydrogenation to form a CH_x surface species. By further hydrogenation this species is converted to methane, which desorbs from the surface and terminates the reaction sequence. In contrast, by insertion of molecularly adsorbed CO an acylic or enol-type species is formed. This species can be converted to ethanol by hydrogenation or function as intermediate for the formation of C_{2+} oxygenates by further CO insertion/hydrogenation. In the case of CO_2, dissociative adsorption leading to molecularly adsorbed CO is assumed [162].

TABLE 6.2
Comparison of Selected Catalysts for the Conversion of Syngas to Ethanol

	Experimental Conditions						Selectivity S_i (%)				
Catalyst	T (K)	p (MPa)	H_2:CO	Flow	X_{CO} (%)	CO_2	MeOH	EtOH	Oxy	HC	Reference
1%Rh/V$_2$O$_5$	493	0.1	1	n.a.	4.5	6.0	6.2	37.2		50.5	[101]
2%Rh–2.5%Fe/TiO$_2$	567	2	1	8,000[a]	17.7			23.7	19.1	57.2	[97]
5%Rh–5%Mn/SBA15	573	1	2	9,000[a]	21.9	23.8	1.2	12.9	0.1	61.6	[94]
5%Rh–2.5%Fe/SBA15	573	1	2	9,000[a]	19.5	18.7	2.9	20.6	2.9	54.8	[99]
5%Rh–5%FeO$_x$/SiO$_2$	523	2	2	8,000[a]	12.4	3.5	18.0	42.0		31.2	[98]
K–Co–Mo/AC	603	5	2	4,800[b]	14.3		9.3	8.1	4.4	53.4	[111]
K/Co/β–Mo$_2$C	573	8	1	2,000[b]	51.0	Omitted	11.1	15.5	13.0	60.4	[110]
K/Ni/β–Mo$_2$C	573	8	1	2,000[b]	73.0	50.9	6.0	9.4	7.9	25.8	[109]
La$_{0.2}$MoCo$_{0.1}$K$_{0.6}$	603	3	2	2,225[b]	17.2	Omitted	29.6	40.3	5.5	24.8	[113]
K–Ni–Mn/MoS$_2$	588	9.5	2	6,000[b]	17.8	14.3	32.0	23.0	14.7	16.0	[114]
K–Co/MoS$_2$/clay	573	13.8	1.1	2,000[b]	30.5	Omitted	17.9	36.0	23.2	22.9	[112]
K–Ni/MoS$_2$	573	8	1	2,500[b]	34.1	Omitted	8.3	21.2	33.2	37.3	[107]
K–Ni/MoS$_2$	643	9.1	1	7,000[a]	16.7	Omitted	23.8	32.3	23.0	20.9	[115]
Cu/ZnO/Cr$_2$O$_3$	583	7.6	0.45	5,300[a]	13.8	Omitted	53.4	6.9	26.9	9.4	[119]
3%Cs–Cu/ZnO/Cr$_2$O$_3$	583	7.6	0.45	5,300[a]	14.5	Omitted	42.7	7.3	36.6	10.6	[119]
0.3%Cs–Cu/ZnO/Al$_2$O$_3$	583	7.6	0.45	5,300[a]	11.5	Omitted	79.9	5.1	8.5	6.5	[119]
2.5%Cs–Cu/ZnO/Al$_2$O$_3$	583	7.6	0.45	5,300[a]	10.7	Omitted	82.3	3.6	8.4	5.7	[119]

Catalyst											
1%K–Cu$_{0.5}$Mg$_5$CeO$_x$	583	4.5	1	6,000^a	9.9	15.6	63.5	2.2	11.0	4.0	[118]
Pd–K–Cu$_{0.5}$Mg$_5$CeO$_x$	583	4.5	1	6,000^a	10.6	19.2	60.0	2.9	7.8	5.1	[118]
Ir–Ru/SiO$_2$	573	5	2	2,000^b	5.6	4.5	4.2	17.2	5.8	64.1	[134]
Li–Ir–Ru/SiO$_2$	573	5	2	2,000^b	6.5	5.7	2.3	21.6	9.9	55.7	[134]
Co–Ir/SiO$_2$	573	5.1	2	2,000^b	n.a.	3.8	23.0	27.0	7.8	38.7	[135]
10%Co–2%Pd/SiO$_2$	543	1	2	24,000^a	8.2	6.4	1.2	2.5	1.7	88.2	[132]
Fe–Cu/SiO$_2$	603	10	1	4,000^a	4.4	9.9			9.8^c	80.3	[127]
Fe–Cu–La–Mo/SiO$_2$	603	10	1	4,000^a	5.3	9.5			16.2^c	77.6	[127]
Fe–Cu–La–Mo/SiO$_2$	603	10	3	4,000^a	11.8	4.9			22.4^c	73.1	[127]
Co–Cu/ZnO/Al$_2$O$_3$	563	4	2	3,000^b	95.0	Omitted	5.7		1.6^d	92.8	[149]
1%K–Co–Cu/ZnO/Al$_2$O$_3$	563	4	2	3,000^b	82.0	Omitted	18.9		4.9^d	76.2	[149]
5%K–Co–Cu/ZnO/Al$_2$O$_3$	563	4	2	3,000^b	47.0	Omitted	14.6		5.7^d	79.1	[149]
Co–Cu (1:3)	543	2	2	18,000^b	<1	27.8	2.8	5.3	35.2	29.1	[152]
Co–Cu (1:24)	543	2	2	18,000^b	<1	48.8	6.6	11.4	16.0	17.3	[152]

EtOH, ethanol; HC, hydrocarbons including CH$_4$; MeOH, methanol; n.a., data not available; Oxy, all other oxygenates including C$_{2+}$ alcohols, ethers, aldehydes, etc.; S$_i$, selectivity of species i; X$_{CO}$, CO conversion.

^{a} Flow given in L kg$_{cat}^{-1}$ h^{-1}.
^{b} Flow given in h^{-1}.
^{c} Including methanol and ethanol.
^{d} Including ethanol.

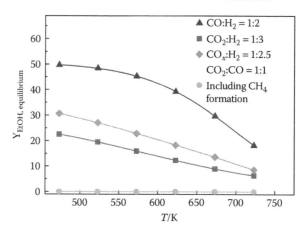

FIGURE 6.7 Equilibrium yields of ethanol applying different syngas compositions with and without consideration of the competitive methanation (calculated using the AspenPlus® software).

The reaction mechanism for the synthesis of higher alcohols on modified Cu-based catalysts is discussed much more controversially. Most mechanistic models assume the formation of the C_2 intermediate by a coupling reaction of two methanol molecules, which was supported by nuclear magnetic resonance (NMR) and isotopic measurements [163]. Formyl was found as an important intermediate, which adsorbs on basic sites on alkali-modified Cu-based catalysts. By nucleophilic attack of the formyl species on formaldehyde the C_2 intermediate is formed [116,118,163]. Both formyl and formaldehyde are assumed to be formed from methanol [164]. Elliott and Pannella [165] describe the formation of ethanol based on an adsorbed C_1 species on the catalyst surface, which can originate from both synthesis gas $(CO + H_2)$ and methanol. This C_1 intermediate is not further specified by the authors, but it is likely assumed to be either formyl or formaldehyde [90]. Recently, Wang and coworkers [166] published an alternative mechanism based on density functional theory (DFT) calculation. They propose that CO is successively hydrogenated to adsorbed H_3CO* species, which either forms methanol with $H*$ or is dissociated to CH_3* and $OH*$. Ethanol formation then occurs by insertion of CO into the CH_3* species followed by hydrogenation.

Concerning the influence of CO_2 in the syngas, a promoting [167] as well as an inhibiting effect on the formation of higher alcohols was reported [118]. However, most publications on higher alcohol synthesis focus on CO_2-free synthesis gas.

6.2.4 ETHANOL PRODUCTION FROM SYNTHESIS GAS

In the 1980s, IFP built a pilot plant for mixed alcohols from synthesis gas in Chiba, Japan. They used a Cu–Co-based catalyst with different transition and alkali metals, which produced a mixture of C_1–C_6 alcohols with methanol being the major component. The plant had an overall production capacity of about 670 t per year. However, IFP has not pursued this work [168,169]. Between 1982 and 1987, a 15,000 t/y pilot plant was launched in Italy by the joint cooperation of Snamprogetti, Enichem, and Haldor Topsøe (SEHT). The facility produced a mixture of alcohols based on a

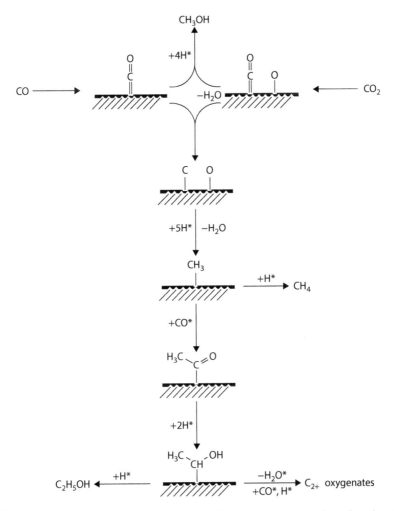

FIGURE 6.8 Schematic illustration of the simplified reaction mechanisms for ethanol formation on Rh-, FT-, and Mo-based catalysts [107,130,160,161].

modified methanol synthesis catalyst. Although they were able to sell their product as gasoline blend, the work was discontinued [86,169]. The Lurgi–Octamix process also used a methanol synthesis catalyst, but with somewhat different promoters compared to the system applied in the SEHT process. Lurgi built a pilot plant in Juelich, Germany, in 1990 with a capacity of about 2 t/d. However, Lurgi indicated that they no longer pursued this technology [86,169]. Dow used a MoS_2-based catalyst that showed good productivities for linear C_2–C_5 alcohols, but also produced significant amounts of CO_2, methanol, and hydrocarbons [86]. This bench scale technology was planned to be scaled up to a 2 t/d pilot plant by Power Energy Fuels, Inc. (PEFI) called the Ecalene™ mixed alcohol process, but further and more recent information is not available [86,170]. Range Fuels launched its first commercial-scale ethanol facility in 2007. It provided the conversion of biomass-derived synthesis gas to a mixture of aliphatic alcohols over an Mo-based catalyst with a capacity of about 30,000 t/y [171,172]. However, the plant failed and had to be shut down in 2011. In addition to the direct conversion of synthesis gas to ethanol and higher alcohols, indirect consecutive processes were recently described, which for instance involve the downstream conversion of methanol [173,174] or the formation of higher alcohols from previously formed DME and synthesis gas [175].

Although no commercial plant has been established yet, the demand of an industrially feasible process for the direct conversion of synthesis gas to ethanol and higher alcohols is still high. Much more sophisticated research in the fields of catalyst development and reaction engineering is required to achieve this goal.

ACKNOWLEDGMENTS

We gratefully thank the members of the Laboratory of Industrial Chemistry of the Ruhr-University Bochum and the Catalytic Processes group at Fraunhofer UMSICHT for their contribution to this chapter, especially Kevin Kähler, Holger Ruland, and Johan Anton. Our partners from the BETSY project ("*Bio-Et*hanol from *S*ynthesis Gas") funded by the Ministry for Economy, Energy, Building, Habitation and Transportation of North Rhine-Westfalia within the programme Ziel2.NRW financed by the European Union through the European regional development fund (ERDF) are acknowledged for the fruitful collaboration.

REFERENCES

1. A. Mittasch, M. Pier, K. Winkler, BASF (1925) DE415686.
2. C. Lormand, *Ind. End. Chem.* 17 (1925) 430.
3. M.R. Fenske, P.K. Frolich, *Ind. Eng. Chem.* 21 (1929) 1052.
4. J.T. Gallagher, J.M. Kidd, ICI (1965) GB1159035.
5. T.D. Casey, G.M. Chapman, Catalysts&Chemicals, Inc. (1969) GB1286970.
6. A.B. Stiles, DuPont (1973) GB1436773.
7. E.F. Magoon, Shell (1973) US3709919.
8. F.J. Broecker, K.-H. Gründler, L. Marosi, M. Schwarzmann, B. Triebskorn, G. Zirker, BASF (1978) DE2846614.
9. M. Schneider, K. Kochloefl, J. Ladebeck, Süd-Chemie AG (1984) EP0125689.
10. P. Courty, C. Travers, D. Durand, A. Forestière, P. Chaumette, IFP (1986) US4596782.
11. Homepage of the Methanol Institute, http://methanol.org (accessed 09/2013).
12. J.B. Hansen, P.E.H. Nielsen, Methanol Synthesis, in: G. Ertl, H. Knözinger, F. Schüth, J. Weitkamp (eds.), *Handbook of Heterogeneous Catalysis*, Wiley-VCH, Weinheim, 2008.
13. G.A. Olah, *Angew. Chem. Int. Ed.* 44 (2005) 2636.
14. E. Kunkes, M. Behrens, Methanol Chemistry, in: R. Schlögl (ed.), *Chemical Energy Storage*, De Gruyter, Berlin, 2013.
15. H. Wilmer, T. Genger, O. Hinrichsen, *J. Catal.* 215 (2003) 188.
16. M. Kurtz, H. Wilmer, T. Genger, O. Hinrichsen, M. Muhler, *Catal. Lett.* 86 (2003) 77.
17. J.C. Frost, *Nature* 334 (1988) 577.
18. R. Burch, S.E. Golunski, M.S. Spencer, *J. Chem. Soc. Faraday Trans.* 86 (1990) 2683.
19. M.S. Spencer, *Top. Catal.* 8 (1999) 259.
20. T. Fujitani, J. Nakamura, *Appl. Catal. A* 191 (2000) 111.
21. I. Nakamura, H. Nakano, T. Fujitani, T. Uchijama, J. Nakamura, *Surf. Sci.* 402–404 (1998) 92.

22. J. Nakamura, Y. Choi, T. Fujitani, *Top. Catal.* 22 (2003) 277.
23. M. Kurtz, N. Bauer, C. Büscher, H. Wilmer, O. Hinrichsen, R. Becker, S. Rabe et al., *Catal. Lett.* 92 (2004) 49.
24. R. Naumann d'Alnoncourt, M. Kurtz, H. Wilmer, E. Löffler, V. Hagen, J. Shen, M. Muhler, *J. Catal.* 220 (2003) 249.
25. J.-D. Grundwaldt, A.M. Molenbroek, N.-Y. Topsøe, H. Topsøe, B.S. Clausen, *J. Catal.* 194 (2000) 452.
26. N.-Y. Topsøe, *Catal. Today* 113 (2006) 58.
27. M. Behrens, F. Studt, I. Kasatkin, S. Kühl, M. Hävecker, F. Abild-Pedersen, S. Zander et al., *Science* 336 (2012) 893.
28. F. Schüth, M. Hesse, K.K. Unger, Precipitation and Coprecipitation, in: G. Ertl, H. Knözinger, F. Schüth, J. Weitkamp (eds.), *Handbook of Heterogeneous Catalysis*, Wiley-VCH, Weinheim, Germany, 2008.
29. S. Schimpf, M. Muhler, Methanol Catalysts, in: K.P. de Jong (ed.), *Synthesis of Solid Catalysts*, Wiley-VCH, Weinheim, Germany, 2009.
30. S. Kaluza, M. Behrens, N. Schievenhövel, B. Kniep, R. Fischer, R. Schlögl, M. Muhler, *ChemCatChem* 3 (2011) 189.
31. P. Kurr, S. Kaluza, M. Hieke, B. Kniep, M. Muhler, R. Fischer, Süd-Chemie AG (2010) DE102010021792.
32. S. Kaluza, M. Muhler, *Catal. Lett.* 129 (2009) 287.
33. S. Kaluza, M. Muhler, *J. Mater. Chem.* 19 (2009) 3914.
34. G.H. Graaf, P.J.J.M. Sijtsema, E.J. Stamhuis, G.E.H. Joosten, *Chem. Eng. Sci.* 41 (1986) 2883.
35. G. Natta, *Catalysis* 3 (1955) 349.
36. A.Y. Rozovskii, Y.B. Kagan, G.I. Lin, E.V. Slivinskii, S.M. Loktev, L.G. Liberov, A.N. Bashkirov, *Kinet. Catal.* 17 (1976) 1314.
37. J.C.J. Bart, R.P.A. Sneeden, *Catal. Today* 2 (1987) 1.
38. G.C. Chinchen, P.J. Denny, J.R. Jennings, M.S. Spencer, K.C. Waugh, *Appl. Catal.* 36 (1988) 1.
39. K. Klier, *Adv. Catal.* 31 (1982) 243.
40. K. Klier, V. Chatikavanij, R.G. Herman, G.W. Simmons, *J. Catal.* 74 (1982) 343.
41. G. Liu, D. Willcox, M. Garland, H.H. Kung, *J. Catal.* 96 (1985) 251.
42. G.C. Chinchen, P.J. Denny, D.G. Parker, G.D. Short, M.S. Spencer, K.C. Waugh, D.A. Whan, *Am. Chem. Soc. Div. Fuel Chem.* 29 (1984) 178.
43. G.C. Chinchen, P.J. Denny, D.G. Parker, M.S. Spencer, D.A. Whan, *Appl. Catal.* 30 (1987) 333.
44. R. Kieffer, E. Ramaroson, A. Deluzarche, Y. Trambouze, *React. Kinet. Catal. Lett.* 16 (1981) 207.
45. K.G. Chanchlani, R.R. Hudgins, P.L. Silveston, *J. Catal.* 136 (1992) 59.
46. M. Sahibzada, I.S. Metcalfe, D. Chadwick, *J. Catal.* 174 (1998) 111.
47. G.H. Graaf, E.J. Stamhuis, A.A.C.M. Beenackers, *Chem. Eng. Sci.* 43 (1988) 3185.
48. T.S. Askgaard, J.K. Nørskov, C.V. Ovesen, P. Stoltze, *J. Catal.* 156 (1995) 229.
49. K.M. Vanden Bussche, G.F. Froment, *J. Catal.* 161 (1996) 1.
50. O.A. Malinovskaya, A.Y. Rozovskii, I.A. Zolotarskii, Y.V. Lender, Y.S. Matros, G.I. Lin, G.V. Dubovich, N.A. Popova, N.V. Savostina, *React. Kinet. Catal. Lett.* 34 (1987) 87.
51. T. Kubota, I. Hayakawa, H. Mabuse, K. Mori, K. Ushikoshi, T. Watanabe, M. Saito, *Appl. Organomet. Chem.* 15 (2001) 121.
52. Y. Yang, J. Evans, J.A. Rodriguez, M.G. White, P. Liu, *Phys. Chem. Chem. Phys.* 12 (2010) 9909.
53. M. Bowker, R.A. Hadden, H. Houghton, J.N.K. Hyland, K.C. Waugh, *J. Catal.* 109 (1988) 263.
54. B.A. Sexton, *Surf. Sci.* 88 (1997) 319.

55. B.E. Hayden, K. Prince, D.P. Woodruff, A.M. Bradshaw, *Surf. Sci.* 133 (1983) 589.
56. P.A. Taylor, P.B. Rasmussen, C.V. Ovesen, P. Stoltze, I. Chorkendorff, *Surf. Sci.* 261 (1992) 191.
57. I. Chorkendorff, P.A. Taylor, P.B. Rasmussen, *J. Vac. Sci. Technol. A* 10 (1992) 2277.
58. S.-I. Fujita, H. Ito, N. Takezawa, *Catal. Lett.* 33 (1995) 67.
59. M. Bowker, H. Houghton, K.C. Waugh, *J. Chem. Soc. Faraday Trans.* 77 (1981) 3023.
60. A. Ueno, T. Onishi, K. Tamaru, *Trans. Faraday Soc.* 67 (1971) 3585.
61. S.G. Neophytides, A.J. Marchi, G.F. Froment, *Appl. Catal. A Gen.* 86 (1992) 45.
62. K. Kaehler, M.C. Holz, M. Rohe, J. Strunk, M. Muhler, *ChemPhysChem* 11 (2010) 2521.
63. J.F. Edwards, G.L. Schrader, *J. Phys. Chem.* 88 (1984) 5620.
64. S.-I. Fujita, M. Usui, E. Ohara, N. Takezawa, *Catal. Lett.* 13 (1992) 349.
65. R. Yang, Y. Fu, Y. Zhang, N. Tsubaki, *J. Catal.* 228 (2004) 23.
66. J.E. Bailie, C.H. Rochester, G.J. Millar, *Catal. Lett.* 31 (1995) 333.
67. V. Sanchez-Escribano, M.A. Larrubia Vargas, E. Finocchio, G. Busca, *Appl. Catal. A Gen.* 316 (2007) 68.
68. K.M. Vanden Bussche, G.F. Froment, *Appl. Catal. A Gen.* 112 (2004) 37.
69. M. Bowker, R.J. Madix, *Surf. Sci.* 95 (1980) 190.
70. A.V. de Carvalho, M.C. Asensio, D.P. Woodruff, *Surf. Sci.* 273 (1992) 381.
71. J.C. Lavalley, J. Saussey, T. Raïs, *J. Mol. Catal.* 17 (1982) 289.
72. J. Saussey, J.-C. Lavalley, J. Lamotte, T. Raïs, *J. Chem. Soc. Chem. Commun.* (1982) 278.
73. J. Saussey, J.C. Lavalley, T. Raïs, A. Chakor-Alami, J.P. Hindermann, A. Kiennemann, *J. Mol. Catal.* 26 (1984) 159.
74. G.A. Vedage, R.G. Herman, K. Klier, *J. Catal.* 95 (1985) 423.
75. T. Wurzel, DGMK conference "Synthesis Gas Chemistry," Dresden, Germany, 2006.
76. J. Haid, U. Koss, *Stud. Surf. Sci. Catal.* 136 (2001) 399.
77. J. Topp-Jørgensen, *Stud. Surf. Sci. Catal.* 36 (1988) 293.
78. F.J. Keil, *Micropor. Mesopor. Mat.* 29 (1999) 49.
79. J. DiPardo, Outlook for Biomass Ethanol Production and Demand, ftp://ftp.eia.doe.gov/pub/pdf/multi.fuel/biomass.pdf (accessed 10/2013), US Energy Information Administration, Washington, DC.
80. Homepage of the Renewable Fuels Association, http://ethanolrfa.org (accessed 10/2013)
81. M. Wang, C. Saricks, D. Santini, *Effects of Fuel Ethanol Use on Fuel-Cycle Energy and Greenhouse Gas Emission*, ANL/ESD-38, Argonne National Laboratory, Argonne, IL, 1999.
82. R.G. Herman, *Catal. Today* 55 (2000) 233.
83. K. Weissermel, H.-J. Arpe, in: *Alcohols, Industrial Organic Chemistry*, Wiley-VCH, Weinheim, 2003.
84. K. Winnacker, L. Küchler, *Chemische Technik. Prozesse und Produkte*, Whiley-VCH, Weinheim, 2004.
85. Position paper: Change in the Raw Materials Base, German Chemical Society (GDCh), Society for Chemical Engineering and Biotechnology (DECHEMA), German Society for Petroleum and Coal Science and Technology (DGMK), the German Chemical Industry Association (VCI), Frankfurt a.M., Germany, 2010.
86. P.L. Spath, D.C. Dayton, Preliminary Screening—Technical and Economic Assessment of Synthesis Gas to Fuels and Chemicals with Emphasis on the Potential for Biomass-Derived Syngas, NREL/TP-510-34929, National Renewable Energy Laboratory, Golden, CO, 2003.
87. E. Boles, Eta[energie] 01 (2007) 42.
88. J.H. Clark, F.E.I. Deswarte, T.J. Farmer, *Biofuels Bioprod. Bioref.* 3 (2009) 72.
89. B. Digman, H.S. Joo, D.-S. Kim, *Environ. Prog. Sustain. Energy* 28 (2009) 47.
90. J.J. Spivey, A. Egbebi, *Chem. Soc. Rev.* 36 (2007) 1514.
91. V. Subramani, S.K. Gangwal, *Energy Fuels* 22 (2008) 814.

92. M.A. Gerber, M. Gray, J.F. White, D.J. Stevens, *Evaluation of Promoters for Rhodium-Based Catalysts for Mixed Alcohol Synthesis*, Pacific Northwest National Laboratory and US Department of Energy, USA, 2008.
93. H. Kato, M. Nakashima, Y. Mori, T. Mori, T. Hattori, Y. Murakami, *Res. Chem. Intermed.* 21 (1995) 115.
94. G. Chen, X. Zhang, C.-Y. Guo, G. Yuan, *C. R. Chimie* 13 (2010) 1384.
95. A.B. Boffa, C. Lin, A.T. Bell, G.A. Somorjai, *Catal. Lett.* 27 (1994) 243.
96. F. Li, D. Jiang, X.C. Zeng, Z. Chen, *Nanoscale* 4 (2012) 1123.
97. M.A. Haider, M.R. Gogate, R.J. Davis, *J. Catal.* 261 (2009) 9.
98. J. Wang, Q. Zhang, Y. Wang, *Catal. Today* 171 (2011) 257.
99. G. Chen, C.-Y. Guo, Z. Huang, G. Yuan, *Chem. Eng. Res. Des.* 89 (2011) 249.
100. W.M.H. Sachtler, M. Ichikawa, *J. Phys. Chem.* 90 (1986) 4752.
101. P. Gronchi, E. Tempesti, C. Mazzocchia, *Appl. Catal. A Gen.* 120 (1994) 115.
102. P.-Z. Lin, D.-B. Liang, H.-Y. Luo, C.-H. Xu, H.-W. Zhou, S.-Y. Huang, L.-W. Lin, *Appl. Catal. A Gen.* 131 (1995) 207.
103. S.C. Chuang, J.G. Goodwin Jr., I. Wender, *J. Catal.* 95 (1985) 435.
104. J.R. Katzer, A.W. Sleight, P. Gajardo, J.B. Michel, E.F. Gleason, S. McMillan, *Faraday Discuss. Chem. Soc.* 72 (1981) 121.
105. S. Trautmann, M. Baerns, *J. Catal.* 150 (1994) 335.
106. http://www.degussa-goldhandel.de/de/rhodium_neu.aspx, September 2013.
107. D. Li, C. Yang, W. Li, Y. Sun, B. Zhong, *Top. Catal.* 32 (2005) 233.
108. A. Muramatsu, T. Tatsumi, H. Tominaga, *Bull. Chem. Soc. Jpn.* 60 (1987) 3157.
109. M. Xiang, D. Li, W. Li, B. Zhong, Y. Sun, *Catal. Commun.* 8 (2007) 513.
110. M. Xiang, D. Li, W. Li, B. Zhong, Y. Sun, *Catal. Commun.* 8 (2007) 503.
111. Z. Li, Y. Fu, J. Bao, M. Jiang, T. Hu, T. Liu, Y. Xie, *Appl. Catal. A Gen.* 220 (2001) 21.
112. J. Iranmahboob, H. Toghiani, D.O. Hill, *Appl. Catal. A Gen.* 247 (2003) 207.
113. Y. Yang, Y. Wang, S. Liu, Q. Song, Z. Xie, Z. Gao, *Catal. Lett.* 127 (2009) 448.
114. H. Qi, D. Li, C. Yang, Y. Ma, W. Li, Y. Sun, B. Zhong, *Catal. Commun.* 4 (2003) 339.
115. R. Andersson, M. Boutonnet, S. Järås, *Fuel* (2013) doi: 10.1016/j.fuel.2013.07.057.
116. K.J. Smith, R.B. Anderson, *Can. J. Chem. Eng.* 61 (1983) 40.
117. G.A. Vedage, P.B. Himelfarb, G.W. Simmons, K. Klier, *ACS Symp. Ser.* 279 (1985) 295.
118. A.-M. Hilmen, M. Xu, M.J.L. Gines, E. Iglesia, *Appl. Catal. A Gen.* 169 (1998) 355.
119. J.G. Nunan, R.G. Herman, K. Klier, *J. Catal.* 116 (1989) 222.
120. E.M. Calverley, K.J. Smith, *J. Catal.* 130 (1991) 616.
121. J.M. Campoa-Martín, J.L.G. Fierro, A. Guerrero-Ruiz, R.G. Herman, K. Klier, *J. Catal.* 163 (1996) 418.
122. K. Klier, A. Beretta, Q. Sun, O.C. Feeley, R.G. Herman, *Catal. Today* 36 (1997) 3.
123. P. Forzatti, E. Tronconi, I. Pasquon, *Catal. Rev.—Sci. Eng.* 33 (1991) 109.
124. M.J.L. Gines, E. Iglesia, *J. Catal.* 176 (1998) 155.
125. B.H. Davis, *Top. Catal.* 32 (2005) 143.
126. M.E. Dry, *J. Chem. Technol. Biotechnol.* 77 (2001) 43.
127. A. Razzaghi, J.-P. Hindermann, A. Kiennemann, *Appl. Catal.* 13 (1984) 193.
128. M. Inoue, T. Miyake, Y. Takegami, T. Inui, *Appl. Catal.* 11 (1984) 103.
129. K. Fujimoto, T. Oba, *Appl. Catal.* 13 (1985) 289.
130. S.A. Hedrick, S.S.C. Chuang, A. Pant, A.G. Dastidar, *Catal. Today* 55 (2000) 247.
131. V.R. Surisetty, J. Kozinski, A.K. Dalai, *Int. J. Chem. React. Eng.* 9 (2011) A50.
132. N. Kumar, M.L. Smith, J.J. Spivey, *J. Catal.* 289 (2012) 218.
133. S. Sartipi, J.E. van Dijk, J. Gascon, F. Kapteijn, *Appl. Catal. A Gen.* 456 (2013) 11.
134. H. Hamada, Y. Kuwahara, Y. Kintaichi, T. Ito, K. Wakabayashi, H. Iijima, K. Sano, *Chem. Lett.* (1984) 1611.
135. Y. Kintaichi, Y. Kuwahara, H. Hamada, T. Ito, K. Wakabayashi, *Chem. Lett.* (1985) 1305.

136. Y. Kintaichi, T. Ito, H. Hamada, H. Nagata, K. Wakabayashi, Gakkaishi, S. *J. Jpn. Pet. Inst.* 41 (1998) 66.
137. L. Guczi, G. Boskovic, E. Kiss, *Catal. Rev.* 52 (2010) 133
138. K. Takeuchi, T. Matsuzaki, T.-A. Hanaoka, H. Arakawa, Y. Sugi, K. Wei, *J. Mol. Catal.* 55 (1989) 361.
139. A. Sugier, E. Freund, IFP (1978) US4122110.
140. A. Sugier, E. Freund, IFP (1981) US4291126.
141. P. Courty, P. Chaumette, D. Durand, C. Verdon, IFP (1988) US4780481.
142. P. Courty, D. Durand, A. Sugier, E. Freund, IFP (1982) DE3310540.
143. A. Sugier, E. Freund, J.-F. Le Page, IFP (1080) DE3012900.
144. A. Sugier, E. Freund, IFP (1978) DE2748097.
145. P. Courty, D. Durand, E. Freund, A. Sugier, *J. Mol. Catal.* 17 (1982) 241.
146. A. Kiennemann, P. Chaumette, B. Ernst, J. Saussey, J.C. Lavalley, *Stud. Surf. Sci. Catal.* 107 (1997) 55.
147. S. Velu, K. Suzuki, S. Hashimoto, N. Satoh, F. Ohashi, S. Tomura, *J. Mater. Chem.* 11 (2001) 2049.
148. N. Tien-Thao, M.H. Zahedi-Niaki, H. Alamdari, S. Kaliaguine, *J. Catal.* 245 (2007) 348.
149. I. Boz, *Catal. Lett.* 87 (2003) 187.
150. N. Tien-Thao, H. Alamdari, S. Kaliaguine, *J. Solid State Chem.* 181 (2008) 2006.
151. V. Mahdavi, M.H. Peyrovi, M. Islami, J.Y. Mehr, *Appl. Catal. A Gen.* 281 (2005) 259.
152. N.D. Subramanian, G. Balaji, C.S.S.R. Kumar, J.J. Spivey, *Catal. Today* 147 (2009) 100.
153. P. Buess, R.F.I. Caers, A. Frennet, E. Ghenne, C. Hubert, N. Kruse, ExxonMobil (2003) US20030036573.
154. Y. Xiang, V. Chitry, P. Liddicoat, P. Felfer, J. Cairney, S. Ringer, N. Kruse, *J. Am. Chem. Soc.* 135 (2013) 7114.
155. Y. Xiang, V. Chitry, N. Kruse, *Catal. Lett.* 143 (2013) 936.
156. X. Mo, Y.-T. Tsai, J. Gao, D. Mao, J.G. Goodwin Jr., *J. Catal.* 285 (2012) 208.
157. X. Dong, X.-L. Liang, H.-Y. Li, G.-D. Lin, P. Zhang, H.-B. Zhang, *Catal. Today* 147 (2009) 158.
158. L. Shi, W. Chu, S. Deng, *J. Nat. Gas Chem.* 20 (2011) 48.
159. S. Mawson, M.S. McCutchen, P.K. Lim, G.W. Roberts, *Energy Fuels* 7 (1993) 257.
160. M. Ichikawa, T. Fukushima, *J. Chem. Soc. Chem. Commun.* (1985) 321.
161. A. Takeuchi, J.R. Katzer, *J. Phys. Chem.* 86 (1982) 2438.
162. M.F.H. van Tol, A. Gielbert, B.E. Nieuwenhuys, *Appl. Surf. Sci.* 67 (1993) 166.
163. J.G. Nunen, C.E. Bogdan, K. Klier, K.J. Smith, C.-W. Young, R.G. Herman, *J. Catal.* 113 (1988) 410.
164. J.R. Fox, F.A. Pesa, B.S. Curatolo, *J. Catal.* 90 (1984) 127.
165. D.J. Elliott, F. Pannella, *J. Catal.* 114 (1988) 90.
166. R. Zhang, G. Wang, B. Wang, *J. Catal.* 305 (2013) 238.
167. D.J. Elliott, *J. Catal.* 111 (1988) 445.
168. P. Courty, D. Durand, A. Sugier, E. Freund, IFP (1987) US4659742.
169. K. Ibsen, Equipment Design and Cost Estimation for Small Modular Biomass Systems, Synthesis Gas Cleanup, and Oxygen Separation Equipment; Task 9: Mixed Alcohols From Syngas—State of Technology, NREL/SR-510-39947, National Renewable Energy Laboratory, Golden, CO, 2006.
170. G.R. Jackson, D. Mahajan, PowerEnerCat, Inc. (2001) US6248796.
171. R.C. Stites, J. Hohman, Range Fuels, Inc. (2009) US2009318573.
172. Range Fuels Commercial-Scale Biorefinery, US Department of Energy, http://energy .gov/eere/office-energy-efficiency-renewable-energy (accessed 10/2013).
173. Enerkem homepage, http://www.enerkem.com/en/home.html (accessed 10/2013).
174. Maverick Biofuels homepage, http://www.maverickbiofuels.com/ (accessed 10/2013).
175. D. Wang, G. Yang, Q. Ma, Y. Yoneyama, Y. Tan, Y. Han, N. Tsubaki, *Fuel* 109 (2013) 54.

7 Steam Reforming

Karin Föttinger

CONTENTS

7.1   Introduction 193
7.2   Steam Reforming of Monofunctional Alcohols 195
    7.2.1   Methanol Steam Reforming.. 195
        7.2.1.1   Cu Catalysts 196
        7.2.1.2   Pd-Based Catalysts.. 197
    7.2.2   Ethanol Steam Reforming 203
7.3   Steam Reforming of Polyols: Glycerol Reforming....................................207
7.4   Perspectives 208
References...208

This chapter compiles recent progress in steam-reforming reactions for H_2 production focusing in particular on alcohols as feedstock. Advances in understanding the surface chemistry and reaction pathways are discussed.

7.1   INTRODUCTION

H_2 production is of enormous importance to meet future energy demands. It is a fast-growing field in particular with the development of proton exchange membrane (PEM) fuel cells. Ideally, utilizing feedstock from plants would result in carbon-neutral sustainable H_2 production technologies with net zero emission of CO_2. However, currently the majority of the hydrogen is produced from natural gas. Therefore, the utilization of biomass and bio-derived liquids such as bioethanol is an enormously growing topic of research and strong efforts are dedicated to move toward sustainable resources. A variety of bio-liquid feedstock such as sugars, alcohols such as ethanol, bio-oils, and cellulose from nonedible plants may potentially be exploited.

In this respect, alcohols as a feed for H_2 production by steam reforming (SR) have received increasing attention in the recent past. Simple molecules like methanol and ethanol as well as complex polyfunctional molecules (e.g., sugars, lignin, glycerol) are the focus of current research. Although methanol is currently mainly produced from syngas obtained from fossil sources, its production from sustainable resources such as syngas obtained from biomass, wood, and straw is feasible. Methanol is of interest in particular for mobile nonstationary applications, such as automotive and portable devices [1,2] due to the fact that it is converted to hydrogen under very mild conditions.

SR of methanol and ethanol is widely studied and a number of reviews have been published on this process, for example, [1,3–7]. This chapter is thus not intended to be an extensive and comprehensive review but focuses on certain aspects such as mechanistic understanding and identification of active centers.

Generally, SR of alcohols is an endothermic reaction requiring supply of energy. Alternative processes generating hydrogen from alcohols include auto-thermal reforming (ATR) (also called oxidative SR) and partial oxidation (POX). The latter reaction uses oxygen instead of water vapor for the H_2 production reaction; the former adds a small amount of oxygen to the SR feed and thus combines SR and POX reactions. The main advantage is that POX is exothermic and therefore does not consume energy, which allows for reactor designs that are more compact, owing to absence of external heating. The main drawback is, however, the lower hydrogen yields and thus efficiency. In addition to the reforming processes performed at high temperatures the reforming processes performed at high temperatures, an additional low-temperature pathway has opened up for some liquid compounds derived from biomass (e.g., ethanol, glycerol), so-called aqueous-phase reforming (APR). This approach has received more and more interest as it has the potential to greatly improve process efficiency and hydrogen yield and, on top of that, decrease costs. Figure 7.1 shows an overview of possible routes to hydrogen production.

The chapter's main focus is on SR, but it will also address APR. Section 7.2 is dedicated to monofunctional alcohols, describing methanol steam reforming (MSR) in Section 7.2.1 and ethanol steam reforming (ESR) in Section 7.2.2. Methanol and ethanol are the most promising and widely studied monofunctional

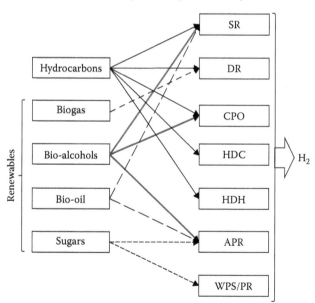

FIGURE 7.1 Overview on feedstock and process alternatives for hydrogen production. APR, aqueous phase reforming; CPO, catalytic partial oxidation; DR, dry reforming; HDC, hydrocarbons decomposition; HDH, hydrocarbons dehydrogenation; SR, steam reforming; WPS/PR, water photo-splitting and photo-reforming. (From Dal Santo, V. et al., *Catal. Today*, 197, 190, 2012. With permission.)

alcohols and will therefore be discussed here. Many aspects of ESR (active catalysts, mechanisms and reaction pathways, deactivation issues) are also relevant for SR of propanol and higher alcohols. Section 7.3 discusses glycerol steam reforming (GSR) as an exemplary case study for the utilization of polyfunctional alcohols.

7.2 STEAM REFORMING OF MONOFUNCTIONAL ALCOHOLS

7.2.1 METHANOL STEAM REFORMING

Due to the difficulties in storing and transporting hydrogen for future energy and fuel supply, methanol has emerged as a promising compound to chemically store hydrogen. It is transformed to hydrogen in a catalytic SR unit "onboard" upstream of, for example, a PEM fuel cell. Methanol, as a liquid fuel, offers numerous advantages: it is easy to store, transport, and handle, and it has a high H:C ratio and requires only mild reaction conditions with typical reaction temperatures of 473–600 K. Therefore, it is especially interesting for use in decentralized and mobile applications [1]. As a further advantage, methanol is already produced in large quantities worldwide, as it is one of the major base chemicals, and the process of industrial methanol synthesis is well established. The syngas needed for methanol synthesis can be produced not only from fossil, but also from renewable sources or even from CO_2 hydrogenation.

Besides MSR, methanol decomposition (MDC) and reverse water–gas shift [(R)WGS] can occur under reaction conditions.

$$CH_3OH + H_2O \rightarrow CO_2 + 3H_2 \quad \text{methanol steam reforming (MSR)} \qquad (7.1)$$

$$CH_3OH \rightarrow CO + 2H_2 \quad \text{methanol decomposition (MDC)} \qquad (7.2)$$

$$CO_2 + H_2 \rightarrow CO + H_2O \quad \text{reverse water–gas shift [(R)WGS]} \qquad (7.3)$$

The major challenge in the SR process is selectivity, avoiding production of the unwanted by-product CO, which acts as a strong poison for the Pt-based electrocatalyst in PEM fuel cells. Although the resistance of the fuel cell anode electrocatalyst against poisoning has been increased by employing Pt-alloys (e.g., with Ru or Co), the required limit of CO concentration in the H_2 feed is still highly demanding (<10–20 ppm) [9]. Potential sources of CO are MDC and (R)WGS. Current state-of-the-art SR catalysts have not fulfilled these stringent requirements on selectivity, which makes an additional cleanup step necessary for removing CO. Thus, a great deal of research activity is going on to solve this issue in catalytic MSR.

The catalysts that are active and selective in MSR are similar to those applied in methanol synthesis. Predominantly, Cu-based catalysts have been used due to their superior performance [1,2]. Analogies to methanol synthesis via CO_2 hydrogenation are obvious, since MSR is the reverse reaction, thus likely involving similar or the same reaction intermediates and active sites. Pd/ZnO has appeared as a promising candidate, potentially able to substitute Cu, which has the drawback of low long-term stability due to sintering; besides, Cu is pyrophoric once it is reduced [2,10,11]. For potential application, long-term stability is crucial. In a direct comparison, Conant et al. [12] observed constant conversion for long operation times (60 h) on $Pd/ZnO/Al_2O_3$

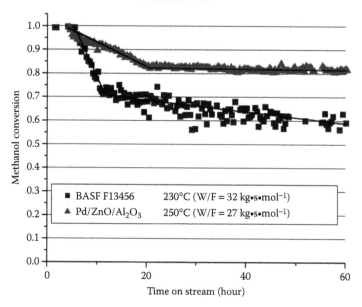

FIGURE 7.2 Comparison between a commercial Cu-based catalyst (BASF F13456) and Pd/ZnO/Al$_2$O$_3$ MSR catalysts: catalytic activity over 60 h. The reaction temperature was 523 K for the Pd-based catalyst and 503 K for the Cu-based catalyst. (From Conant, T. et al., *J. Catal.*, 257, 64, 2008. With permission.)

after an initial activity loss of 17%, while conversion over Cu/ZnO/Al$_2$O$_3$ dropped by 40% within 60 h under the same conditions, with a fast initial drop followed by continuous long-term deactivation (Figure 7.2). Moreover, the Pd catalyst could be fully regenerated by oxidation–reduction cycles, which was not the case for Cu. Both Cu- and Pd-based catalysts will be discussed here.

7.2.1.1 Cu Catalysts

Cu-based catalysts are widely used in C$_1$ chemistry, in particular in methanol synthesis and low-temperature WGS. The identical catalysts are active and selective in MSR. Therefore, in many studies the "classical" methanol synthesis catalyst has been applied as reforming catalyst. The commercially applied state-of-the-art catalyst for the production of methanol from syngas (mixture of CO, CO$_2$, and H$_2$) is Cu/ZnO/Al$_2$O$_3$, with a typical composition of 50% and 75% CuO, between 10% and 30% ZnO, and 5–10 mol% Al$_2$O$_3$, forming porous aggregates of highly dispersed Cu and ZnO [13]. Cu/ZnO is also active without alumina, but addition of alumina enhances the activity and stability. The preparation and composition of Cu/ZnO/Al$_2$O$_3$ catalysts have been empirically optimized. Although methanol synthesis is a large-scale process, there are controversial issues not solved yet, such as the nature of active sites, the role of ZnO, and the reaction mechanism [14]. The same holds for Cu catalysts in MSR.

Despite a large number of investigations, the nature of the active centers catalyzing methanol synthesis and SR is still under discussion. Different copper species have been proposed, including Cu^0 [15], Cu$^+$ dispersed in or on ZnO [16], the Cu–ZnO interface [17], and Cu/Zn alloys [18]. Microstructural properties such as structural disorder, defects, and strain of copper crystallites were proposed to play an important role [19], with ZnO inducing such microstructural disorder. Although the Cu surface area is one of the important parameters determining the catalytic activity, no linear correlation is observed [18], but strongly different intrinsic activities indicate that not all Cu sites are equivalently active. Especially, the nature of sites and state of the catalyst under actual relevant reaction conditions is highly debated. Dynamic and reversible morphology changes were observed in the Cu/ZnO system. While rather spherical particles with a smaller interface to the ZnO prevail under oxidizing conditions (i.e., a high content of CO_2 and water), more disc-like particles with a larger interface to ZnO were found under reducing conditions [20]. The dynamic adjustment of the catalyst to the respective reaction conditions emphasizes the importance of *in situ* characterization of the active state under reaction atmosphere. In an inverse model catalyst study on CuZn near-surface alloys, Rameshan et al. [21] identified an optimum bifunctional catalyst state for high activity and CO_2 selectivity, which consists of bimetallic $Cu(Zn)^0$ regions favoring methanol dehydrogenation to formaldehyde and redox-active $Cu(Zn)^0$-Zn(ox) sites [i.e., a surface alloy covered by a thin wetting layer of interfacial Zn(ox)], assisting in water activation and thus providing hydroxide or oxygen for further oxidation of HCHO to CO_2.

The reaction mechanism of MSR has been controversially debated as well, with main open questions concerning the nature of the involved reaction intermediates and surface species. The two most commonly discussed reaction mechanisms proceed (1) via methyl formate [22,23], which is then hydrolyzed to adsorbed formate or formic acid, or (2) over methylenebisoxy directly to adsorbed formate [24]. In both cases, the adsorbed formate/formic acid decomposes primarily to CO_2 and H_2. An early view of the MSR reaction proceeding via methanol decomposition to CO and H_2 followed by WGS has been discarded due to CO concentrations in the product stream being lower than the equilibrium concentration [25]. A comprehensive microkinetic model was established by Frank et al. [26], describing in detail the elementary steps occurring on the surface of the catalyst (Figure 7.3). Briefly, the reaction is considered to proceed via methoxy and formate species identified by infrared (IR) spectroscopy with dehydrogenation of methoxy as the rate-limiting step. This model involves both the aforementioned pathways (1) and (2), as well as two kinds of sites. Site A would adsorb all carbon-containing species, while site B would be responsible for H adsorption, in agreement with a previous suggestion of Peppley et al. [27]. Based on the available kinetic data they could, however, not determine whether the methyl formate pathway (1) or the methylenebisoxy route (2) predominates.

Since possible reaction sequences proposed for Cu MSR catalysts are basically similar to those suggested to occur on PdZn, the common mechanistic details are also relevant for Pd-based catalysts, although only few mechanistic studies have been published for these systems [24,29,30].

7.2.1.2 Pd-Based Catalysts

Motivated by the drawback of low long-term stability of nanoparticulate Cu, new catalysts based on noble metals emerged in the middle of the 1990s, which are less prone to sintering. Iwasa and coworkers were the first to report that Pd supported on certain reducible oxides, namely ZnO, Ga_2O_3, and In_2O_3, has exceptionally high

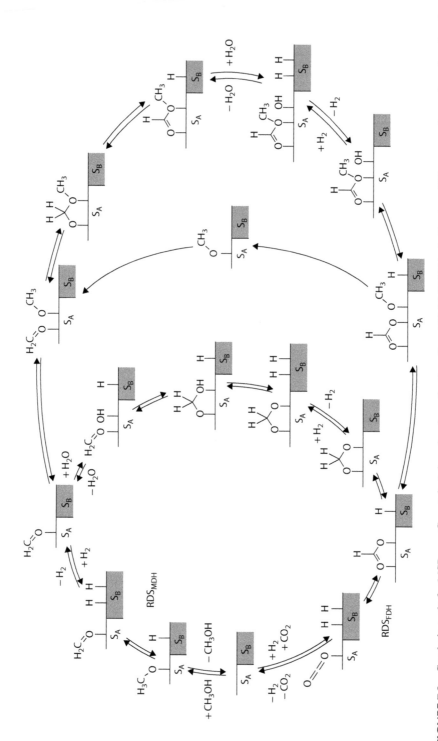

FIGURE 7.3 Catalytic cycle for MSR over Cu catalysts based on investigations from [22,24,27,28] including two different reactive surface sites A and B. (From Frank, B. et al., *J. Catal.*, 246, 177, 2007. With permission.)

activity and selectivity for MSR [24,31–33]. Since then, a number of studies have been carried out, most of them dealing with the Pd/ZnO system, while far less work has been devoted to Pd supported on Ga_2O_3 and In_2O_3. Recently published reviews on Pd/ZnO catalysts describe their selected potential applications, including MSR, and the observed correlations of catalytic performance with important properties [5,34]. The related Pd/Ga_2O_3 system was recently reviewed in [35,36].

Typically, Pd on inert supports catalyzes methanol decomposition to CO and H_2 (MDC). However, when supported on ZnO, Ga_2O_3, and In_2O_3, a completely different selectivity resembling that of Cu was observed [24,31–33,37]. Iwasa and coworkers connected the catalytic properties to alloy/intermetallic compound (IMC) formation of Pd with reduced Zn, Ga, and In upon reduction at >573 K, based on ex situ XRD and XPS [24,31–33]. In order to explain the modified reactivity, they proposed a different reaction pathway on PdZn as compared to Pd [24,33]. While methanol is rapidly dehydrogenated to CO and H_2 via formaldehyde (HCHO) on Pd, the intermediate HCHO would react with water to CO_2 and H_2 on PdZn, most likely via HCOOH, in a similar pathway to that probably proceeding over Cu catalysts.

From surface science, it is known that HCHO adsorbs in a different geometry on Cu [on-top $\eta^1(O)$] [38,39] as compared to Pd, where a bridging $\eta^2(C,O)$ configuration is preferred [40]. Iwasa and coworkers [24,33] suggested an η^1 adsorption of intermediate formaldehyde on PdZn (similar to Cu), which is less prone to C–H (and also C–O) bond cleavage than the bridging η^2 species on Pd. This proposed different adsorption geometry of formaldehyde on PdZn would thus be a plausible cause of the different reactivity as compared to Pd. However, according to density functional theory (DFT) calculations by Lim et al. [41], the η^2 geometry is the most stable configuration of formaldehyde on PdZn surfaces as well, with the C bonding to Pd and O interacting with Zn. Experimental proof of η^1 adsorption of HCHO on PdZn under actual reaction conditions is lacking up to now.

Instead of altering the bonding configuration of intermediates, the role of Zn may be the modification of reaction barriers, as suggested by Neyman and coworkers [41–43] as well as Huang and Chen [44]. DFT calculations indicated an increased barrier for C–H bond cleavage in adsorbed methoxy and formaldehyde on PdZn surfaces as compared to Pd leading to a stabilization of intermediate formaldehyde and thus to a higher probability of further reaction with hydroxyls to formates, which subsequently decompose to CO_2 and H_2. The fast dehydrogenation of formaldehyde on Pd is therefore considerably slowed down on PdZn. Activation energies for C–H bond breaking in methoxy and formaldehyde were calculated to be only 33 and 38 kJ mol^{-1} for Pd(111) but 113 and 64 kJ mol^{-1} on PdZn(111), respectively [41,43]. Based on X-ray photoelectron spectroscopy (XPS) and UV photoelectron spectroscopy (UPS) measurements, Tsai et al. [45] and Bayer et al. [46] suggested that the reactivity of PdZn is governed by its strongly modified electronic structure being similar to Cu but different to Pd, in agreement with DFT calculations. The alteration of the electronic structure upon alloy formation considerably modifies the reaction barriers for methanol decomposition as compared to Pd and might explain the similar catalytic properties of PdZn and Cu.

The electronic properties were studied in detail on ultra-high vacuum model systems by Rameshan et al. [47,48]. By utilizing in situ XPS at mbar pressures of methanol/water, they showed that the Cu-like electronic structure is also present under MSR conditions. A comparison of monolayer PdZn versus multilayer PdZn surface alloys revealed that the "thickness" of the PdZn surface alloy matters. Interestingly, the properties of the topmost PdZn surface monolayer were different irrespective of whether the second (subsurface) layer comprised PdZn or pure Pd [47,48]. Since a surface monolayer of PdZn exhibits a higher thermal stability than "deeper" (second, third, etc.) PdZn layers, the transformation of multi- to monolayer occurred upon annealing at higher temperatures (>550 K) [49]. The subsurface layers deplete quickly in Zn due to Zn diffusion into the Pd bulk resulting in the formation of a PdZn monolayer (on top of Pd substrate). This change in subsurface coordination induces a strong change in electronic properties of the surface atoms (Figure 7.4) and consequently in their catalytic performance [47]. Although the surface composition itself remains unchanged, the change in coordination environment with almost only Pd in the sub-surface layer leads to a modification from a Cu-like density of states (DOS) on mul-tilayer alloys toward a Pd-like DOS on monolayer alloys (Figure 7.4, valence band spectra). This modification considerably influences the catalytic behavior with high MSR selectivity on multi- and low-MSR selectivity on monolayer surface alloys. In addition to the altered electronic properties, geometric changes were observed

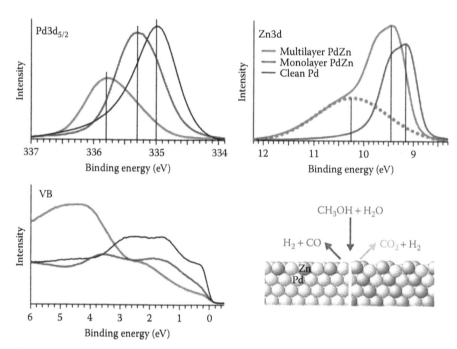

FIGURE 7.4 Ambient pressure-XPS spectra [Pd3d, Zn3d, and valence-band (VB) regions] acquired *in situ* during MSR on the PdZn 1:1 multilayer (red traces) and monolayer alloy (blue traces). For comparison, the corresponding "pure" Pd spectra are added (black traces). The oxidized ZnOH component is highlighted by the dotted red line (middle panel). To obtain equal information depth for all spectra, the Pd3d spectra were recorded with 650 eV photon energy, and the Zn3d and valence-band regions at 120 eV. Reaction conditions: 0.12 mbar methanol, 0.24 mbar water, 553 K. (From Rameshan, C. et al., *Angew. Chem. Int. Ed.*, 49, 3224, 2010. With permission.)

with a change in corrugation from Zn-out (multilayer) to Pd-out (monolayer) [48,49], which is in agreement with DFT calculations [50,51].

While there is general agreement on the importance of PdZn alloy formation, open questions concern the role of support and nature of the intermetallic phase (ordered intermetallic versus random alloy phases). In an attempt to disentangle the catalytic properties of the IMC/alloy phase from those of the ZnO support, unsupported alloyed or intermetallic particles were studied by Halevi et al. [52,53] and Friedrich et al. [54] following different synthesis approaches. Halevi et al. compared the reactivity of unsupported α-PdZn [53] and β-phase PdZn [52] prepared by spray pyrolysis to figure out the importance of the nature of the IMC/alloy. While nearly 100% selectivity to MSR was found on the ordered 1:1 tetragonal PdZn β phase, α-PdZn, a solid solution of Zn in fcc Pd, was 100% selective to CO, behaving similar to Pd. The authors thus concluded that the formation of the tetragonal 1:1 intermetallic compound is required to obtain a selective MSR catalyst. However, they found significant amounts of ZnO at the surface of the β-PdZn particles after reduction at 523 K, which makes it difficult to exclude the presence of ZnO under reaction conditions.

Using metallurgically prepared single-phase intermetallic PdZn compounds with different bulk compositions (Pd_xZn_{100-x}, $x = 46.8 - 59.1$), Friedrich et al. [54] detected a strong connection between the composition and the catalytic performance in MSR. While selectivity in the desired CO_2 and H_2 formation was low on Pd-rich samples, substantially higher selectivity was observed on Zn-rich samples, which was explained by the existence of oxidized Zn species at the surface of the Zn-rich samples allowing for an easier adsorption and dissociation of methanol and water as on pure metallic surfaces. An important role of the oxide support was also suggested by Smith et al. [55] based on DFT calculations, which showed very low or practically no activation barriers of methanol and water dissociation on the polar Zn-terminated ZnO(0001) surface, whereas the dissociation of methanol and water was highly activated on flat, defect-free PdZn surfaces. Therefore, the reaction might occur at the metal–oxide interface, with the initial dissociation of methanol and water occurring on the oxide and the metal being involved in further dehydrogenation steps [55].

We have studied thoroughly the structural, adsorption, and reaction properties of wet-chemically prepared ZnO-supported PdZn [56,57] and β-Ga_2O_3-supported Pd_2Ga nanoparticles [30,37,56,58]. A range of in situ techniques have been combined, including vibrational spectroscopy (FTIR), XPS, and XAS for a fundamental understanding of active phases present under MSR conditions, their stability, and mechanistic details. The dynamic adaptation of ZnO-supported Pd nanoparticles to the reaction environment, that is, when exposed to methanol/water, and its impact on reactivity were demonstrated in [57]. Quick-EXAFS was utilized to investigate the formation of a PdZn alloy in methanol/water by following the structural and electronic changes. The near-edge region is shown in Figure 7.5. The same structural changes occur upon reduction of Pd/ZnO in H_2. In parallel, the selectivity continuously changes from MDC (characteristic for metallic Pd) to MSR on PdZn (Figure 7.5). These time-resolved in situ XAS measurements represent a direct proof of PdZn formation under MSR conditions leading to reactivity different from that of metallic Pd [57]. The structural and electronic properties resemble those of the tetragonal Pd:Zn IMC with a 1:1 stoichiometry, which is the thermodynamically stable phase in the compositional range of around 50 at% [5].

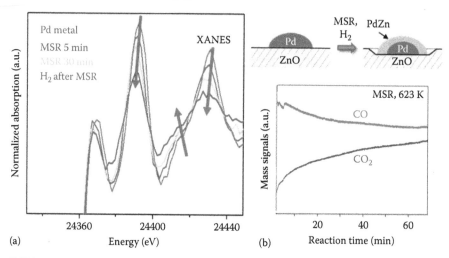

FIGURE 7.5   (a) Pd K edge XANES spectra obtained upon exposure of a 7.5 wt% Pd/ZnO catalyst to MSR conditions at 623 K without pre-reduction ($p_{CH_3OH} = p_{H_2O} = 20$ mbar). The arrows illustrate the changes due to *in situ* formation of PdZn. (b) Corresponding mass spectrometry traces of the products CO and CO_2, which are representative for the selectivity to MDC and MSR, respectively. (From Föttinger, K. et al., *J. Phys. Chem. Lett.*, 2, 428, 2011. With permission.)

In contrast to the tetragonal PdZn, which has a rather broad compositional range of existence (from 37 to 56 at % Pd), Pd and Ga form a number of IMCs of different stoichiometries and structures with a narrow and more defined range of composition [59,60]. Once more, we have utilized *in situ* spectroscopy to identify structure and composition of the active phase(s) present under MSR reaction conditions. In good accordance, XRD, EXAFS, and XPS detect formation of Pd_2Ga under reaction conditions (the H_2 enables for reduction and IMC formation), as well as upon reduction in the relevant temperature range of around 573–673 K [30,37,58].

The modification of adsorption sites upon alloying was assessed by FTIR spectroscopy of CO adsorption and occurs for Pd_2Ga similar as for PdZn. In contrast to Pd/ZnO, PdZn intermetallic nanoparticles exhibit only on-top adsorption of CO with vibrational frequencies around 2070 cm^{-1}. CO adsorbs on the Pd atoms only whereas adsorption on bridge and hollow sites, which is typically observed on Pd metal, is absent on the intermetallic PdZn nanoparticles. The CO adsorption spectra are in excellent agreement with PM-IRAS spectra of CO adsorbed on model PdZn/Pd(111) surface alloys, as shown by Weilach et al. [61]. In addition, a clear red shift in the stretch-vibration frequency of on-top CO by about 20 cm^{-1} was observed on ZnO-supported PdZn [57]. The red shift reflects a change in electronic properties compared to metallic Pd due to charge transfer from Zn to Pd, as predicted by DFT calculations [62]. A change in the electronic structure of PdZn as compared to Pd has likely an influence on the adsorption strength of reactants and intermediates, thereby modifying the catalytic properties. Accordingly, DFT calculations [41,43,62] suggest changes in the reaction barriers for the undesired methanol decomposition on PdZn. In particular, the calculated reaction barriers for the dehydrogenation of the intermediate formaldehyde on PdZn and Cu surfaces are similar, whereas those on Pd are much lower (thus CH_2O decomposes to CO before it reacts with H_2O).

A detailed mechanistic study was performed on Pd_2Ga/Ga_2O_3 utilizing steady state and concentration modulation *in situ* FTIR spectroscopy. By this approach, Haghofer et al. [30] could show that the reduced Ga_2O_3 surface plays an important role in the selective MSR reaction mechanism. Strongly enhanced formation of surface formates was detected on Pd_2Ga/Ga_2O_3, attributed to reactive oxygen sites in the Ga_2O_3 surface formed during IMC formation by reduction. The reaction sequence was suggested to predominantly proceed on the Ga_2O_3 modified by the high-temperature reduction— likely at the bimetal–oxide interface—and is promoted by the presence of the intermetallic particles [30]. Bonivardi and coworkers arrived at similar conclusions concerning the role of the oxide support in methanol synthesis on Pd/Ga_2O_3 catalysts [63].

A very important aspect for potential application is the stability of the PdZn and Pd_2Ga intermetallic particles. Importantly, both PdZn and Pd_2Ga are partially unstable under MSR reaction conditions [37,56]. *In situ* IR spectroscopy revealed the presence of characteristic bands of CO adsorbed on bridge and hollow sites on metallic Pd in addition to on-top CO on intermetallic Pd under MSR reaction conditions. The instability of the intermetallic surfaces under MSR conditions has been attributed to the presence of CO. In room temperature CO adsorption experiments, it was detected that prolonged exposure of the intermetallic PdZn and Pd_2Ga surfaces to CO led to the appearance of increasing amounts of bridge-bonded CO in the IR spectra, characteristic of metallic Pd (Figure 7.6). As a result of the CO-induced partial degradation, domains of metallic Pd are formed at the surface under reaction conditions, which then produce even more CO via MDC.

IMC surface degradation is more pronounced at lower temperatures, whereas a larger fraction of the surface is in an alloyed state at higher reaction temperatures (Figure 7.6). A possible explanation is that H_2 produced in MSR can faster and more efficiently regenerate the IMC at higher temperatures [56].

Overall, important insights have been gained from the powerful combination of *in situ* spectroscopy, computational chemistry, and surface science, and the broad range of materials that have been studied, including powder catalysts, single crystal–based model systems, and unsupported bulk IMCs.

7.2.2 ETHANOL STEAM REFORMING

Ethanol is an interesting candidate as a source for hydrogen production by the SR process and has thus been widely studied. Bioethanol can be produced by fermentation from sugarcane, corn, etc., but also from agricultural waste and wood (so-called second-generation bioethanol). Several reviews [4,6,7,64,65] have summarized the status and progress on ESR. In contrast to MSR, ESR requires the cleavage of the ethanolic C–C bond. Consequently, deactivation by carbon deposition is a frequently observed problem yet to be solved. Improving catalyst stability is therefore currently the greatest challenge for practical application.

Besides ESR, APR of ethanol has attracted attention [66,67]. Due to the fact that APR of ethanol is carried out at much lower reaction temperatures, undesired

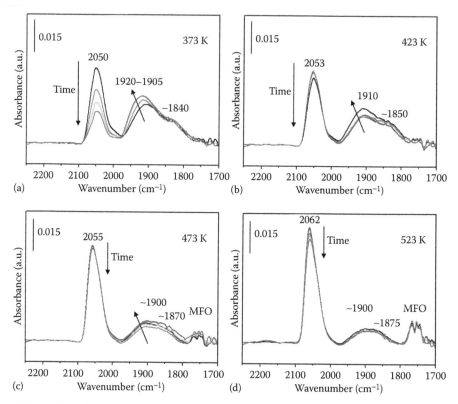

FIGURE 7.6 *In situ* FTIR spectra during methanol decomposition over Pd$_2$Ga/Ga$_2$O$_3$ at different reaction temperatures after reduction at 673 K. The CO stretch vibration range is displayed for spectra recorded every 5 min in 1 vol% methanol in He. (From Föttinger, K., *Catal. Today*, 208, 106, 2013. With permission.)

decomposition reactions are less favorable and further conversion of CO by WGS is enhanced. Generally, the same reactions and products were observed for APR as for ESR.

ESR is described by the following equation:

$$C_2H_5OH + 3H_2O \rightarrow 2CO_2 + 6H_2 \quad \text{ethanol steam reforming (ESR)} \quad (7.4)$$

Several other reactions can occur under the conditions of ESR. Compared to methanol conversions, the reaction network of ethanol is considerably more complex exhibiting a variety of products and side reactions.

$$C_2H_5OH \rightarrow CO + CH_4 + H_2 \quad \text{ethanol decomposition (EDC)} \quad (7.5)$$

$$C_2H_5OH \rightarrow C_2H_4O + H_2 \quad \text{dehydrogenation to acetaldehyde} \quad (7.6)$$

$$C_2H_4O \rightarrow CO + CH_4 \quad \text{acetaldehyde decomposition} \quad (7.7)$$

$$C_2H_5OH \rightarrow C_2H_4 + H_2O \quad \text{dehydration to ethylene} \qquad (7.8)$$

$$2C_2H_5OH + H_2O \rightarrow CH_3COCH_3 + CO_2 + 4H_2 \quad \text{formation of acetone} \qquad (7.9)$$

In addition, (R)WGS, methane steam reforming, methane dry reforming, as well as methanation typically occur under reaction conditions. Some side products, in particular ethylene and acetone, can react further forming coke precursors and coke. For example, ethylene polymerizes over acidic sites, and acetone can undergo condensation reactions.

Higher hydrogen yields are typically obtained at high reaction temperatures (>673 K), while by-products such as acetaldehyde, ethylene, and acetone are often detected at lower reaction temperatures due to kinetic limitation of the reforming reactions. The exothermic methanation reactions of CO and CO_2 contribute to methane formation at lower temperatures, whereas reforming of produced methane is favored at higher temperatures. Thus, undesired methane yields typically decrease with increase in reaction temperature.

The product distribution strongly depends on the reaction conditions and the catalyst used. When using supported catalysts, besides the active phase, the support may interact with ethanol as well and influence the product distribution. Active catalysts comprise noble metals (Pt, Pd, Rh, Ru) and base metals (Ni, Co, Cu); however, metal oxides (e.g., ZnO, Fe_2O_3, CeO_2, La_2O_3) also have been shown to be active; see, for example, [4,6,7,64,65,68] and references therein. In a direct comparison of noble metal catalysts, Rh has shown the best performance in terms of hydrogen selectivity [69] which was attributed to the high reforming ability of Rh. Comparing a range of noble and base metals Ni and Rh on Al_2O_3 [70] and Ni, Co, and Rh on MgO [71], respectively, showed the best H_2 selectivity in ESR. However, also metal dispersion and the nature of the support affect the catalyst performance. Overall, Ni and Co are the most promising metal catalysts due to their low costs and good performance.

A range of intermediates and surface species are involved in the complex reaction network of ethanol on metal and oxide surfaces. Figure 7.7 displays the proposed reaction networks on transition metal surfaces as well as the main oxygenate intermediates from ethanol, while Figure 7.8 summarizes the reactions proposed to occur on oxide surfaces. Intermediates observed on transition metal surfaces include alkoxides, η^1 and η^2 aldehydes, acyl compounds, and carboxylates [4,72].

Similar to methanol, ethanol interacts with active metal surfaces by dissociative adsorption forming an ethoxide. Further dehydrogenation produces acetaldehyde, which is a key intermediate in ESR and can adsorb in different configurations (η^1 and η^2) influencing the stability/reactivity, in analogy to MSR. The aldehyde intermediate can then either desorb, or react further by C–O bond cleavage and hydrogenation to alkanes or via dehydrogenation to acetyl, followed by C–C bond cleavage. The latter reaction is more favorable on metal surfaces and produces CO and CHx. On Rh surfaces, a somewhat different mechanism has been observed via oxametallacycle species as intermediates [72].

On oxides and supported metal catalysts, a higher concentration of oxidized intermediates such as acetates and carbonates are detected. Ethanol adsorbs as ethoxy species on oxide surfaces and can further dehydrogenate to acetaldehyde, which is

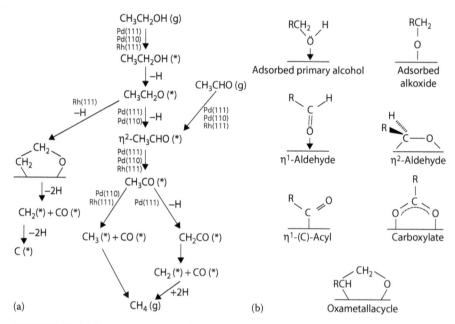

FIGURE 7.7 (a) Reactions pathways of ethanol on transition metal surfaces. (b) Main oxygenated intermediates from ethanol on transition metals. (From Mavrikakis, M. and Barteau, M. A., *J. Mol. Catal. A Chem.*, 131, 135, 1998. With permission.)

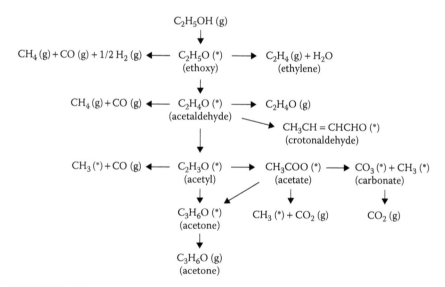

FIGURE 7.8 Reactions pathways of ethanol on oxide surfaces. (From Mattos, L. V. et al., *Chem. Rev.*, 112, 4094, 2012. With permission.)

favored in the presence of OH groups. Again, acetaldehyde is a key intermediate and can either desorb, dehydrogenate further to acetyl [73], undergo aldolization [74], or react with surface O or OH to acetate [75]. Acetate species can either decompose to CH_4 and CO or oxidize further to carbonates [4,76].

As in MSR, water activation occurs on the oxide surface. De Lima et al. [77] studied the reaction mechanism of ESR on $Pt/CeZrO_2$ by DRIFTS under reaction conditions. They found that steam promoted the forward decomposition of acetate, which was formed by support-induced oxidation of acetaldehyde with surface OH, to carbonate and methane. This is in good accordance with the water-promoted formate reaction to CO_2 during MSR that was detected by Haghofer et al. [30]. In summary, the synergy between metal and oxide and an important role of the metal–oxide interface are crucial for ESR as well as for MSR activity and selectivity.

In addition, as mentioned Section 7.1, side reactions can occur on acidic and basic sites on the support, such as the dehydration of ethanol to ethylene taking place on acid sites or aldolization. This can lead to catalyst deactivation by carbonaceous deposits, which is a major challenge in ESR. Also, the Boudouard reaction and decomposition of hydrocarbons to carbon contribute to coke formation and activity loss.

$$2CO \rightarrow CO_2 + C \quad \text{Boudouard reaction} \qquad (7.10)$$

Different strategies were followed in order to overcome the frequently observed deactivation and to improve catalyst stability by preventing carbon formation, such as optimization of the reaction conditions (high steam:ethanol ratio, addition of small amounts of oxygen to the feed, high reaction temperatures) and modification of the catalyst (e.g., choice of the support oxide, use of bimetallic catalysts, adding promoters that modify the support acidity). These approaches and relevant references are described in detail in the review by Mattos et al. [4].

7.3 STEAM REFORMING OF POLYOLS: GLYCEROL REFORMING

Glycerol with its three hydroxyl groups is discussed as an exemplary and prominent example in this section. Besides methanol and ethanol, glycerol is an interesting candidate as a source of hydrogen produced by SR. It is formed in large amounts as a by-product in biodiesel production and can also be obtained from renewable-source processes such as by glucose fermentation. The crude glycerol stream contains a range of contaminants and other compounds including water. SR is therefore an interesting means of further conversion as no more expensive separation or refining steps are necessary, compared to other routes of glycerol utilization. On top of that, it is nontoxic, nonflammable, and biodegradable. Reviews on GSR were published recently, for example, [65,78–80].

The information gained on GSR can be transferred to other polyols such as sugars. Generally, selectivity is a challenge due to the fact that the more complicated the reactant molecule is and the more functional groups it has, the more reaction pathways and products are possible. However, in general, the reactions occurring are similar to those of ethanol, and many aspects of GSR are analogous to ESR. GSR is described by the following equation:

$$C_3H_8O_3 + 3H_2O \rightarrow 3CO_2 + 7H_2 \qquad (7.11)$$

Besides GSR, APR is highly interesting. GSR with water vapor in gas phase requires high temperatures (usually around 800°C), while APR is carried out in liquid phase at lower temperatures (approximately 200°C–250°C) and elevated pressures (typically 20 bar). Many aspects of GSR and APR are similar to ethanol reforming reactions, including mechanistic aspects and the nature of catalytically active materials. The same catalysts are active in glycerol reforming reactions like in ESR, that is, particularly Ni, Ru, Co, and Pt [65,78,80].

The major challenges and problems to be solved in GSR are by-product formation (predominantly CO and methane), catalyst deactivation, and high energy consumption due to the high reaction temperatures required. In APR, the main drawbacks are the high-pressure requirement, lower H_2 selectivity, and production of alkanes as by-products.

7.4 PERSPECTIVES

In summary, SR of methanol and ethanol has been investigated in detail and significant understanding has been reached especially for MSR. In particular, the combination of studies on a broad range of materials with DFT calculations has enabled detailed insights into the processes and chemistry occurring at catalysts surfaces. Studies on polyfunctional alcohols as feedstock are receiving increasing interest recently, which will certainly boost thorough understanding of these reactions in the near future.

The main challenges currently remaining in alcohol SR with respect to commercial application are catalyst stability and the purity of the produced hydrogen. Purity is a major issue for hydrogen intended for use in PEM fuel cells. Hydrogen production processes must therefore either produce high-purity hydrogen or additional purification processes are required, which is less favorable especially for mobile applications. This demands highly selective and stable catalysts and optimized reaction conditions. A molecular-level understanding of alcohol reforming reactions is an important prerequisite for fulfilling these demands and can be reached by the combination of *in situ*/operando spectroscopic approaches and computational chemistry, as has been successfully demonstrated for MSR.

REFERENCES

1. D.R. Palo, R.A. Dagle, and J.D. Holladay, *Chemical Reviews*, 107 (2007) 3992.
2. D.L. Trimm and Z.I. Onsan, *Catalysis Reviews*, 43 (2001) 31.
3. S. Sá, H. Silva, L. Brandão, J.M. Sousa, and A. Mendes, *Applied Catalysis B: Environmental*, 99 (2010) 43.
4. L.V. Mattos, G. Jacobs, B.H. Davis, and F.B. Noronha, *Chemical Reviews*, 112 (2012) 4094.
5. M. Armbrüster, M. Behrens, K. Föttinger, M. Friedrich, É. Gaudry, S.K. Matam, and H.R. Sharma, *Catalysis Reviews*, 55 (2013) 289.
6. P.D. Vaidya and A.E. Rodrigues, *Chemical Engineering Journal*, 117 (2006) 39.
7. A. Haryanto, S. Fernando, N. Murali, and S. Adhikari, *Energy & Fuels*, 19 (2005) 2098.

8. V. Dal Santo, A. Gallo, A. Naldoni, M. Guidotti, and R. Psaro, *Catalysis Today*, 197 (2012) 190.
9. X. Cheng, Z. Shi, N. Glass, L. Zhang, J. Zhang, D. Song, Z.-S. Liu, H. Wang, and J. Shen, *Journal of Power Sources*, 165 (2007) 739.
10. K.D. Jung, O.S. Joe, S.H. Han, S.J. Uhm, and I.J. Chung, *Catalysis Letters*, 35 (1995) 303.
11. C. Zhang, Z. Yuan, N. Liu, S. Wang, and S. Wang, *Fuel Cells*, 6 (2006) 466.
12. T. Conant, A.M. Karim, V. Lebarbier, Y. Wang, F. Girgsdies, R. Schlögl, and A. Datye, *Journal of Catalysis*, 257 (2008) 64.
13. M. Behrens, *Journal of Catalysis*, 267 (2009) 24.
14. J.B. Hansen and P.E. Hojlund Nielsen, in G. Ertl, H. Knözinger, F. Schüth, and J. Weitkamp (Editors), *Handbook of Heterogeneous Catalysis*, 2nd ed., Wiley-VCH, Weinheim, Germany, 2008, p. 2920.
15. H. Kobayashi, N. Takezawa, M. Shimokawabe, and K. Takahashi, *Studies in Surface Science and Catalysis*, 16 (1983) 697.
16. K. Klier, *Advances in Catalysis*, 31 (1982) 243.
17. J.C. Frost, *Nature*, 334 (1988) 577.
18. M.M. Günter, T. Ressler, R.E. Jentoft, and B. Bems, *Journal of Catalysis*, 203 (2001) 133.
19. B.L. Kniep, T. Ressler, A. Rabis, F. Girgsdies, M. Baenitz, F. Steglich, and R. Schlögl, *Angewandte Chemie International Edition*, 43 (2004) 112.
20. P.L. Hansen, J.B. Wagner, S. Helveg, J.R. Rostrup-Nielsen, B.S. Clausen, and H. Topsøe, *Science*, 295 (2002) 2053.
21. C. Rameshan, W. Stadlmayr, S. Penner, H. Lorenz, N. Memmel, M. Hävecker, R. Blume, D. et al., *Angewandte Chemie International Edition*, 51 (2012) 3002.
22. C.J. Jiang, D.L. Trimm, M.S. Wainwright, and N.W. Cant, *Applied Catalysis A: General*, 93 (1993) 245.
23. K. Takahashi, N. Takezawa, and H. Kobayashi, *Applied Catalysis*, 2 (1982) 363.
24. N. Takezawa and N. Iwasa, *Catalysis Today*, 36 (1997) 45.
25. E. Santacesaria and S. Carrá, *Applied Catalysis*, 5 (1983) 345.
26. B. Frank, F.C. Jentoft, H. Soerijanto, J. Kröhnert, R. Schlögl, and R. Schomäcker, *Journal of Catalysis*, 246 (2007) 177.
27. B.A. Peppley, J.C. Amphlett, L.M. Kearns, and R.F. Mann, *Applied Catalysis A: General*, 179 (1999) 31.
28. B.A. Peppley, J.C. Amphlett, L.M. Kearns, and R.F. Mann, *Applied Catalysis A: General*, 179 (1999) 21.
29. S.E. Collins, M.A. Baltanás, and A.L. Bonivardi, *Applied Catalysis A: General*, 295 (2005) 126.
30. A. Haghofer, D. Ferri, K. Föttinger, and G. Rupprechter, *ACS Catalysis*, 2 (2012) 2305.
31. N. Iwasa, S. Masuda, N. Ogawa, and N. Takezawa, *Applied Catalysis A: General*, 125 (1995) 145.
32. N. Iwasa, T. Mayanagi, N. Ogawa, K. Sakata, and N. Takezawa, *Catalysis Letters*, 54 (1998) 119.
33. N. Iwasa and N. Takezawa, *Topics in Catalysis*, 22 (2003) 215.
34. K. Föttinger, in J.J. Spivey (Editor), *Catalysis*, Vol. 25, The Royal Society of Chemistry, Cambridge, 2013, p. 77.
35. H. Lorenz, C. Rameshan, T. Bielz, N. Memmel, W. Stadlmayr, L. Mayr, Q. Zhao, S. Soisuwan, B. Klötzer, and S. Penner, *ChemCatChem*, 5 (2013) 1273.
36. M. Armbrüster, M. Behrens, F. Cinquini, K. Föttinger, Y. Grin, A. Haghofer, B. Klötzer, A. et al., *ChemCatChem*, 4 (2012) 1048.
37. A. Haghofer, K. Föttinger, F. Girgsdies, D. Teschner, A. Knop-Gericke, R. Schlögl, and G. Rupprechter, *Journal of Catalysis*, 286 (2012) 13.
38. B.A. Sexton, A.E. Hughes, and N.R. Avert, *Surface Science*, 155 (1985) 366.
39. D.B. Clarke, D.-K. Lee, M.J. Sandoval, and A.T. Bell, *Journal of Catalysis*, 150 (1994) 81.

40. J.L. Davis and M.A. Barteau, *Surface Science*, 235 (1990) 235.
41. K.H. Lim, Z.X. Chen, K.M. Neyman, and N. Rösch, *Journal of Physical Chemistry B*, 110 (2006) 14890.
42. Z.X. Chen, K.H. Lim, K.M. Neyman, and N. Rösch, *Journal of Physical Chemistry B*, 109 (2005) 4568.
43. Z.X. Chen, K.M. Neyman, K.H. Lim, and N. Rösch, *Langmuir*, 20 (2004) 8068.
44. Y. Huang and Z.-X. Chen, *Langmuir*, 26 (2010) 10796.
45. A.P. Tsai, S. Kameoka, and Y. Ishii, *Journal of the Physical Society of Japan*, 73 (2004) 3270.
46. A. Bayer, K. Flechtner, R. Denecke, H.P. Steinrück, K.M. Neyman, and N. Rösch, *Surface Science*, 600 (2006) 78.
47. C. Rameshan, W. Stadlmayr, C. Weilach, S. Penner, H. Lorenz, M. Hävecker, R. Blume et al., *Angewandte Chemie International Edition*, 49 (2010) 3224.
48. C. Rameshan, C. Weilach, W. Stadlmayr, S. Penner, H. Lorenz, M. Hävecker, R. Blume et al., *Journal of Catalysis*, 276 (2010) 101.
49. W. Stadlmayr, C. Rameshan, C. Weilach, H. Lorenz, M. Hävecker, R. Blume, T. Rocha et al., *Journal of Physical Chemistry C*, 114 (2010) 10850.
50. H.P. Koch, I. Bako, G. Weirum, M. Kratzer, and R. Schennach, *Surface Science*, 604 (2010) 926.
51. G. Weirum, M. Kratzer, H.P. Koch, A. Tamtögl, J. Killmann, I. Bako, A. Winkler, S. Surnev, F.P. Netzer, and R. Schennach, *Journal of Physical Chemistry C*, 113 (2009) 9788.
52. B. Halevi, E.J. Peterson, A. DeLaRiva, E. Jeroro, V.M. Lebarbier, Y. Wang, J.M. Vohs et al., *Journal of Physical Chemistry C*, 114 (2010) 17181.
53. B. Halevi, E.J. Peterson, A. Roy, A. DeLariva, E. Jeroro, F. Gao, Y. Wang et al., *Journal of Catalysis*, 291 (2012) 44.
54. M. Friedrich, D. Teschner, A. Knop-Gericke, and M. Armbrüster, *Journal of Catalysis*, 285 (2012) 41.
55. G.K. Smith, S. Lin, W. Lai, A. Datye, D. Xie, and H. Guo, *Surface Science*, 605 (2011) 750.
56. K. Föttinger, *Catalysis Today*, 208 (2013) 106.
57. K. Föttinger, J.A. van Bokhoven, M. Nachtegaal, and G. Rupprechter, *The Journal of Physical Chemistry Letters*, 2 (2011) 428.
58. A. Haghofer, K. Föttinger, M. Nachtegaal, M. Armbrüster, and G. Rupprechter, *Journal of Physical Chemistry C*, 116 (2012) 21816.
59. H. Okamoto, *Journal of Phase Equilibria and Diffusion*, 29 (2008) 466.
60. H. Okamoto, in T.B. Massalski (Editor), *Binary Alloy Phase Diagrams*, 2nd ed., ASM International, Materials Park, OH, 1990, p. 3068.
61. C. Weilach, S.M. Kozlov, H.H. Holzapfel, K. Föttinger, K.M. Neyman, and G. Rupprechter, *Journal of Physical Chemistry C*, 116 (2012) 18768.
62. K.M. Neyman, K.H. Lim, Z.X. Chen, L.V. Moskaleva, A. Bayer, A. Reindl, D. Borgmann, R. Denecke, H.P. Steinrück, and N. Rösch, *Physical Chemistry Chemical Physics*, 9 (2007) 3470.
63. S.E. Collins, J.J. Delgado, C. Mira, J.J. Calvino, S. Bernal, D.L. Chiavassa, M.A. Baltanás, and A.L. Bonivardi, *Journal of Catalysis*, 292 (2012) 90.
64. M. Ni, D.Y.C. Leung, and M.K.H. Leung, *International Journal of Hydrogen Energy*, 32 (2007) 3238.
65. P.R.d.l. Piscina and N. Homs, *Chemical Society Reviews*, 37 (2008) 2459.
66. I.O. Cruz, N.F.P. Ribeiro, D.A.G. Aranda, and M.M.V.M. Souza, *Catalysis Communications*, 9 (2008) 2606.
67. A.V. Tokarev, A.V. Kirilin, E.V. Murzina, K. Eränen, L.M. Kustov, D.Y. Murzin, and J.P. Mikkola, *International Journal of Hydrogen Energy*, 35 (2010) 12642.
68. J. Llorca, P.R.d.l. Piscina, J. Sales, and N. Homs, *Chemical Communications*, (2001) 641.

69. D.K. Liguras, D.I. Kondarides, and X.E. Verykios, *Applied Catalysis B: Environmental*, 43 (2003) 345.

70. F. Auprêtre, C. Descorme, and D. Duprez, *Catalysis Communications*, 3 (2002) 263.

71. F. Frusteri, S. Freni, L. Spadaro, V. Chiodo, G. Bonura, S. Donato, and S. Cavallaro, *Catalysis Communications*, 5 (2004) 611.

72. M. Mavrikakis and M.A. Barteau, *Journal of Molecular Catalysis A: Chemical*, 131 (1998) 135.

73. R. Shekhar, M.A. Barteau, R.V. Plank, and J.M. Vohs, *Journal of Physical Chemistry B*, 101 (1997) 7939.

74. A. Yee, S.J. Morrison, and H. Idriss, *Journal of Catalysis*, 191 (2000) 30.

75. M. Dömök, M. Tóth, J. Raskó, and A. Erdőhelyi, *Applied Catalysis B: Environmental*, 69 (2007) 262.

76. A. Erdőhelyi, J. Raskó, T. Kecskés, M. Tóth, M. Dömök, and K. Baán, *Catalysis Today*, 116 (2006) 367.

77. S.M. de Lima, I.O. da Cruz, G. Jacobs, B.H. Davis, L.V. Mattos, and F.B. Noronha, *Journal of Catalysis*, 257 (2008) 356.

78. S. Adhikari, S.D. Fernando, and A. Haryanto, *Energy Conversion and Management*, 50 (2009) 2600.

79. A. Behr, J. Eilting, K. Irawadi, J. Leschinski, and F. Lindner, *Green Chemistry*, 10 (2008) 13.

80. P.D. Vaidya and A.E. Rodrigues, *Chemical Engineering & Technology*, 32 (2009) 1463.

8 Biomass to Liquid Biofuels via Heterogeneous Catalysis

Michael Stöcker and Roman Tschentscher

CONTENTS

8.1 Introduction 214
8.2 Biofuels: Definition, Political Impact, and Current Technology 216
8.3 The Concept of Biorefineries (Second- and Third-Generation Biofuels) 217
8.4 Use of Biofuels Today 218
8.5 The Harvest of Feedstock and Biomass Conversion versus
    Combustion 219
8.6 Pretreatment of Biomass 220
    8.6.1 Drying 221
    8.6.2 Mechanical Treatments 221
    8.6.3 Steam/CO$_2$ Explosion 221
    8.6.4 Swelling 221
8.7 Composition of Biomass 221
    8.7.1 Energy Storage Molecules 221
        8.7.1.1 Saccharides 221
        8.7.1.2 Sucrose 222
        8.7.1.3 Starch 222
    8.7.2 Lignocellulosic Materials (Main Components
          of Wooden-Based Biomass) 222
        8.7.2.1 Cellulose 223
        8.7.2.2 Hemicellulose 224
        8.7.2.3 Lignin 224
    8.7.3 Ash 225
8.8 Biomass Platforms 226
    8.8.1 Conversion of Biomass 226
        8.8.1.1 Vegetable Oils/Animal Fats 226
        8.8.1.2 Biomass Conversion via Bio-Oils 228
        8.8.1.3 Polysaccharides 235
8.9 Outlook 247
References 248

Biomass as renewable energy resource for the production of transportation fuels has received increasing interest due to the declining availability of fossil-based feedstocks. In addition, the use of biomass as renewable energy resource becomes quite essential, taking into account the focus on global warming, CO_2 emission, increased competition, and a secure energy supply as well as less consumption of fossil-based fuels. Furthermore, political incentives have contributed to an enhanced focus on biomass in general as an alternative and sustainable feedstock with respect to a prospective production of biofuels (fuels obtained from the processing of biomass), last but not least due to their carbon-neutral balance and abundant availability. However, their low energy density and heating value as well as their high oxygen content require substantial efforts to arrive at components resembling hydrocarbon-based fuel fractions. Finally, heterogeneous catalysis plays a vital role with respect to the conversion of biomass to fuels, beneficial for an enhanced selectivity as well as a safe and environmentally benign processing. In this chapter, multiple biomass resources are presented with respect to a variety of fuels, however, with focus on the catalytic conversion of different types of biomass to liquid biofuels.

8.1   INTRODUCTION

The production of transportation fuels based on crude oil must be seen on the background of diminishing petroleum resources. Therefore, a strong focus has been on the research and development of clean technologies, enabling the production of transportation fuels based on renewable energy resources. Vehicles powered by electricity, solar energy, hydrogen fuel cells, and biofuels are all being actively researched on in order to reduce the dependence on crude oil as energy source. However, these new technologies require significant improvements to be economically and technically viable. Based on this scenario, liquid biofuels produced from renewable biomass are unique in their similarity to the currently preferred fuel sources, avoiding major changes to the engines used and the transportation infrastructure in place. As a consequence, bioethanol and biodiesel are already used commercially as blending components for fossil-based fuels [1].

The first-generation type of biofuels, bioethanol and biodiesel, are produced mainly from either agricultural crops like plants/cereals containing sugar or vegetable oils, respectively. Lignocellulosic biomass is one of the main sources for the production of second-generation biofuels, with synfuels as the main target products. The raw materials applied for the manufacture of second-generation biofuels do not compete with the food market directly, but can be in competition for the use of available land. However, before the utilization of lignocellulosic biomass can be economically viable, the challenges related to the pretreatment and hydrolysis must be solved in order to improve the suitability of this feedstock. Finally, algae crops have been considered as a suitable source of feedstocks for biofuels, and consequently, these marine-derived fuels have been termed as third-generation biofuels [1].

Currently, fossil-based energy resources like petroleum, coal, and natural gas are responsible for about three quarters of the world's primary energy consumption. Alternatives to fossil-based energy resources are nuclear power, hydropower, solar energy, fuel cells, biomass, etc. representing currently about one-quarter of the world's primary energy consumption. Biomass represents an abundant and carbon neutral renewable

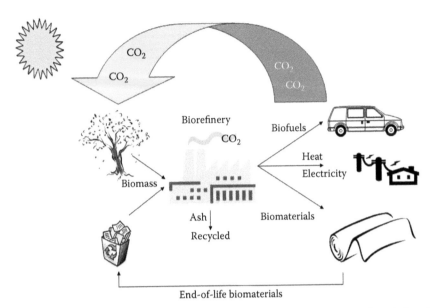

FIGURE 8.1 Integrated biorefinery concept for sustainable technologies. (Ragauskas, A. J. et al., *Science*, 311, 484–489, 2006. Reproduced by permission of Science.)

energy resource for the production of biofuels [2]. In addition, energy production from biomass has the advantage to form lower amounts of greenhouse gases compared to the conversion of fossil fuels, since the carbon dioxide generated during the energy conversion is consumed during subsequent biomass regrowth (see Figure 8.1) [3].

Moving the world market dependence away from fossil-based energy resources to renewable alternatives, like biomass resources, can be regarded as an important contribution to the establishment of favorable framework conditions for the global climate and a sustainable industrial market situation [3]. In this respect, the today's production and application of first-generation biofuels, like biodiesel and bioethanol, are steps in the right direction; however, the second and third generation of biofuels will be based on biomass resources and will be processed from integrated biorefineries, covering the production of biofuels, heat, and electricity as well as biomaterials (Figure 8.1).

Currently, multiple biomass resources are applied to produce a variety of fuels, chemicals, and energy products. Resources could be from trade and industry, forestry and agriculture, as shown in Figure 8.2. Processing covers biological, thermal, and/or chemical conversion, as well as mechanical treatment in order to obtain solid, liquid, or gaseous fuels and/or valuable chemicals. Besides fuels or valuable chemicals, heat and electricity can be generated as energy products as well.

The challenges regarding the processing of different types of biomass will be highlighted along with the application of heterogeneous catalysts for the biomass-to-fuels track within the biorefinery concept. New challenges related to the catalytic conversion of biomass, like mechanistic understanding of the complex reactions taking place, the catalyst and process developments as well as the product pattern to be envisaged will be discussed. Finally, the current situation with respect to upgrading of

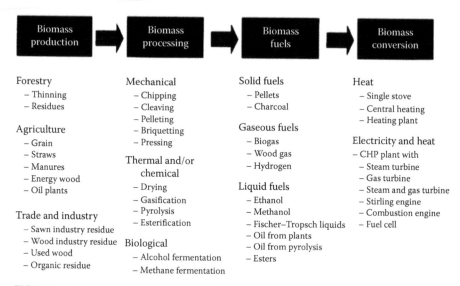

FIGURE 8.2   Renewable energy vs. fossil based: From multiple biomass resources to a variety of fuels and energy products. (From Stöcker, M.: Biofuels and biomass-to-liquid fuels in the biorefinery: Catalytic conversion of lignocellulosic biomass using porous materials. *Angew. Chem. Int. Ed.* 2008. 47. 9200–9211. Copyright Wiley-VCH Verlag GmbH & Co. KGaA. Reproduced with permission.)

the process technology (pilot and commercial units) will be addressed. The focus of this chapter is concentrated on the catalytic conversion of biomass to liquid biofuels.

8.2   BIOFUELS: DEFINITION, POLITICAL IMPACT, AND CURRENT TECHNOLOGY

Liquid biofuels produced from biomass represent an abundant and carbon neutral renewable energy resource for the production of transportation fuels. The CO_2 emission is comparable to the consumption during photosynthesis. However, most of the first-generation biofuels have a number of disadvantages, like the competition for water and land as well as with food for humans and livestock, limited greenhouse gas emission reduction, the high production cost, and, finally, the potentially negative impact on biodiversity [4]. In order to secure the future energy supply and to handle the global warming due to greenhouse effects, energy from renewable sources must be increased relative to the use of fossil-based fuels.

There is, in addition, a strong political focus on renewable biofuel alternatives [4]:

1. The United Nation's climate panel requests the greenhouse gas emission to be reduced with 50%–80% by 2050 (global warming decrease).
2. The Renewable Energy Directive of the European Union (EU) requests the use of 10% biofuels of the total automotive fuel consumption by 2020.
3. The US Renewable Fuel Standard requires around 30% biofuels of the transport pool by 2022.

However, the use of agricultural areas for growing of bioenergy plants will compete with the production of food and livestock, and in order to reach the EU Commission's goal of 10% within 2020, a need of up to 13% of the EU's total agricultural area is estimated. This means there is a strong political dimension in relation to the future application of bioenergy plants.

Technologies for processing second- and third-generation biofuels are still under development with focus on the utilization of lignocellulosic and seaweed-based biomass.

At present, the production of the first-generation bioethanol is mainly based on carbohydrate-rich plants containing the easily digestible saccharides, such as sugar and starch. These are maize, sugarcane, wheat, barley, potato, wood, corn, or sugar beet, which are processed via hydrolysis and fermentation in order to obtain bioethanol. The hydrolysis of cellulose is an energy-intensive and complex production process. So far, no commercial lignocellulosic or algae-based processes are available. The first-generation biodiesel [fatty acid methyl esters (FAME)] production is based on a very simple process: transesterification of vegetable oils from rapeseeds [rapeseed methyl ester (RME)], soy beans, sunflower, palm oil, etc. or animal fats like slaughterhouse waste, fish oil, etc. [2].

8.3   THE CONCEPT OF BIOREFINERIES (SECOND- AND THIRD-GENERATION BIOFUELS)

Whereas the first generation of biofuels is based on well-established technologies, the development of processes related to the production of second- and third-generation biofuels is still in the stage of research and development. Technologies for the gasification of biomass to synthesis gas (syngas = CO and hydrogen) have been developed. Syngas, also available via pyrolysis of lignocellulosic biomass, can be used to produce biomethanol or Fischer–Tropsch liquids (FTLs) by applying well-known technologies (see Figure 8.3). Lignocellulosic biomass can be converted to bio-oil via fast pyrolysis, besides the complete gasification to syngas. The bio-oil can be gasified in a second step to syngas (which can be processed further to Fischer–Tropsch products or others like methanol with subsequent production of olefins and/or gasoline) or can be separated to give phenolic compounds and/or carbohydrate fractions. The phenolic compounds might be used for the production of phenolic resins, whereas the carbohydrate fractions are catalytically transformed to hydrogen.

In general, a modern biorefinery aims to parallel a crude oil refinery where an abundant raw material consisting of, for example, renewable lignin, cellulose, and hemicellulose, enters the biorefinery and is converted through a sequence of process steps to a variety of products covering biofuels, valuable chemicals, heat, and electricity. However, one should not underestimate the logistic challenges concerning running a modern biorefinery, especially with respect to gather sufficient biomass, for example, to operate a Fischer–Tropsch unit from an economic point of view. Furthermore, carbohydrates from biomass represent a viable route to esters, carboxylic acids, and alcohols, which are regio- and stereochemically pure. Expensive chiral catalysts or advanced synthesis routes, which are normally

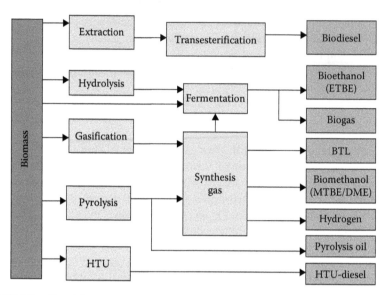

FIGURE 8.3 Simplified process flow within a biorefinery. (From Stöcker, M.: Biofuels and biomass-to-liquid fuels in the biorefinery: Catalytic conversion of lignocellulosic biomass using porous materials. *Angew. Chem. Int. Ed.* 2008. 47. 9200–9211. Copyright Wiley-VCH Verlag GmbH & Co. KGaA. Reproduced with permission.)

requested to selectively introduce chemical functionality in dedicated compounds, can be avoided [3].

8.4 USE OF BIOFUELS TODAY

Concerning the use of low admixture fuel: up to 5 vol% in common gasoline or diesel is allowed and this can be used on common vehicles without any form of adaption (E5—Europe). With respect to high admixture fuel, which means more than 5 vol%, the following compositions are currently available in the countries listed below:

Gasoline
Brazil: up to 25% EtOH (E25)
USA: 10% EtOH/90% gasoline (E10)
Sweden: 85% EtOH/15% gasoline (E85)

Diesel
5–30 vol% are common: B5, B10, and B30—FAME, E95—for diesel engines
However, the vehicles applying this type of fuel must be adapted.

Biobutanol may have a potential as a more suitable gasoline range biofuel than bioethanol. The use of biobutanol would increase the octane number in the gasoline pool, compared to bioethanol. Still, it has a comparable energy density. In addition, biobutanol has a lower vapor pressure and a lower water solubility, which simplifies handling procedures and the infrastructure [5].

8.5 THE HARVEST OF FEEDSTOCK AND BIOMASS CONVERSION VERSUS COMBUSTION

The cost for the feedstock strongly depends on its origin. The feedstock can be the waste products of various industries producing high-value materials, such as paper and wood, but also from agricultural residue and food waste. Typical examples are straw from corn, pulp waste, etc. In those cases, the costs for agriculture and harvest are covered by the main product price and the actual costs consist of collection, packaging, and transportation only.

In other instances such as switchgrass or sugarcane for ethanol production the entire costs including agriculture have to be considered. Figure 8.4 shows a world map and the main crops for biofuel production.

One can see that the biomass sources are highly diversified, especially in the tropical regions. The crop of choice depends on various factors such as irradiation time and density, temperatures, rainfall, season length, and soil quality and is often closely related to the farming traditions in certain regions. Polar regions only allow the cultivation of softwood. In arid regions, sweet sorghum and switchgrass are preferred due to the low demand of water. The current research aims to develop crops producing good yields while being very adapted and less demanding.

Table 8.1 shows typical energy costs for harvesting of various feedstocks and their productivity. Important is a fast growth under low cultivation costs, such as

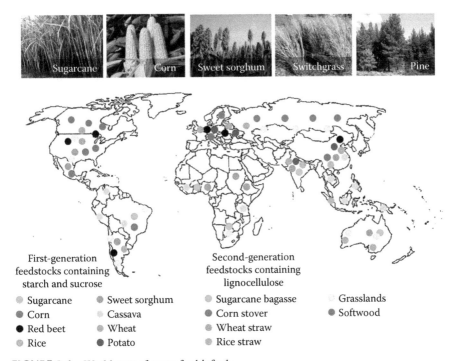

FIGURE 8.4 World map of crops for biofuels.

TABLE 8.1

Higher Heating Value and Energy Input for Biomass Cultivation and Collection

Feedstock	Higher Heating Value (%)	Energy Input (GJ/ha)	Productivity (dry tons/acre)
Wheat straw[a]	17.4	15–17 (Wheat)	1–2
Corn stover[a]	17.9–18.4	16–27 (Corn)	3–4.5
Bagasse[a]	18.4–19.4	15–17 (Sugarcane)	5–7
Switchgrass	18.6–19.0	2–7	2–8

Source: Zhu, J. Y. and Zhuang, X. S., *Prog. Energy Combust. Sci.*, *38*, 583–598, 2012. With permission.

[a] Waste products from processing of edible parts and sucrose and starch production.

water and fertilizer consumption. Further a high yield per acre of land is important, especially in highly populated areas, but also regarding the fuel consumption for cultivation and harvest. Using pure lignocellulosic feedstocks, one of the hot candidates is switch grass and similar crops that can be cultivated in pasture and grass lands of North and South America, central Asia, and Africa. The important aspect is their low consumption of fertilizer and water.

The conversion of lignocellulosic feedstocks is always in competition to the direct combustion. With emerging technologies, such as electro cars and fuel cells, combustion processes are a reasonable alternative. Various corn straw incinerators have been constructed feeding small communities with energy. As an example, bagasse from sugar production is commonly combusted to provide energy for the sugarcane mills, refinery operations, and provide energy for the local communities. Furthermore, the prices of feedstocks can dramatically alternate depending on the length of the harvest season.

Maritime feedstocks, such as algae and seaweed, having much higher water contents than lignocellulose are more promising for conversion to fuels and chemicals. The high costs for drying make a use for burning not feasible.

8.6 PRETREATMENT OF BIOMASS

Prior to any thermochemical or biochemical treatment, the biomass has to be crushed into particles that can be processed. For the production of vegetable oils, sucrose, and starch, these processes have a long tradition and can be performed at low costs.

For other feedstocks originating from waste, such as bagasse and straw, the pretreatment is already done during the harvest or prior to the extraction process of the saccharides. If wood is directly converted it has to be chopped or milled to smaller particles. The main target is the increase of the surface area and a good accessibility of the pore structure, which is always a trade-off with regard to the pretreatment costs as well as the removal of impurities, such as ash and salts.

8.6.1 DRYING

The drying step is energetically very costly and should be avoided if possible. For various routes, such as pyrolysis, however, a low water content is necessary. Also the energy input for milling can be strongly reduced if the biomass water content is low.

8.6.2 MECHANICAL TREATMENTS

Mechanical treatments are very energy-intensive processes, but an essential part of the pretreatment. Especially going to particle sizes of less than 1 mm, the costs explode [7]. The treatment type and energy input strongly depend on the mechanical properties of the feedstock. The latter strongly increases with the moisture content of biomass [7]. Feedstocks, such as corn stove, straw, and switchgrass, can easily be milled than wood [7]. Especially for hardwood the costs for milling are significant and can be comparable to the energy content of the produced ethanol.

8.6.3 STEAM/CO$_2$ EXPLOSION

Using steam explosion the biomass is heated to around 180°C–200°C under elevated pressure. Then the pressure is released rapidly resulting in an expansion of the gas and a destruction of the fibril structure. This process takes only a few minutes or seconds, the water in the biomass is commonly sufficient. The process lacks, however, of high costs for pressurization. Nevertheless, it is a standard process of biomass pretreatment, especially because there are no chemicals involved. It can result in amounts of char far beyond 5 wt% [8] and does not necessarily increase the hydrolysis rate [8]. It is, therefore, done in the first step of ethanol production from lignocellulose. Using CO_2, an acidic environment is created, resulting in rather mild hydrolysis prior to expansion.

8.6.4 SWELLING

Immersion of biomass in water results in swelling of the pore structure and a better accessibility. The energy costs are low; however, large storage tanks have to be used if natural resources, such as rivers and lakes are not available. In some instances the biomass pretreatment is part of the harvest and extraction process. Sugarcane, for instance, is milled and washed from sucrose resulting in wet bagasse with a particle size in the millimeter range and can directly be fed to the hydrolysis unit.

8.7 COMPOSITION OF BIOMASS

8.7.1 ENERGY STORAGE MOLECULES

8.7.1.1 Saccharides

Storage saccharide, such as sucrose and starch, are commonly digestible by human and are considered as first-generation biofuel feedstocks. Various crops contain large amounts of those easy accessible storage saccharides. These are commonly stored in the amyloplasts of the cells. Sugarcane contains more than 15 wt% of sucrose and the maize kernel contains about 64wt% of starch.

8.7.1.2 Sucrose

The production of sucrose, a dimer of fructose and glucose, is mainly done from the crops sugar beet and sugarcane. Sugarcane is a fast growing grass type with a harvest period of around 180 days per year. The sugarcane producers are traditionally found in South America, in particular in Brazil, but also the northeast parts of Australia are providing good conditions for sugarcane growth. Australia has launched a large research program to utilize the sugarcane. In a standard process, sucrose is milled into smaller particles, then pressed, and the remaining solid parts are washed with water. This leaves a bagasse with a sugar content of a few weight percent. During drying a part of the remaining sucrose will convert to ethanol and other fermentation products. In colder climate sugar beet is produced with, however, a much lower productivity compared to sugarcane. Sugar beet waste is fed to animals and is, thereby, used completely.

Currently, the main amount of sucrose is directly fed to fermenters to produce bioethanol. Other ways are the production of polyols or furanes. The routes to those products will be discussed in the following sections.

8.7.1.3 Starch

Starch is mainly produced from corn and potatoes. It is currently the main source of bioethanol in the United States and Europe. Starch does only appear in an amorphous form, due to the coil structure. It swells in water and can even be dissolved in hot water. This makes starch a valuable source for bioethanol production, although, due to the need for an additional hydrolysis step, the conversion is more costly than for sucrose.

8.7.2 LIGNOCELLULOSIC MATERIALS (MAIN COMPONENTS OF WOODEN-BASED BIOMASS)

Lignocellulosic feedstocks originate from nonedible parts of plants and crops, residues, etc. and are, therefore, considered as second-generation biofuel feedstocks. Lignocellulosic biomass consists of various compounds, as shown in Table 8.2.

TABLE 8.2
Biomass Composition of Different Lignocellulosic Feedstocks

Feedstock	Hemicellulose + Cellulose (%)	Lignin (%)	Extractives (%)	Ash (%)
Wheat straw	55.3	16.9	13.0	10.2
Corn stover	59.3–61.5	18.2–20.2	3.3–4.8	11.4–12.5
Switchgrass	55.4–60.5	17.4–20.5	5.7–6.2	5.7–6.2
Bagasse	61.6–68.3	23.1–24.1	1.5–4.4	2.8–4.0

Source: Zhu, J. Y. and Zhuang, X. S., *Prog. Energy Combust. Sci.*, 38, 583–598, 2012. With permission.

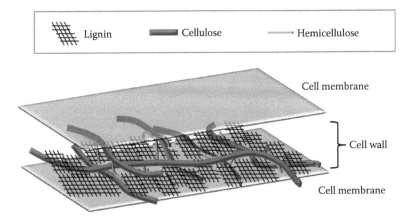

FIGURE 8.5   Cell wall structure of cellulosic plants.

The main compounds are hemicellulose, cellulose, and lignin. Examples of feedstocks highly promising are softwood in the northern hemisphere and switchgrass in arid regions. Bagasse, straw, and other residues from agriculture are also feasible feedstocks as they represent a large part of the plant weight. The density of bagasse and straw is very low. Furthermore, the infrastructure in certain countries is not well developed. This means, the transport costs per kilogram are very high.

Other biomass sources, such as biological waste from households, contain more water. Burning is, therefore, less of an option and the conversion to biofuels does make more sense. Lignocellulose consists of three main components, as shown in Figure 8.5.

The classification is based on the molecular structure, as discussed in the following sections. The separation is, however, less strict. Hydrolysis of hemicellulose using diluted acid often results in the partial hydrolysis of cellulose and the dissolution of acid-soluble lignin. Other ingredients are ash, mainly silica, in amounts of at least a few weight percent and proteins. Table 8.2 shows the typical composition of typical feeds, such as bagasse and straw.

8.7.2.1   Cellulose

Cellulose is the main part of lignocellulosic biomass, contributing by around 40% to the lignocellulose mass. Therefore, the feasibility of biomass conversion via sugars depends on the cellulose conversion. Cellulose consists of linear molecules of D-glucose bound together via beta (1,4)-glycosidic linkages, densely packed to form crystalline and amorphous parts. This makes it difficult for heterogeneous catalysts and even enzymes to access and hydrolyze the molecules. Cellulose is a valuable source for bio-based materials and composites, microfibrillated cellulose used in composites, textiles, paper industry but also medical applications [9].

8.7.2.2 Hemicellulose

Glucose

Galactose

Mannose

Xylose

Arabinose

Glucuronic acid

Is an amorphous and heterogeneous group of branched polysaccharides. The skeleton consists of xylose units, while the branches are a mixture of xylose, glucose, arabinose, and galactose. Hemicellulose is not crystalline and, thus, easier solvated and hydrolyzed. Hemicellulose contributes by 25%–30% to the mass of lignocellulose.

8.7.2.3 Lignin

Lignin has the most complex structure of lignocellulose. It is a rigid network increasing the mechanical stability of the plant. Lignin is a highly complex, 3D polymer of different phenyl–propane units, such as *p*-cumaryl alcohol, coniferyl alcohol, and sinapyl alcohol, bound together via ether and carbon–carbon bounds. Lignin fills the cell wall spaces between cellulose, hemicellulose, and pectin components, causing structural rigidity. The mass amount is about 15%–25%, as shown in Table 8.2.

In addition, small amounts of *extraneous organics* are found in lignocellulosic materials (up to 4%). The given percentages are mean values, since they vary depending on the type of lignocellulosic raw materials used.

8.7.3 ASH

The presence of ash is a major concern not only for combustion but also for conversion. Ash and inorganic residues originate not only from the plant itself but also from soil taken up during the harvest. Silica can lead to abrasion in the milling and extrusion process. Potassium salts generally cause equipment corrosion. Iron from soil, from equipment abrasion, such as sugarcane mills and other sources, is a catalyst poison for noble metals.

8.8 BIOMASS PLATFORMS

Living systems are able to convert a large number of sources to a broad range of compounds needed for the cell growth. This transformation occurs mainly via key metabolites for energy, redox reactions, and carbon, so-called bowties. This system can easily be copied to the production of biofuels and chemicals, as Sanders et al. pointed out [10]. Looking at the composition of biomass, the most obvious choice are storage compounds, such as oils, starch, and fats, but similarly promising are structural compounds, such as cellulose, hemicellulose, and lignin, as they form the main mass of a plant and do not directly compete with the food chain.

The chapter is divided according to the different biomass platforms, consisting of vegetable oils/animal fats, lignin/bio-oils, and polysaccharides. The focus is on the catalytic conversion of biomass to liquid biofuels, with respect to the materials used, the process technologies applied, as well as the results obtained on lab-, pilot-, and industrial scale, when appropriate.

8.8.1 Conversion of Biomass

8.8.1.1 Vegetable Oils/Animal Fats

Triglyceride esters of fatty acids (e.g., FAME) are usually termed as vegetable oils or animal fats, which can be converted into biodiesel via extraction and transesterification with methanol or ethanol. Glycerol is formed as by-product. The fatty acids usually processed contain 16 or 18 carbon atoms. Triglyceride esters can be transformed into liquid biofuels much easier than wooden-based biomass since they contain less oxygen and represent a high energy level already. These esters can be used as diesel fuel as such, but their high viscosity and low volatility as well as injector coking and other disadvantages hinder their direct application in diesel engines [4,11].

Vegetable oils are thermally not stable, and they can be converted without applying catalysts upon rapid heating in the absence of air. However, the use of zeolites [like ZSM-5 (structure code: MFI), ultra-stable Y zeolite (USY, structure code: FAU), and beta-zeolite (structure code: BEA zeolite)] can improve the product yield substantially due to cracking and pyrolysis of vegetable oils [11]. On the other hand, coprocessing of vegetable oils with crude oil in existing petroleum refineries is an option for future biofuel production, contributing to a reduction of the energy use–related greenhouse gas emissions. This can be achieved by either mixing of vegetable oils with crude oil and coprocessing in the atmospheric distillation or mixing with the middle-distillate fraction from the atmospheric distillation as feed for further processing. Alternatively, vegetable oil can be used together with fossil-based feedstocks for fluidized catalytic cracking (FCC) processing [12].

The major concern with respect to biodiesel feedstocks is the competition with food and animal feed sources. Therefore, nonedible oils should preferably be used as feedstocks for the manufacture of biodiesel, for example, seaweed-based oils, waste vegetable oils, which are locally available at low cost. The quality and properties of the manufactured biodiesel depends strongly on the carbon chain lengths of the fatty acid, the degree of unsaturation of the applied ester, the content of impurities and moisture, as well as the presence of other functionalities of the ester carbon chain [4].

Concerning the transesterification of vegetable oils, both base and acidic catalysts are in operation; however, base-type catalysts show a much higher activity than acidic systems. The commercial production of biodiesel is currently performed by either sodium hydroxide or sodium methylate, classical homogeneous alkaline catalysts. However, one of the main challenges with respect to the transesterification of vegetable oils with methanol or ethanol is to substitute the liquid base catalysts with heterogeneous systems in order to arrive at higher purity products. Therefore, heterogeneous catalysts (including micro- and mesoporous materials) have been developed for the transesterification of vegetable oils to allow easy recovery of the catalysts, simplified purification steps and to achieve high resistance to catalyst deactivation. Furthermore, another request is to develop robust catalysts to handle vegetable oils with high impurity contents, like used cooking oils or plant oils with a high content of free acid groups [13]. An interesting comparison of both homogeneous and heterogeneous catalytic systems has been reported for the transesterification of triacetin as model compound. At 60°C and a methanol to triacetin ratio of 6:1, the following sequence with respect to the catalytic activity was observed: NaOH > H_2SO_4 > ETS-10 (Na, K) > Amberlyst-15 > sulfated zirconia > Nafion > MgO ~ tungstated zirconia > supported phosphoric acid > H-BEA > ETS (H) [4].

Large pore microporous materials (zeolites FAU, BEA) and mordenite (structure code: MOR) have been used as catalysts for the transesterification of fatty acids. However, to a large extent, these conversions mainly occur on the external surface of the catalyst particles, due to the processing of voluminous substrate molecules which limit the internal molecular diffusions. Compared to the catalytic performances of sodium hydroxide or sodium methylate, the activity of zeolites has been lower than the homogeneous catalysts. So far, no microporous-based catalysts have been considered for the commercial production of biodiesel [4].

A number of companies have developed process technologies for the complete hydrogenation of triglycerides to hydrocarbons, in this way bypassing the coproduction of glycerol (e.g., Neste Oil, BP, Petrobras, Haldor Topsøe, Axens, UOP-Eni, and Conoco-Phillips). As an example, the UOP-Eni Ecofining TM process is based on catalytic hydrodeoxygenation (HDO), decarboxylation (DCO), and isomerization reactions to produce an isoparaffin-rich diesel ("green diesel"). This process is flexible with respect to the feedstocks used, such as inedible feeds, cooking oils, animal fats. Amorphous oxides, mixture of oxides (silica-alumina, sulfated zirconia), zeolites (FAU, MOR, BEA, MFI), zeotype materials (SAPO-11, SAPO-31, SAPO-41), and mesoporous materials (MCM-41, amorphous silica-alumina with controlled mesoporous pore architecture) have been applied for this purpose [4].

A different approach concerning the processing of vegetable oils is the hydrotreatment of vegetable oils to saturated hydrocarbons, either in dedicated units, like the NExBTL process of Neste Oil, or in the hydrotreatment units of petroleum refineries, like the H-BIO process of Petrobras. In principle, the hydrotreatment process is better adapted than the transesterification to produce biodiesel from low-purity vegetable oils and can be conducted in existing petroleum refineries [13].

8.8.1.2 Biomass Conversion via Bio-Oils

8.8.1.2.1 Pyrolysis of Lignocellulosic Components

A detailed description of the catalytic pyrolysis of biomass is given in Chapter 9, and this part of the current chapter will only address a few general aspects related to the pyrolysis of lignocellulosic components. Cellulose, hemicellulose, and lignin pyrolyze or degrade at different rates and by various mechanisms and pathways. The direct use of the lignocellulosic material as chemical feedstock faces problems due to their complex structure and the difficulty to separate their components in an economically feasible way. As we change from a crude oil refinery to biorefinery, the item of separation technology will be different. In petroleum refinery, distillation is the main separation operation due to the fact that we are handling volatile compounds. For valuable chemicals derived from lignocellulosic biomass solvent–based extraction of lignin, chromatography or membrane separation are the main choices due to the nonvolatile nature of most biomass components. Removal of the polysaccharide fractions can be done by stepwise hydrolysis, as discussed in Section 8.8.1.3.

Prior to the pyrolysis of the different components of wooden-based biomass, their separation into the three major components is performed by, for example, steam splitting. Alternatively, aqueous-phase reforming (APR) can be applied, in which the lignocellulosic biomass first undergoes treatment to produce an aqueous solution of sugars or polyols [1].

The *pyrolysis of cellulose* is started at quite low temperatures (around 50°C); however, the thermal degradation proceeds through two types of reaction: a gradual decomposition at low temperatures and a rapid volatilization at higher temperatures. The initial decomposition reactions cover hydrolysis, oxidation, depolymerization, dehydration, and DCO. The *thermal decomposition of hemicellulose* occurs more readily than that of cellulose. The pyrolysis starts at 100°C during heating for 48 hours, and steaming at high temperatures for a short time depolymerizes hemicellulose. Hemicelluloses contain more moisture than lignin, however, their thermal decomposition proceeds at lower temperature compared to lignin degradation [14].

The lignin component of the wooden-based biomass must be addressed strongly with respect to increase the competitiveness of the biorefinery concept. *Lignin decomposes* over a wider temperature range compared to cellulose and hemicellulose. Currently, residual lignin from paper pulping is burned off in order to generate heat and electricity; however, *lignin pyrolysis* at temperatures between 250°C and 600°C has demonstrated the potential of formation of low-molecular weight and valuable feedstocks for further processing. These results suggest that the application of shape-selective cracking catalysts would allow running the process at lower temperatures, providing at the same time an improved product distribution pattern. For example, lignin has been depolymerized by a base-catalyzed treatment into a bio-oil consisting of low-molecular-weight phenolic compounds [15]. The formation of phenolics is based on the cracking of the phenyl–propane units of the macromolecular lattice of lignin. The phenolic fraction can be separated in order to arrive at phenolic resins, or the bio-oil can be subjected to a hydroprocessing step, which yields a mixture of alkylbenzenes useful as a potential liquid biofuel [3,14].

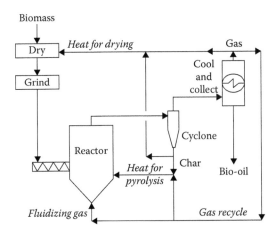

FIGURE 8.6 Flow sheet of bio-oil production from lignin. (From Stöcker, M.: Biofuels and biomass-to-liquid fuels in the biorefinery: Catalytic conversion of lignocellulosic biomass using porous materials. *Angew. Chem. Int. Ed.* 2008. 47. 9200–9211. Copyright Wiley-VCH Verlag GmbH & Co. KGaA. Reproduced with permission.)

Figure 8.6 shows a flow sheet of the bio-oil production from lignin. Besides the upgrading of bio-oil via separation of both phenolics and motor fuel components, the carbohydrate fraction can be converted to hydrogen via catalytic steam reforming. In addition, the gasification of bio-oil to syngas opens the entire route to conventional Fischer–Tropsch processing and the indirect pathway via methanol to olefins and gasoline. The liquid fraction of the pyrolysis products consists of two phases: an aqueous phase containing a wide variety of organo-oxygen compounds of low-molecular weight and a nonaqueous phase containing insoluble organics (mainly aromatics) of high-molecular weight. This phase is called bio-oil and is the product of greatest interest [14].

8.8.1.2.2 Application of Heterogeneous Catalysts for the Conversion of Wooden-Based (Lignocellulosic) Biomass

The conversion of biomass to first-generation biofuels applies only a small part of the biomass available for processing, and the greenhouse gas emission reduction is not optimal. However, the application of wooden- or seaweed-based biomass with respect to the production of second- or third-generation biofuels is more expensive since considerable investment costs are requested. The conversion of wooden-based biomass in order to produce biofuels via heterogeneous catalysis has received strong attention during the last years. Biomass pyrolysis turns out to be the preferred conversion among the different thermochemical processes available [16–18]. Pyrolysis is an appropriate process for the conversion of large amounts of wooden-based biomass to bio-oil. Bio-oil can easily be upgraded to biofuels and chemicals. However, some important bio-oil characteristics, like high water and oxygen content, corrosiveness, increased instability, immiscibility with crude oil–based fuels, high acidity, high viscosity, and low calorific value, cause technical challenges. Therefore, improvement of the bio-oil quality is a prerequisite before upgrading to biofuels can be envisaged [19]. Currently, there are a number of processes introduced for the thermochemical conversion of wooden-based biomass to biofuels and/or chemicals; however, usually these processes end up with a low-quality bio-oil, not suitable for processing to valuable products.

In general, the following three alternatives are currently of interest for the conversion of wooden-based biomass to biofuels:

1. Biomass-to-liquids (BTLs) with subsequent refining of the obtained bio-oil
2. Gasification of biomass followed by catalytic upgrading of the products
3. Separation of sugars with subsequent catalytic conversion

The literature available reports about the production of bio-oil with improved quality; however, the disfavor of additional coke and water formation has to be taken into account. In addition, less amounts of organic phases are produced [20,21].

So far, four main tracks have been studied with respect to the upgrading of bio-oil with improved quality:

1. FCC: $C_6H_8O_4 \rightarrow C_{4.5}H_6 + H_2O + 1.5CO_2$
2. DCO: $C_6H_8O_4 \rightarrow C_4H_8 + 2CO_2$
3. HDO: $C_6H_8O_4 + 4H_2 \rightarrow C_6H_8 + 4H_2O$
4. Hydrotreating (HT): $C_6H_8O_4 + 7H_2 \rightarrow C_6H_{14} + 4H_2O$

Concerning FCC, the preferred catalysts to be applied are based on ZSM-5 (structure code: MFI) and zeolite Y (structure code: FAU). So far, upgrading of bio-oil via FCC has been investigated only via using fixed-bed equipment within the temperature range of 340°C–500°C [2].

The state of the art in HDO is based on HT using NiMo- and CoMo-sulfided catalysts processed at about 400°C and under high pressure of hydrogen. However, due to the questionable availability of hydrogen in refineries, processing of bio-oil via HDO might not be the final solution. Complete DCO might be the best upgrading route for bio-oil, since hydrocarbons are produced and hydrogen is not requested [22]. ZSM-5 and USY zeolites have been used for this process. DCO of the organic acids leads to a bio-oil with improved quality since fewer acids are present and the obtained bio-oil is less corrosive, more stable, and has higher energy content. However, large formation of coke during this process requires new catalysts for a deeper DCO of bio-oil in order to make this technology acceptable from an economic point of view, since mainly bio-oils obtained via thermal conversion have been used so far. Anyway, catalytically improved bio-oils should overcome the aforementioned problems.

Novel developments with respect to the improvement of the bio-oil quality, using heterogeneous catalysis, should be given high priority. This will allow the production of improved bio-oil in order to upgrade these bio-oils within modern refinery streams. The overriding aim should be that pure bio-oils could be co-processed with hydrocarbon fractions (like vacuum gas oils) in a conventional refinery, giving the bio-oil the role of feeds or co-feeds in petroleum streams.

In order to meet these challenges, novel mono-/bifunctional catalysts such as zeolites, mesoporous materials with uniform pore size distribution (MCM-41, MSU, SBA-15), micro/mesoporous hybrid materials doped with noble and transition metals and base catalysts should be investigated. The main task of the novel catalysts should be to selectively prefer the decarboxylation reactions producing high-quality bio-oil with low amounts of oxygen and water. Following this approach, a reduced formation of undesirable oxygenated compounds (such as alcohols, ketons, acids, and carbonyls) is envisaged, since these compounds are known to be detrimental for the direct use or further coprocessing of the formed high-quality bio-oil. The application of catalysts doped with noble metals would be a benefit with respect to the promotion of reactions of bio-oil regarding oxygen removal and ring-opening, at the same time with a minimum consumption of hydrogen.

Furthermore, the hydrothermal stability of the novel catalysts must be improved and this can be studied by the successive addition of water to the dry feedstock. The resistance to deactivation and the catalyst behavior upon regeneration must be investigated in order to tune new catalyst formulations. This includes the controlled formation of appropriate catalyst particles as well, and tailoring the catalytic properties of the novel catalysts, like porosity, acidity, basicity, metal-support interactions, etc., should be part of the entire picture.

The *in situ* catalytic upgrading of bio-oil via biomass pyrolysis using novel catalysts will take place in the pyrolysis reactor. This means, we have a direct contact of the solid biomass with the solid catalyst, in comparison with a crude oil refinery, where usually liquid or vapor phases are in contact with the solid catalysts. This requests a proper tuning of the catalytic behavior in order to arrive at suitable structure–property relationships.

The bio-oil with improved quality could then be upgraded via FCC and/or HT, which means we can use the catalytically produced bio-oil as a blend in FCC and/or in HT. The direct processing of high-quality bio-oil and the use of bio-oil as co-feed with hydrocarbon fractions should be investigated both with respect to novel and commercially available catalysts [2].

The behavior of H-ZSM-5 (a zeolite or microporous material with the structure code MFI, see Figure 8.7) has been studied in connection with the pyrolysis of wooden-based biomass [2]. The acidic sites of this catalyst are conducting the performance of this zeolite in the conversion of lignocellulosic feedstocks. This occurs via a carbonium ion mechanism, promoting deoxygenation, and DCO of the bio-oil components, as well as cracking, alkylation, isomerization, cyclization, oligomerization, and aromatization. However, as a consequence of the catalytic conversion, tar and coke were formed as undesirable by-products as well. The regeneration of the deactivated catalyst by coke burn-off at 500°C in air reduced the effectiveness of the zeolite in the catalytic conversion of lignocellulosic biomass to bio-oil and the further processing to aromatic products. H-ZSM-5 activated at 500°C revealed mainly Brønsted acid sites; however, the formation of Lewis acid sites was observed at higher temperatures resulting in dehydroxylation reactions. The best bio-oil quality was obtained at 450°C using this catalyst. Even fast pyrolysis of vegetable biomass in a fluidized-bed reactor and using Ni-H-ZSM-5 as a catalyst has been successfully performed [2].

However, recent focus with respect to catalyst development devoted to the conversion of lignocellulosic materials has been concentrated on mesoporous materials with a uniform pore size distribution, like the MCM-41 (Mobil Composition of Matter) or MSU (Michigan State University). These materials are relatively new, and their pore diameter can be tailored within the range of 2–10 nm, allowing to process large organic molecules represented in the wooden-based biomass feedstocks (see Figure 8.8).

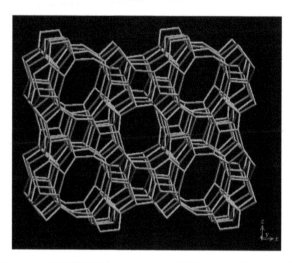

FIGURE 8.7   Structure of H-ZSM-5 (structure code: MFI, three-dimensional channel system with 10-membered rings, pore diameter: 5.1 × 5.6 Å). (From Stöcker, M.: Biofuels and biomass-to-liquid fuels in the biorefinery: Catalytic conversion of lignocellulosic biomass using porous materials. *Angew. Chem. Int. Ed.* 2008. 47. 9200–9211, 2008. Copyright Wiley-VCH Verlag GmbH & Co. KGaA. Reproduced with permission.)

The concept for using mesoporous materials for the catalytic conversion of wooden-based biomass has been adapted from FCC (within conventional crude oil refining), where the large residue molecules are first cracked to gas-oil components in the macro- and mesopores, before the final gasoline or propylene (LPG) will be formed during cracking performed in the micropores of the zeolite Y or ZSM-5, respectively (see Figure 8.9). The very large pores of mesoporous materials are able to process large lignocellulosic molecules; even so the chemistry is different since we are dealing with carbohydrates concerning wooden-based biomass conversion, whereas hydrocarbons are processed in conventional FCC.

The mesoporous materials can be applied as such; however, attempts have been made with respect to the improvement of the hydrothermal stability of these catalysts by steam treatment (since these systems must tolerate certain amounts of water) or by the introduction of noble and/or transition metals in order to enhance the oxygen removal capacity and/or promote decarboxylation reactions for the benefit of getting high-quality bio-oils [2].

So far, the application of Al-MCM-41 materials as catalysts for the pyrolysis of wooden-based biomass appears to be promising, especially regarding the improvement of the quality of the obtained bio-oil via increased formation of phenols. Even the siliceous MCM-41 material was active in biomass pyrolysis, producing high amounts of total liquid products via the enhanced thermal cracking of lignocellulosic biomass due to the high surface area of the mesopores. A fine-tuning between acidity and porosity of the MCM-41-based materials seems to be a prerequisite for the improvement of the bio-oil quality as well as the enhanced product selectivity [21].

Future studies have to focus on the acidic properties of the mesoporous materials (type, strength, density, and number of acid sites) and on the pore architecture in

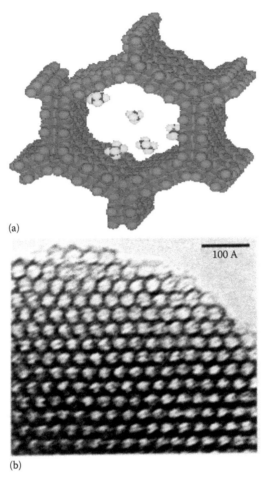

(a)

(b)

FIGURE 8.8 Schematic structure of MCM-41 consisting of silicon, aluminum, and oxygen units. Small molecules are shown in the pores (a). Hexagonal array of the one-dimensional structure of MCM-41 as revealed by high-resolution transmission electron microscopy (HRTEM) (b). (From Stöcker, M.: Biofuels and biomass-to-liquid fuels in the biorefinery: Catalytic conversion of lignocellulosic biomass using porous materials. *Angew. Chem. Int. Ed.* 2008. 47. 9200–9211, 2008. Copyright Wiley-VCH Verlag GmbH & Co. KGaA. Reproduced with permission.)

order to get information about the structure–property relationships when applying mesoporous materials as catalysts in the conversion of lignocellulosic biomass [20].

8.8.1.2.3 Application of Catalysts for the Conversion of Biomass Other than Wood-Based Liquid Biofuels

The one-pot conversion of algal (and lignocellulosic) biomass into a liquid fuel [2,5-dimethylfuran (DMF)] has been reported by De et al. [24], applying a multicomponent catalytic system comprising N, N-dimethylacetamide and CH_3SO_3, Ru/C, and formic acid. The synthesis of DMF from all substrates was carried out under mild reaction

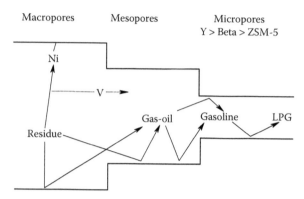

FIGURE 8.9 Conceptual pore architecture design of an FCC catalyst. (Reproduced from *Stud. Surf. Sci. Catal.*, 85, Sie, S. T., Past, present and future role of microporous catalysts in the petroleum industry, 587–631, Copyright 1994, with permission from Elsevier.)

conditions. The reaction progressed via 5-hydroxymethylfurfural (HMF) in the first step followed by hydrogenation and hydrogenolysis of HMF with the Ru/C catalyst.

Kim et al. applied Pd catalysts supported on different metal oxides for the production of gasoline-range branched hydrocarbons from butanal. Among the prepared catalysts, Pd/zirconia showed a complete butanal conversion by the formation of C_7–C_9 branched hydrocarbons with 75% yield [25].

The production of liquid biofuels from wet organic waste matter was reported by Hammerschmidt et al. [26], applying potassium carbonate combined with zirconia in a continuous one-step process under hydrothermal and near-critical water conditions.

Hierarchical macroporous–mesoporous SBA-15 sulfonic acid catalysts were used for the biofuel synthesis and demonstrated for the case of transesterification of the bulky glyceryl trioctanoate and the esterification of long-chain palmitic acid. The enhanced mass transport properties of this type of hierarchical architecture paves the way to a broad application of those bimodal catalysts with respect to catalytic biomass conversion [27].

8.8.1.2.4 Applications of Bio-Oil

High-quality bio-oil can be used to obtain biofuels and/or valuable chemicals following the flow sheet shown in Figure 8.3. Gasification of bio-oil to syngas opens the conventional routes to Fischer–Tropsch products and methanol. From methanol, olefins and gasoline can be obtained. Separation of bio-oil can reveal routes to phenolics and—via the carbohydrate fractions—to hydrogen. However, alternatively, the bio-oil can be used to produce heat and electricity, as shown in Figure 8.10.

The bio-oil with improved quality can be upgraded via FCC and/or HT, which means we can use the catalytically produced bio-oil as a blend in FCC and/or in HT. The direct processing of high-quality bio-oil and the use of bio-oil as co-feed with hydrocarbon fractions will contribute to a decreased use of fossil-based energy sources. The overriding aim should be that pure bio-oils could be co-processed with hydrocarbon fractions (like vacuum gas oils) in a conventional refinery, giving the bio-oil the role of feeds or co-feeds in petroleum streams [22].

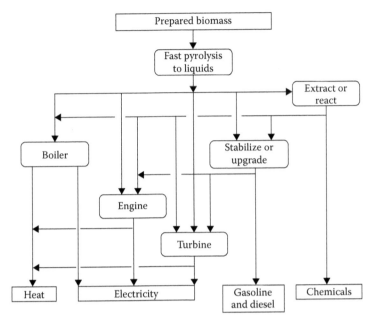

FIGURE 8.10 Applications of bio-oil. (From Stöcker, M.: Biofuels and biomass-to-liquid fuels in the biorefinery: Catalytic conversion of lignocellulosic biomass using porous materials. *Angew. Chem. Int. Ed.* 2008. 47. 9200–9211. Copyright Wiley-VCH Verlag GmbH & Co. KGaA. Reproduced with permission.)

8.8.1.3 Polysaccharides

The North American National Renewable Energy Laboratory (NREL) report for the US Department of Energy (DoE) on chemicals from biomass lists the 15 most important key compounds to chemicals and fuels from biomass [28], which were later reevaluated by Bozell and Petersen [29]. Among those, glucose, xylose, and their derivatives, such as sorbitol, xylitol, are listed. The route via the hydrolysis of cellulose and hemicellulose to monomers is, therefore, a key step. Additionally, for high-value compounds and energy wise it makes more sense to use the chemical structure of the biomass building blocks. Saccharides can even be used as a H_2 source for the reduction of alcohols and the hydrogenation of olefins [30].

In any case, the production of chemicals from biomass is often only economically feasible if in parallel fuels are produced from these crops. The production of chemicals, as commodities and as fine chemicals, allows higher margins, but only as long as the market is not saturated. As an example for Germany, the market of fuels from biomass could replace about 20%–25% of the fossil fuels. If all this biomass would be used for the production of chemicals, their price would drop significantly making the production not economic.

8.8.1.3.1 Polysaccharide Hydrolysis

Carbohydrates from biomass are present as oligo- and polymers. Their hydrolysis to smaller units is an essential step for any further conversion into chemicals and fuels.

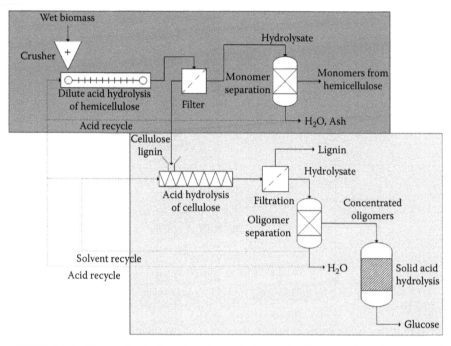

FIGURE 8.11 Consecutive hydrolysis of hemicellulose and cellulose resulting in mono- and oligomeric saccharides.

Storage saccharides, such as sucrose and starch, can be dissolved in water, filtered, and, if needed, hydrolyzed to monomers.

In the bioethanol industry, these feedstocks are commonly used directly for fermentation. However, the main part of the plant consists of water insoluble structural saccharides, such as cellulose and hemicellulose. There have been many attempts to use solid acid catalysts at this stage; however, the use of solid acid catalysts results in a poor contact of acid groups and hemicellulose/cellulose embedded in the lignin matrix resulting in poor hydrolysis rates. Any use of shear forces to break the structure during the hydrolysis would also damage the catalyst. Additionally, dealing with biomass as feedstock, one has to consider the presence of ash in significant quantities up to 12 wt%. Ions dissolved would replace the protons of the solid catalyst requiring the need for regeneration. This would be an expensive way of *in situ* production of homogeneous acid catalysts. On top of that, hydrolysis of hemicellulose and/or cellulose results in solid products like lignocellulose or lignin, as shown in Figure 8.11. Especially, the lignin particles after cellulose hydrolysis are in the nanometer size range. They block the solid catalyst pores and need to be burnt off the catalyst.

8.8.1.3.2 Sucrose Hydrolysis

Sucrose is currently the main source of bioethanol. Apart from that, it can be hydrolyzed at very mild conditions. Weak acidic catalysts and even hydrogenation catalysts such as supported noble metal catalyst or Raney-Ni are sufficient. Hydrogenolysis of sucrose to sorbitol and mannitol can easily be done at temperatures below 120°C and pressures of less than 80 bar. At low hydrogen pressures, the isomerization between glucose and fructose occurs.

8.8.1.3.3 Starch Hydrolysis

Starch is stored in the amyloplasts as a coiled structure, which prevents the formation of dense packings or crystalline forms. It can be dissolved in hot water, which makes it easily accessible to solid catalysts. Currently, starch is industrially hydrolyzed using amylases, which are significantly less expensive than cellulases. However, attempts have been made to use solid acids. Yamaguchi and Hara (2010) achieved 70% yield of glucose at 100°C–150°C using carbon-based solid acids [31–33]. Small amounts of the glucose dehydration products levoglucoson, levulinic acid (LA), and formic acid have been detected.

8.8.1.3.4 Structural Saccharides Hydrolysis

These saccharides make up the largest part of the plant. They are present in a dense, partly crystalline structure embedded in the lignin matrix of the cell walls. The most feasible way is, therefore, to use homogeneous acids in the initial stage of hydrolysis. This is common practice also in ethanol production to improve the digestibility of the feedstock. Figure 8.11 shows a sequential hydrolysis of hemicellulose and cellulose. This has several advantages, which we will discuss in Sections 8.8.1.3.5 and 8.8.1.3.6.

8.8.1.3.5 Hemicellulose Hydrolysis

Hemicellulose consists of a mixture of pentoses, mainly xylose making up the main strain of the molecules. Further arabinose, glucose, and galactose are found in the branches. These sugars would not be stable under the conditions required for cellulose hydrolysis. Any simultaneous hydrolysis of cellulose and hemicellulose results in large amounts of humins. Hemicellulose hydrolyzes at rather mild conditions. HCl concentrations of less than 5% and temperatures of less than 90°C are sufficient to produce monomers with high yields. The large amount of water results in a low concentration of oligomers. Under these conditions, the hydrolysis of cellulose is very slow, resulting in a good separation of hemicellulose hydrolysates and cellulose.

By converting the waste from sucrose-containing plants, such as bagasse, one can also convert remaining sucrose. Fructose is very unstable, even at those mild conditions. The stability of sugars is rated as

Fructose < Arabinose, Xylose < Galactose, Glucose

Therefore, before the hemicellulose hydrolysis, the washing of sucrose should be as thoroughly as possible within the economic constraints. Residues from starch production will not be an issue. Starch will eventually hydrolyze to glucose, which is stable under hemicellulose hydrolysis conditions.

The hemicellulose hydrolysis does not require strong shear forces. The larger the lignocellulosic particles after hemicellulose hydrolysis, the lower the costs for filtration. On the other hand, the pores are opened by the removal of hemicellulose. This allows a good access of cellulose in the next hydrolysis step and is more cost effective than steam explosion, especially using hard woods [8]. Additionally the ash is removed together with the hemicellulose hydrolysate. To separate the mono- and oligomers and to recycle the acid, various methods, such as solvent swing adsorption and chromatographic methods, extraction, and crystallization, have been developed [34,35]. Neutralization is currently applied but produces large amounts of salts and is, thus, not a serious option for a large-scale biofuel production. Additionally, the sugars content in the hydrolysate is less than 10% and needs to be purified for further processing.

During hydrolysis arabinose and galactose appear first, as they are easily accessible in the hemicellulose branches. The xylose–xylose bonds are broken earlier than the xylose–glucose and finally the glucose–glucose bonds. Especially, arabinose and xylose are chemically less stable than glucose. At conditions required for cellulose hydrolysis the C5 saccharides already degrade to humins.

8.8.1.3.6 Cellulose Hydrolysis

In the hydrolysis using industrial acids such as HCl or H_2SO_4, we need to distinguish between low- and high-concentrated acid hydrolysis. Typical acids with a low pKa value, such as HCl, H_2SO_4, *para*-toluenesulfonic acid, and H_3PO_4, are used [36]. Drawbacks of diluted hydrolysis are that crystalline segments of cellulose are broken up slowly. This means that glucose produced from amorphous parts can degrade, while crystalline cellulose is still hydrolyzed up resulting in poor yields [37]. This can, however, be reduced using other pretreatment steps, such as biomass wetting/swelling or milling. Cellulose more effectively decrystallizes in concentrated acid and the hydrolysis becomes more homogeneous in nature, although cellulose is still not soluble in those media. The demands on reactor and equipment material are, however, high, requiring acid stable construction materials. As the hydrolysis is performed under low pressure, the use of polymer equipment or linings could be a low-cost alternative.

For the hydrolysis and degradation rate, a severity factor is often used, which depends exponentially on the acid concentration, the temperature, and the reaction time. The yield of monomers in dependence of this factor shows commonly an optimum [38]. This is a simple way, but does not entirely reflect the heterogeneous nature of the process.

For dissolved sugars, the activation energy of hydrolysis is higher than the degradation activation energy of the monomers. This means, one should first work at rather mild conditions to form soluble oligomers, which are then converted to monomers at higher temperatures. Cellulose hydrolysis is significantly enhanced if a solvent is used. Ionic liquids have been tested extensively, however, they suffer from high viscosity and poor recyclability. Especially, the separation of the contaminants from those solvents has not been tackled yet. The low viscosity is especially important for the separation of the fine lignin particles. Low-cost solvents are molten salts, such as zinc chloride. Sub-coordinated with water molecules, they bind to the OH groups of cellulose and are able to dissolve amounts of up to 10%. They have been used in the fiber production and processes for sugars separation and solvent recycling have been developed [39].

Once the cellulose has been hydrolyzed to oligomers, these can be separated by adsorption and a highly concentrated saccharide stream can be hydrolyzed over a solid acid catalyst. In general, any solids with acidic functions could be used for oligomer hydrolysis. Solid catalysts and the applied hydrolysis conditions are listed in Table 8.3.

It is important that the catalyst particle shows a hierarchical pore structure of macro- and mesopores, as the diffusion of oligomers is slow and glucose should be

TABLE 8.3
Solid Acid Catalysts Systems for Cellulose Hydrolysis

Catalyst	Feedstock	Conditions	Yield	Reference
		Sulfonated Materials		
Sulfonated resins	Cellohexaose	90°C, 1 h	8% Glucose	[40]
(Amberlyst, Dowex)	Cellulose	150°C, 24 h	26% Glucose	[41]
Sulfonated carbons	Cellohexaose	90°C, 1 h	8% Glucose	[40]
	Cellulose	100°C, 3 h	4% Glucose	[31]
	Cellulose	150°C, 24 h	70% Glucose & Oligomers	[42]
	Cellulose	150°C, 24 h	75% Glucose	[43]
			41% Glucose	[41]
Sulfonated zirconia on SBA15	Cellobiose	160°C, 1.5 h	57% Glucose	[44]
		Porous Solid Acids		
H-Mordenite	Cellulose	150°C, 24 h	7% Glucose	[41]
Zirconium phosphate	Cellulose	150°C, 5 h	30% Glucose	[37]
	Cellobiose	150°C, 2 h	97% Glucose	[37]
H-Beta, H-ZSM5	Cellulose	150°C, 24 h	27% Oligomers	[41]
			9%–11% Glucose	
		Supported Noble Metals		
Ru/AC	Cellulose	230°C, 3 h	30% Glucose	[32]

transported rapidly out of the pores. Typically shaped zeolites provide these properties as well as tailored sulfonated resins.

Other means of cellulose hydrolysis have been reported. The research group of Schüth used ball milling of cellulose under acid atmosphere achieving high sugar yield [45,46]. However, small amounts of water need to be added for hydrolysis. Additionally, ball milling is a very costly process. This cannot be a reliable source for biofuels. Further, remaining lignin has to be removed from the milling device. Other means of shear stress are less energy intensive. The use of microwaves has been explored to disrupt the hydroxyl bonds. But, apart from the provided heat, it is still under debate whether microwaves have a chemical benefit.

8.8.1.3.7   Enzymatic Hydrolysis

After pre-hydrolysis and removal of lignin, the use of immobilized enzymes can be a serious alternative to solid acid catalysts. The immobilization of enzymes on polymers, silica, and other supports has been investigated [47]. A purified stream of oligomers would also strongly improve the long-term stability of the enzymes. Currently, the high enzyme preparation costs are a hurdle for an economically feasible process [38]. Depending on the throughput they so far are continuously added to the process or produced *in situ*.

Enzymatic hydrolysis compared to acid hydrolysis has the advantage of high selectivity to sugars, the medium is less aggressive and, due to the internal mass transfer limitations, can selectively hydrolyze the amorphous fraction only [10]. Still recalcitrance of lignocellulosic substrates and also low-cost pretreatment methods are an issue [38].

Cellulases in solution are bulky molecules. Their hydrolysis activity depends strongly on the accessibility of the polysaccharides External mass transfer to the biomass particle surface and internal mass transfer within the pores plays a large role [48]. As a result, only the external part of cellulose is accessible for digestion. Additional cellulose emerges at the inner parts of the biomass particle. The well-known shrinking core model can illustrate this. The hydrolysis increases significantly with smaller particle size, which in turn adds the costs of pretreatment. Product inhibition can further be an obstacle; cellobiose and glucose can reduce the enzyme activity [48]. Lignin and other non-hydrolytic parts for the cellulases can bind the enzymes. This effect increases with increasing the relative surface of nondigestible surface during the hydrolysis [48]. Further, shear can degrade the enzymes; the agitation should, therefore, kept low. As a result, only a small portion of the enzymes retain their original catalytic activity making the process very slow.

8.8.1.3.8   Further Conversions

The main target of further transformation steps is to reduce the oxygen content of the monomers. By that also the water solubility is reduced. This improves the separation but also increases the energy density of the biofuel.

8.8.1.3.8.1   Furfural Route   In this route, the dehydration step is done first to produce HMF from hexitols or furfural from pentitols. Van Putten et al. and Dutta et al. published good reviews on the state of the art in HMF production [49,50]. One major issue is the low stability of fructose and even lower stability of intermediates and

HMF under dehydration conditions giving rise to degradation to soluble polymeric species and insoluble humins, but also to further dehydration to levulinic acid (LA). Therefore, to achieve high selectivity one has to work at low HMF concentrations, which increases the process costs.

An option is the adsorption. Another one is the direct conversion to more stable compounds, such as LA, discussed in Section 8.8.1.3.8.2. Currently, the extraction using biphasic systems is widely applied. Methyl isobutyl ketone or butanol are used as solvents [51]. NaCl in addition improves the extraction of HMF into the organic phase [51,52]. Operation at conversion levels of less than 35% gives selectivities higher than 95% [49]. As catalysts, sulfonated resins, zirconia, titania, and vanadyl phosphates as well as H-BEA zeolites have been applied at temperatures ranging from 80°C to 200°C. Carlini et al. applied niobium oxide and niobic acid at low temperatures of 100°C [53]. The same research group reports the use of niobium phosphate salts at similar conditions giving 100% selectivity at 30% conversion within 30 minutes [50]. The use of organic solvents, such as DMSO, will strongly increase the yield. Values higher than 95% have been reported for various catalysts within 2 hours at temperatures of less than 120°C. The removal of water increases, however, the process costs significantly.

The upgrade to fuels is done in different directions. One route is the production of DMF from HMF and methyl furan from furfural [51,52]. These compounds have a high energy density, for DMF it is 40% higher than ethanol. Further hydrogenation will ultimately yield alkanes. The process is illustrated in Figure 8.12. The conversion is achieved by hydrogenolysis of the dissolved HMF over a Cu-catalyst. $CuCrO_4$ shows good yields to furanes [31]. The group of Dumesic achieved DMF yields of 61%, but also found that small amounts of chlorine can strongly deactivate the catalyst by sintering of Cu [32,33,51,52]. To improve the stability against chlorine they used a CuRu/C catalyst and achieved DMF yields of 71% for purified HMF streams and 61%

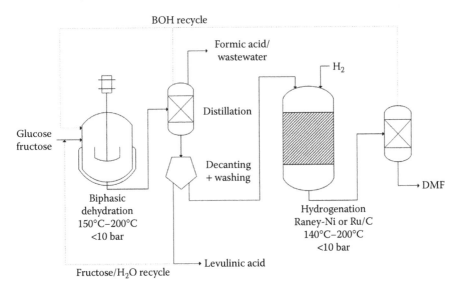

FIGURE 8.12 Process steps and conditions for DMF and levulinic acid production using butanol (BOH) as solvent for HMF extraction.

for streams with Cl⁻ impurities. A techno-economic evaluation of HMF production from fructose with subsequent hydrogenolysis to DMF has been published [51,52]. The sensitivity analysis showed that by far the most important parameters are the yields of HMF and DMF. In particular, a better partition of HMF between aqueous and water phase could have a significant impact on the process economics.

Alternatively, Xing et al. suggested the aldol condensation of furfural with acetone and subsequent dimerization and hydroxydeoxygenation to hydrogenated dimers and tridecane [54]. Drawbacks are the costs for the reactants hydrogen and acetone.

8.8.1.3.8.2 Levulinic Acid Route LA is another platform chemical with a wide range of applications in the pharmaceutical industry, for plasticizers or other additives [30]. It is currently produced via HMF (Figure 8.12) by dehydration applying H_2SO_4 or HCl, at temperatures above 150°C [55]. Raspolli Galletti et al. have reported the production of LA using diluted HCl, where a first biomass pretreatment step at 80°C–100°C opened up the biomass fibers and a second hydrolysis step at 170°C–200°C was applied. Using giant reed as feedstock gave LA yields of more than 80% of the theoretical value [56]. Preferred are, however, solid acid catalysts, such as ion-exchange resins (IERs) and zeolites. Weingarten et al. used zirconium and tin phosphates for the conversion of fructose at temperatures of 160°C [57]. Lewis acid sites catalyze the isomerization from glucose to fructose (zirconium/phosphates ratio of 1), but also the formation of degradation products, such as humins, while HMF and LA yields increase with the amount of Brønsted acid sites.

The LA selectivity is, thus, a function of the ratio between the relative concentrations of Brønsted and Lewis acid sites. High yields of LA were obtained using catalysts with zirconium phosphates having P/Zr ratios of 2 and 3. Especially a ratio of 2 shows a high surface area and a high surface concentration of hydroxyl groups and gives LA yields of 45% [57]. As a by-product, formic acid is produced, in a molar yield similar or slightly higher than LA due to side reactions of dehydration intermediates [55].

From LA various routes lead to potential biofuels, such as ethyl levulinate or ethyl valeroacetone via gamma-valeroacetone (gVL), as indicated in Figure 8.13 [58,59].

Pilot and commercial plants have been developed following route 1, the so-called biofine process producing LA esters using bioethanol, methanol, or mixed alcohols [60]. Levulinates have been extensively tested in gasoline and diesel engines. They show a high miscibility with fossil fuels and can thus be pipelined. The components are certified as fuel additives exceeding ASTM D-975 diesel standards and having higher miles per gallon (MPG) value than ethanol [60]. Ethyl levulinate is especially interesting as cold flow improver for biodiesel increasing the cloud point, pour, and cold filter plugging points [61]. It can be produced from LA and similarly from pentoses-based furfuryl alcohol, as indicated in Figure 8.13.

LA can also be used to produce methyltetrahydrofuran (MTHF), which is an oxygenated fuel extender for gasoline and can further be used as a solvent replacing THF [61]. Hydrogenation in the liquid phase at 200°C–250°C using PdRe/C catalysts gives yields up to 90% [62]. Upare et al. report the gas-phase hydrocyclization over a nickel-promoted copper/silica catalyst with a high copper loading resulting in MTHF yields of more than 90% over 100 hours of operation [63]. The pentane yield is 10%. Over time pentanol is formed at the expense of MTHF.

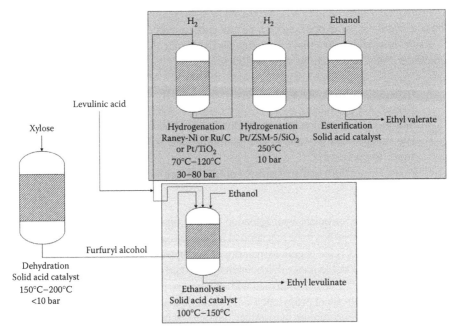

FIGURE 8.13 Process steps and conditions for xylose and levulinic acid conversion to ethyl valerate and ethyl levulinate.

To produce gVL, LA is commonly separated from the acid environment and then hydrogenated over a standard Ru/C catalyst. Raspolli Galletti et al. have reported the hydrogenation of LA under mild conditions using a supported Ru catalyst in combination with a solid acid catalyst [56]. Temperatures of 70°C–100°C and a hydrogen pressure of 30 bar have been applied. A fast conversion of the intermediate HMF is required, as it tends to polymerize to humins. Both steps can be combined to a one-pot route; however, the reaction mixture has to be neutralized or the HCl removed prior to the hydrogenation as it has a strong poisoning effect on the Ru catalyst. The addition of a solid acid catalyst in the hydrogenation step, such as niobium oxide or phosphate, favors the initial activation of the carbonyl group and the subsequent esterification producing gamma-lactone. These acids show strong Brønsted acid sites and medium-strong Lewis acid surface sites and retain their activity at high temperatures [56]. Reuse showed only slight decrease of the yields from 81% to 76% over three cycles.

Formic acid, the by-product of the LA production, is a commodity chemical with a large market. Deng et al. followed a different approach instead of separating it from LA, it can also be used as hydrogen source. They first dehydrated a glucose solution to LA and formic acid. After neutralization, they employed $RuCl_3$ and pyridine or NEt_3 as bases at 150°C resulting in a gVL yield of 83% and an overall yield of 48% based on glucose within 6 hours [55]. Vacuum distillation of the reaction products and recycling of the catalyst resulted in comparable yields. A reduced water content and elevated CO_2 pressures improve the yield, suggesting that CO_2 produced from the formic acid decomposition has an important role in the reaction.

From gVL valeric acid (VA) is produced by hydrogenolysis. It is assumed that first the acid-catalyzed ring opening is done and subsequently the hydrogenation.

Lange et al. did an extensive test program to balance the hydrogenation and acidic functions [59]. Their findings were as follows:

- With a too high metal-to-zeolite ratio, MTHF, pentanal/pentanol, and alkanes are produced.
- At a too low metal loading, the formation of pentenoic acid is favored.
- The shape selectivity and acidic strength of the zeolite have been found of minor importance.
- Pt and Pd are both active, while Rh produced significant amounts of gaseous by-products.

In order to achieve sufficient hydrogenation rates temperatures of 250°C needs to be employed. The catalyst has to be regenerated frequently by H_2 strips or air flow at 400°C. An important aspect is the chemical integrity of the support. It has been found that SiO_2, TiO_2, and ZrO_2 are stable under this combination of carboxylic acid and high temperature. The obtained VA is then esterified with ethylene or propylene glycol using IERs.

A one-step synthesis of propylene valerate can be done by reaction of the produced VA and the gVL conversion by-product pentanol with a selectivity of 20%–50%. Unreacted VA can be recycled. The volatility–ignition properties of the valerates are adjusted by the alkyl chain length to make it compatible to gasoline or diesel applications. They have good energy densities and polarities. These compounds have passed all tests for biofuels including oxidation stability, fouling tendency, corrosion lubricity, and water affinity [59].

An alternative route is the conversion of gVL to liquid alkenes for blending in gasoline, diesel, or jet fuels [64]. In this process, first gVL is decarboxylated to butene and CO_2, which does not require any H_2 to be consumed. Using silica/alumina mixed oxides, the operation at 375°C–400°C and elevated pressures up to 36 bar complete gVL conversions and butene yields of 96% have been obtained. After separation of water, butene was oligomerized on HZSM-5 or Amberlyst 70 at 170°C–230°C and somewhat lower pressures to minimize cracking and favor alkene coupling. Amberlyst was found to be more active allowing a lower operation temperature and C_8–C_{16} yields of 72% have been reached. The authors also suggest a one-pot method without water separation utilizing the formation of CO_2 to keep the pressure high. Inhibition of the acidic catalyst by water has to be compensated by increasing the catalyst load. Over 90 hours, a stable overall yield to C_{8+} alkenes has been achieved.

8.8.1.3.8.3 Isosorbide Route   In this route, the monomeric hydrolysis products are first hydrogenated using a standard supported noble metal catalyst and then subsequently dehydrated to form heterocyclic rings (see Figure 8.14).

The first step is an industrial standard operation to form polyols; hydrogen pressures of up to 80 bar and temperatures around 120°C result in yields close to 100%. As catalysts, Ru/C or Raney-nickel is applied. The latter is significantly less expensive but it also shows higher leaching rates. In this context, it is especially important to remove any gluconic acid, as it is a chelating agent. Furthermore, it is important to note that the

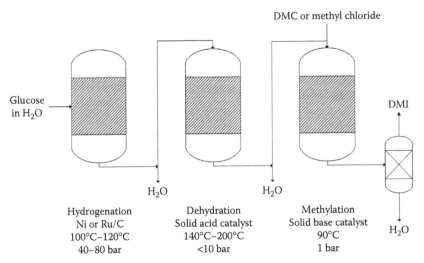

FIGURE 8.14 Process steps for isosorbide and dimethyl isosorbide (DMI) production from glucose.

concentration of hydrogen in the liquid phase has to be sufficient; otherwise, the isomerization reaction to fructose becomes dominant. The glucose hydrogenation product sorbitol is an important platform chemical. It is more stable than glucose. In the same way, fructose is hydrogenated to mannitol and sorbitol. As mentioned earlier, the hydrolysis of sucrose and hydrogenation can be done in one-step using noble metal catalysts. For conversion of more stable oligomers, such as cellobiose, an additional acidic function has to be introduced. Using an acid medium is less of an option. Sulfuric and hydrochloric acids have a strong effect on catalyst leaching and poisoning. In principle, the hydrogenation and acidic function can be introduced separately, but ideally the metallic phase is dispersed on a solid acid catalyst, such as zeolites or sulfonated resins [50]. Examples are bifunctional catalysts such as Pt/ZSM-5, which can also be regenerated easily by burning off the humins formed. This oxidation treatment also re-disperses the catalyst.

The dehydration is performed at temperatures between 140°C and 200°C using solid acid catalysts [50]. Especially sulfonated resins show good conversions and selectivities to dehydrogenation products above 90%, which can stand temperatures up to 160°C. Keeping the water level low improves the conversion and improves the long-term mechanical stability by reducing the swelling. Pentiols and hexiols form one ring under splitting of one water molecule. From those products only monodehydrohexitols are able to form a double ring under removal of one additional water molecule. From all reactants, intermediates, and products, humins and soluble oligomers or precipitates can be formed, which can block the catalyst pores. So, similar to the hydrolysis, the key step is the combination of macro- and mesopores.

For fuel production, a final methylation step is necessary resulting in dimethyl isosorbide (DMI) from isosorbide, dimethyl isomannide from mannitol, and monomethyl dehydroxylitol from xylitol, which show excellent diesel properties [65]. Alternatively, isosorbide dinitrate can be produced, which added in amounts of a few percent strongly increase the fuel octane number. For DMI production, methyl chloride [66], dimethyl carbonate (DMC) [67], or methanol/dimethyl ether [65,67] can, in principle, be used as reactants.

The etherification using methanol or DME produces water. To achieve acceptable yields, one needs to work under nearly water-free conditions, which is a significant cost factor. Solid acid catalysts or solid base catalysts can both be used. An alternative is to use methyl chloride or DMC in combination with a basic catalyst.

The use of those reactants is, though, significantly more costly than the use of methanol. Tundo et al. reported a nearly 100% yield by reaction of DMC with isosorbide in DMF using NaOMe as homogeneous catalyst and using mild conditions of 90°C and 1 bar [67].

For ethers with longer chain length, alkenes can be used as reactants instead, as shown by Rose et al. with yields of more than 60% using a sulfonated resin as catalyst [68]. This is, however, more of an interest for chemicals production than for fuel production.

The feature of the isosorbide route is that the less stable sugars, such as glucose and fructose, are rapidly converted to polyols. Further, moving from left to right in Figure 8.14, the products become less polar, which is beneficial for separation from the aqueous phase by extraction or adsorption. Additionally, only one mole of hydrogen is consumed in addition to two moles of methanol. Several companies have produced isosorbide in pilot plant scale, for instance Roquette using starch as feedstock and Avantiums XYX program to produce polymers from isosorbide and BiCHEM Technology targeting for diesel additives.

8.8.1.3.8.4 Lignin Hydrogenolysis

The lignin fraction contains the highest energy of all three main compounds of lignocellulosic biomass. Lignin has a lower density of polar groups compared to cellulose and is not soluble in water and most solvents.

Regarding the lignin separation from polysaccharides, there are currently two options. The first option is to selectively remove lignin from the feedstock. This can be done using basic media in combination with sulfates and sulfides (Kraft process) or sulfites (Sulfite pulping). Kraft pulping is the most common process in the paper industry. This would have the advantage that one is left with ash-free hemicellulose and cellulose fraction. The Na-lignin from the Kraft process can be used to produce polymers, but is often burned to partly recover the process costs. Lignosulfonates from the sulfite pulping can be used as plasticizers in the cement industry, as dispersers of dyes and for tanning leather.

A complete lignin removal can, however, not be achieved. Commonly around 10% of the lignin fraction remains in solid phase. Additionally due to the severe process conditions, cellulose and hemicellulose partially degrade. To address those issues, various Organosolv processes have been developed utilizing methanol, acetic acid, and other organic acids as solvent, as shown in Figure 8.15 [69,70]. Challenges are still the recycling of the solvent.

The other option is to hydrolyze the saccharides and recover the lignin as residue, as shown in Figure 8.11. Lignin can then be dried and burned to provide energy for the saccharides conversion process or depolymerized and further converted to fuels and chemicals, such as phenols.

Wang et al. reported the direct hydrogenation of lignin to aliphates [71]. However, a significant amount of hydrogen is required, which is reflected on process costs. Commonly, the produced lignin is first depolymerized over a silica-alumina catalyst

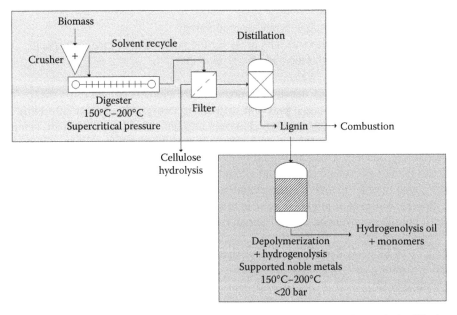

FIGURE 8.15 Process scheme for Organosolv process and subsequent hydrogenolysis of lignin.

and butanol as extraction phase. For efficient lignin depolymerization, hydrogen and a metal function are required. The challenge is to obtain a high hydrogenolysis function, while the hydrogenation of the phenol rings should be minimized.

Torr et al. dispersed wood particles in a dioxane/water mixture. Using Pd/C catalysts they achieved a yield of hydrogenolysis oil of 80% and dihydroconiferyl alcohol of 20% based on the lignin fraction [72]. For the subsequent cracking step of the phenols, Yoshikawa et al. applied a Zr/Al-doped FeOx catalyst giving a total phenol yield of 7% [73]. Addition of steam increased the yields to 17% [74].

Maersk has launched two projects aiming to develop marine fuels based on lignin. The first of these is in conjunction with Progression Industry, a spin-off company of Eindhoven University of Technology, with the challenge of developing a lignin-based fuel capable of meeting strict performance, price, and sustainability standards. The second is co-funded by the Danish National Advanced Technology Foundation and involves Maersk, DONG Energy, and several other companies and academic institutions, and will look at several sources of biofuel including lignin.

8.9 OUTLOOK

The use of carbon dioxide neutral and renewable biomass for the production of fuels is a vital alternative to fossil-based energy resources. However, a number of major obstacles for an economic feasible production of biofuels via the BTL process have to be challenged. These are as follows:

- High investment costs
- Low volumetric energy density of biomass

- Infrastructure
- Limitations in the photosynthesis productivity
- Availability of cultivable land areas for the production of bioenergy plants in competition with food and feed production

Especially the low energy density of wooden-based biomass (which is the cheapest and most abundant biomass) makes it difficult to convert this source into biofuels. The pyrolysis of lignocellulosic biomass into bio-oil of high quality by using well-suited catalysts is still the main challenge within a modern lignocellulosic biorefinery concept. Once high-quality bio-oil is achieved, the subsequent use as direct feed or as co-feed in conventional refinery processes like FCC or HT can be envisaged in order to arrive at diesel and/or at gasoline. HT requires high-pressure hydrogen; however, it might be feasible that this hydrogen demand can be satisfied from biomass conversion as well (catalytic steam reforming of the carbohydrate fraction, see Figure 8.3). There are a number of alternatives available for the utilization of biomass-derived feedstocks in a crude oil refinery and our society will move toward a sustainable economy as we continue to develop biofuels production technology [2].

In a time were focus is on global warming, carbon dioxide emission, increased competition, secure energy supply, less consumption of fossil-based fuels, etc., the use of renewable energy resources becomes essential. Biomass is one of these renewable resources! New challenges related to the catalytic conversion of wooden-based biomass still exist and have to be addressed in the future research and development within this field:

1. Mechanistic understanding of the catalytic conversion of lignocellulosic biomass to bio-oil, including structure–property relationships and product–pattern distribution
2. Catalyst development with respect to porosity, acidity, basicity, metal–support interactions, controlled formation of appropriate catalyst particles, improved hydrothermal stability and resistance to catalyst deactivation
3. Process conditions and large-scale production

The chosen approach of biorefinery based on wooden biomass does not compete with traditional crude oil refining but will represent a complementary to petroleum refining.

REFERENCES

1. Alonso, D. M., Bond, J. Q., and Dumesic, J. A. Catalytic conversion of biomass to biofuels. *Green Chemistry*, *12* (2010) 1493–1513.
2. Stöcker, M. Biofuels and biomass-to-liquid fuels in the biorefinery: Catalytic conversion of lignocellulosic biomass using porous materials. *Angewandte Chemie International Edition*, *47* (2008) 9200–9211.

3. Ragauskas, A. J., Williams, C. K., Davison, B. H., Britovsek, G., Cairney, J., Eckert, C. A., and Tschaplinski, T. The path forward for biofuels. *Science*, *311* (2006) 484–489.

4. Perego, C., and Bosetti, A. Biomass to fuels: The role of zeolite and mesoporous materials. *Microporous and Mesoporous Materials*, *144* (2011) 28–39.

5. Cascone, R. Biofuels: What is beyond ethanol and biodiesel? *Hydrocarbon Processing*, *86* (2007) 95–109.

6. Zhu, J. Y., and Zhuang, X. S. Conceptual net energy output for biofuel production from lignocellulosic biomass through biorefining. *Progress in Energy and Combustion Science*, *38* (2012) 583–598.

7. Barakat, A., De Vries, H., and Rouau, X. Dry fractionation process as an important step in current and future lignocellulose biorefineries: A review. *Bioresource Technology*, (2013) doi: http://dx.doi.org/10.1016/j.biortech.2013.01.169.

8. Jacquet, N., Vanderghem, C., Danthine, S., Quiévy, N., Blecker, C., Devaux, J., and Paquot, M. Influence of steam explosion on physicochemical properties and hydrolysis rate of pure cellulose fibers. *Bioresource Technology*, *121* (2012) 221–227.

9. Lavoine, N., Desloges, I., Dufresne, A., and Bras, J. Microfibrillated cellulose—Its barrier properties and applications in cellulosic materials: A review. *Carbohydrate Polymers*, *90* (2012) 735–764.

10. Sanders, J. P. M., Clark, J. H., Harmsen, G. J., Heeres, H. J., Heijnen, J. J., Kersten, S. R. A., van Swaaij, W. P. M., and Moulijn, J. A. Chemical engineering and processing: Process intensification process intensification in the future production of base chemicals from biomass. *Chemical Engineering & Processing: Process Intensification*, *51* (2012) 117–136.

11. Huber, G. W., and Corma, A. Synergies between bio- and oil refineries for the production of fuels from biomass. *Angewandte Chemie International Edition*, *46* (2007) 7184–7201.

12. Kaltschmitt, M., Andrée, U., and Majer, S. Koraffination von Pflanzenöl in Mineralölraffinerien—Möglichkeiten und Grenzen. *Erdöl Erdgas Kohle*, *126* (2010) 203–210.

13. Gallezot, P. Catalytic conversion of biomass: Challenges and issues. *ChemSusChem*, *1* (2008) 734–737.

14. Demirbas, A. Mechanisms of liquefaction and pyrolysis reactions of biomass. *Energy Conversion & Management*, *41* (2000) 633–646.

15. Shabtai, J. S., and Zmierczak, W. W. (2001). Process for conversion of lignin to reformulated, partially oxygenated gasoline. US Patent 6,172,272 B1.

16. Bridgwater, A. V., and Peacocke, G. V. C. Fast pyrolysis processes for biomass. *Renewable and Sustainable Energy Reviews*, *4* (2000) 1–73.

17. Yaman, S. Pyrolysis of biomass to produce fuels and chemical feedstocks. *Energy Conversion and Management*, *45* (2004) 651–671.

18. Demirbas, M. F., and Balat, M. Recent advances on the production and utilization trends of bio-fuels: A global perspective. *Energy Conversion and Management*, *47* (2006) 2371–2381.

19. Huber, G. W., Iborra, S., and Corma, A. Synthesis of transportation fuels from biomass: Chemistry, catalysts, and engineering. *Chemical Reviews*, *106* (2006) 4044–4098.

20. Triantafyllidis, K. S., Iliopoulou, E. F., Antonakou, E. V., Lappas, A. A., Wang, H., and Pinnavaia, T. J. Hydrothermally stable mesoporous aluminosilicates (MSU-S) assembled from zeolite seeds as catalysts for biomass pyrolysis. *Microporous and Mesoporous Materials*, *99* (2007) 132–139.

21. Iliopoulou, E. F., Antonakou, E. V., Karakoulia, S. A., Vasalos, I. A., Lappas, A. A., and Triantafyllidis, K. S. Catalytic conversion of biomass pyrolysis products by mesoporous materials: Effect of steam stability and acidity of Al-MCM-41 catalysts. *Chemical Engineering Journal*, *134* (2007) 51–57.

22. Kersten, S. R. A., van Swaaij, W. P. M., Lefferts, L., and Seshan, K. (2007). Options for catalysis in the thermochemical conversion of biomass into fuels. In *Catalysis for Renewables*, Eds. Centi, G. and van Santen, R. A. (pp. 119–145). Wiley-VCH Verlag GmbH, Weinheim, Germany.

23. Sie, S. T. Past, present and future role of microporous catalysts in the petroleum industry. *Studies in Surface Science and Catalysis*, *85* (1994) 587–631.

24. De, S., Dutta, S., and Saha, B. One-pot conversions of lignocellulosic and algal biomass into liquid fuels. *ChemSusChem*, *5* (2012) 1826–1833.

25. Kim, S. M., Lee, M. E., Choi, J.-W., Suh, D. J., and Suh, Y.-W. Conversion of biomass-derived butanal into gasoline-range branched hydrocarbon over Pd-supported catalysts. *Catalysis Communications*, *16* (2011) 108–113.

26. Hammerschmidt, A., Boukis, N., Hauer, E., Galla, U., Dinjus, E., Hitzmann, B., and Nygaard, S. D. Catalytic conversion of waste biomass by hydrothermal treatment. *Fuel*, *90* (2011) 555–562.

27. Dhainaut, J., Dacquin, J.-P., Lee, A. F., and Wilson, K. Hierarchical macroporous-mesoporous SBA-15 sulfonic acid catalysts for biodiesel synthesis. *Green Chemistry*, *12* (2010) 296–303.

28. Werpy, T., and Petersen, G. (2004). Top Value Added Chemicals from Biomass Volume I—Results of Screening for Potential Candidates from Sugars and Synthesis Gas Top Value Added Chemicals from Biomass Volume I: Results of Screening for Potential Candidates. *Technical Report National Renewable Energy Laboratory*. Washington, DC, USA.

29. Bozell, J. J., and Petersen, G. R. Technology development for the production of biobased products from biorefinery carbohydrates—The US Department of Energy's "Top 10" revisited. *Green Chemistry*, *12* (2010) 539.

30. Kobiro, K., Sumoto, K., Okimoto, Y., and Wang, P. Saccharides as new hydrogen sources for one-pot and single-step reduction of alcohols and catalytic hydrogenation of olefins in supercritical water. *The Journal of Supercritical Fluids*, *77* (2013) 63–69.

31. Yamaguchi, D., Kitano, M., Suganuma, S., Nakajima, K., Kato, H., and Hara, M. Hydrolysis of cellulose by a solid acid catalyst under optimal reaction conditions. *Journal of Physical Chemistry*, *113* (2009) 3181–3188.

32. Kobayashi, H., Komanoya, T., Hara, K., and Fukuoka, A. Water-tolerant mesoporous-carbon-supported ruthenium catalysts for the hydrolysis of cellulose to glucose. *ChemSusChem*, *3* (2010) 440–443.

33. Yamaguchi, D., and Hara, M. Starch saccharification by carbon-based solid acid catalyst. *Solid State Sciences*, *12* (2010) 1018–1023.

34. Farone, W. A., and Cuzens, J. E. (1998). Method of producing sugars using strong acid hydrolysis. US Patent 5,726,046.

35. Eyal, A., Vitner, A., and Mali, R. (2011). Methods for the separation of HCl from chloride salt and compositions produced thereby. US Patent Application WO 2011/095977 A1.

36. Morales-delaRosa, S., Campos-Martin, J. M., and Fierro, J. L. G. High glucose yields from the hydrolysis of cellulose dissolved in ionic liquids. *Chemical Engineering Journal*, *181/182* (2012) 538–541.

37. Gliozzi, G., Innorta, A., Mancini, A., Bortolo, R., Perego, C., Ricci, M., and Cavani, F. Zr/P/O catalyst for the direct acid chemo-hydrolysis of non-pretreated microcrystalline cellulose and softwood sawdust. *Applied Catalysis B: Environmental*, (2013) doi: http://dx.doi.org/10.1016/j.apcatb.2012.12.035.

38. Moe, S. T., Janga, K. K., Hertzberg, T., Hägg, M., and Dyrset, N. Saccharification of lignocellulosic biomass for biofuel and biorefinery applications. A renaissance for the concentrated acid hydrolysis? *Energy Procedia*, *20*(1876) (2012) 50–58.

39. Tschentscher, R., Menegassi de Almeida, R., Carucci, J. R. H., Van den Bergh, J., and Moulijn, J. A. (2013). Process for recovering saccharides from cellulose hydrolysis reaction mixture. United States Patent Application WO/2013/110814 A1.

40. Kitano, M., Yamaguchi, D., Suganuma, S., Nakajima, K., Kato, H., Hayashi, S., and Hara, M. Adsorption-enhanced hydrolysis of beta-1,4-glucan on graphene-based amorphous carbon bearing SO_3H, COOH, and OH groups. *Langmuir: The ACS Journal of Surfaces and Colloids, 25* (2009) 5068–5075.

41. Onda, A., Ochi, T., and Yanagisawa, K. Selective hydrolysis of cellulose into glucose over solid acid catalysts. *Green Chemistry, 10* (2008) 1033–1037.

42. Geboers, J. A., Van de Vyver, S., Ooms, R., Op de Beeck, B., Jacobs, P. A., and Sels, B. F. Chemocatalytic conversion of cellulose: Opportunities, advances and pitfalls. *Catalysis Science & Technology, 1* (2011) 714.

43. Onda, A. Selective hydrolysis of cellulose and polysaccharides into sugars by catalytic hydrothermal method using sulfonated activated-carbon. *Journal of the Japan Petroleum Institute, 55* (2012) 73–86.

44. Degirmenci, V., Uner, D., Cinlar, B., Shanks, B. H., Yilmaz, A., Santen, R. A., and Hensen, E. J. M. Sulfated zirconia modified SBA-15 catalysts for cellobiose hydrolysis. *Catalysis Letters, 141* (2010) 33–42.

45. Schüth, F., and Rinaldi, R. (2011). Catalytic conversion of cellulose. In *Proceedings of Catbior Conference Malaga*, Ed. Jones, C. W., Olaris Media, Malaga, (pp. 47–48).

46. Hilgert, J., Meine, N., Rinaldi, R., and Schüth, F. Mechanocatalytic depolymerization of cellulose combined with hydrogenolysis as a highly efficient pathway to sugar alcohols. *Energy & Environmental Science, 6* (2013) 92.

47. Kovalenko, G. A., and Perminova, L. V. Immobilization of glucoamylase by adsorption on carbon supports and its application for heterogeneous hydrolysis of dextrin. *Carbohydrate Research, 343* (2008) 1202–1211.

48. Gan, Q., Allen, S. J., and Taylor, G. Kinetic dynamics in heterogeneous enzymatic hydrolysis of cellulose: An overview, an experimental study and mathematical modelling. *Process Biochemistry, 38* (2003) 1003–1008.

49. Van Putten, R.-J., Van Der Waal, J. C., De Jong, E., Rasrendra, C. B., Heeres, H. J., and De Vries, J. G. Hydroxymethylfurfural, a versatile platform chemical made from renewable resources. *Chemical Reviews, 113* (2013) 1499–1597.

50. Dutta, S., De, S., Saha, B., and Alam, M. I. Advances in conversion of hemicellulosic biomass to furfural and upgrading to biofuels. *Catalysis Science & Technology, 2* (2012) 2025–2036.

51. Kazi, F. K., Patel, A. D., Serrano-Ruiz, J. C., Dumesic, J. A., and Anex, R. P. Techno-economic analysis of dimethylfuran (DMF) and hydroxymethylfurfural (HMF) production from pure fructose in catalytic processes. *Chemical Engineering Journal, 169* (2011) 329–338.

52. Román-Leshkov, Y., Barrett, C. J., Liu, Z. Y., and Dumesic, J. A. Production of dimethylfuran for liquid fuels from biomass-derived carbohydrates. *Nature, 447* (2007) 982–985.

53. Carlini, C., Giuttari, M., Maria, A., Galletti, R., Sbrana, G., Armaroli, T., and Busca, G. Selective saccharides dehydration to 5-hydroxymethyl-2-furaldehyde by heterogeneous niobium catalysts. *Applied Catalysis A: General, 183* (1999) 295–302.

54. Xing, R., Subrahmanyam, A. V., Olcay, H., Qi, W., van Walsum, G. P., Pendse, H., and Huber, G. W. Production of jet and diesel fuel range alkanes from waste hemicellulose-derived aqueous solutions. *Green Chemistry, 12* (2010) 1933–1946.

55. Deng, L., Li, J., Lai, D.-M., Fu, Y., and Guo, Q.-X. Catalytic conversion of biomass-derived carbohydrates into gamma-valerolactone without using an external H_2 supply. *Angewandte Chemie International Edition, 48* (2009) 6529–6532.

56. Raspolli Galletti, A. M., Antonetti, C., Ribechini, E., Colombini, M. P., Nassi o Di Nasso, N., and Bonari, E. From giant reed to levulinic acid and gamma-valerolactone: A high yield catalytic route to valeric biofuels. *Applied Energy, 102* (2013) 157–162.

57. Weingarten, R., Kim, Y. T., Tompsett, G. A., Fernández, A., Han, K. S., Hagaman, E. W., Connor Jr. W. C., Dumesic, J. A., and Huber, G. W. Conversion of glucose into levulinic acid with solid metal(IV) phosphate catalysts. *Journal of Catalysis, 304* (2013) 123–134.

58. Neves, P., Lima, S., Pillinger, M., Rocha, S. M., Rocha, J., and Valente, A. A. Conversion of furfuryl alcohol to ethyl levulinate using porous aluminosilicate acid catalysts. *Catalysis Today*, (2013) doi: http://dx.doi.org/10.1016/j.cattod.2013.04.035.

59. Lange, J.-P., Price, R., Ayoub, P. M., Louis, J., Petrus, L., Clarke, L., and Gosselink, H. Valeric biofuels: A platform of cellulosic transportation fuels. *Angewandte Chemie International Edition, 49* (2010) 4479–4483.

60. Hayes, D. J., Ross, P. J., Hayes, P. M. H. B., and Fitzpatrick, P. S. (2008). The biofine process: Production of levulinic acid, furfural and formic acid from lignocellulosic feedstocks. In *Biorefineries-Industrial Processes and Products: Status Quo and Future Directions* (eds. B. Kamm, P. R. Gruber, and M. Kamm), Wiley-VCH Verlag GmbH, Weinheim, Germany.

61. Joshi, H., Moser, B. R., Toler, J., Smith, W. F., and Walker, T. Ethyl levulinate: A potential bio-based diluent for biodiesel which improves cold flow properties. *Biomass and Bioenergy, 35* (2011) 3262–3266.

62. Elliot, D. C., and Frye, J. G. (1999). Hydrogenated 5-carbon compound and method of making. US Patent 5,883,266.

63. Upare, P. P., Lee, J.-M., Hwang, Y. K., Hwang, D. W., Lee, J.-H., Halligudi, S. B., and Chang, J.-S. Direct hydrocyclization of biomass-derived levulinic acid to 2-methyltetrahydrofuran over nanocomposite copper/silica catalysts. *ChemSusChem, 4* (2011) 1749–1752.

64. Bond, J. Q., Alonso, D. M., Wang, D., West, R. M., and Dumesic, J. A. Integrated catalytic conversion of gamma-valeroacetone to liquid alkenes for transportation fuels. *Science, 327* (2010) 1110–1114.

65. Menegassi de Almeida, R., Nederlof, C., Li, J., Moulijn, J. A., Connor, P. O., and Makkee, M. Cellulosic conversion to isosorbide in a molten salt hydrate media. *ChemSusChem, 3* (2007) 325–328.

66. Fuertes, P., and Wiatz, V. (2008). Method for the etherification of isosorbide in a viscous medium. US Patent Application WO 2009/056722 A2.

67. Tundo, P., Aricò, F., Gauthier, G., Rossi, L., Rosamilia, A. E., Bevinakatti, H. S., and Newman, C. P. Green synthesis of dimethyl isosorbide. *ChemSusChem, 3* (2010) 566–570.

68. Rose, M., Thenert, K., Pfützenreuter, R., and Palkovits, R. Heterogeneously catalysed production of isosorbide tert-butyl ethers. *Catalysis Science & Technology, 3* (2013) 938.

69. Muurinen, E. (2000). *Organosolv pulping—A review and distillation study related to peroxyacid*. Dissertation, University of Oulu, Oulu, Finland.

70. Guay, D. F., and Singsaas, E. I. (2012). Lignin-solvent fuel and method and apparatus for making therof. US Patent Application 2012/0329146 A1.

71. Wang, X., Richter, U., and Rinaldi, R. (2011). Hydrogenolysis of lignin towards energy dense biofuels. In *Proceedings of Catbior Conference Malaga*, Ed. Jones, C. W., Olaris Media, Malaga, (pp. 95–99).

72. Torr, K. M., van de Pas, D. J., Cazeils, E., and Suckling, I. D. Mild hydrogenolysis of in-situ and isolated *Pinus radiata* lignins. *Bioresource Technology, 102* (2011) 7608–7611.

73. Yoshikawa, T., Yagi, T., Shinohara, S., Fukunaga, T., Nakasaka, Y., Tago, T., and Masuda, T. Production of phenols from lignin via depolymerization and catalytic cracking. *Fuel Processing Technology, 108* (2013) 69–75.

74. Yoshikawa, T., Shinohara, S., Yagi, T., Ryumon, N., Nakasaka, Y., Tago, T., and Masuda, T. Production of phenols from lignin-derived slurry liquid using iron oxide catalyst. *Applied Catalysis B: Environmental* (2013) doi: http://dx.doi.org/doi:10.1016/j.apcatb.2013.03.010.

9 Catalytic Pyrolysis of Lignocellulosic Biomass

K. Seshan

CONTENTS

9.1 Introduction 253
9.2 Thermal Pyrolysis..255
9.3 Catalytic Pyrolysis...262
    9.3.1 Need and Scope for Catalysts to Upgrade Bio-Oil...........................262
    9.3.2 Selective Deoxygenation..262
    9.3.3 Solid Acid Catalysts...264
    9.3.4 Alkali Metal–Based Catalysts...266
    9.3.5 Case Study: Catalytic Pyrolysis over Na-Based Catalysts 266
        9.3.5.1 State of the Catalyst 272
9.4 Conclusions...276
Acknowledgments...276
References..276

This chapter reports on the latest developments of biomass catalytic pyrolysis for the production of fuels. The primary focus is on the role of catalysts in the process, namely, their influence in the liquefaction of lignocellulosic biomass.

9.1 INTRODUCTION

Biofuels have gained considerable interest in recent years because of the high crude oil prices, energy security concerns, and potential climate change consequences over the utilization of fossil fuel.[1] This chapter deals with the application of catalysis in the production of liquid biofuels via pyrolysis liquefaction of lignocellulosic biomass. Table 9.1 lists proposed biomass-derived fossil fuel equivalents that can be made via thermal, chemical, and catalytic conversions.[1b]

The Natural Resources Defense Council, USA, has projected that an aggressive approach to produce lignocellulosic biofuels in the US could produce the equivalent of 7.9 million barrels oil in 2050, which is more than 50% of the current US oil use in the transport sector.[2,3] Biofuels should not interfere with the food chain in a direct way. Next to ethical objections to the use of food for fuel, by using exclusively lignocellulosic "waste" biomass (e.g., forestry and agricultural residues, food waste, and energy crops), large quantities of biofuels can be introduced, for example, the EU

TABLE 9.1

Popular Fossil Fuels and Possible Alternatives/Blends That Can Be Derived from Lignocellulosic Biomass

Fossil	Biomass Alternative (100% bio-based replacement and/or blending component for fossil fuel)
Gasoline	Alcohols (C1–C4)
	MTHF (methyltetrahydrofuran)
	MTBE (methyl tertiary-butyl ether)
	Aromatics
	Alkanes
	Levulinic acid esters
	Deoxygenated and refined primary bioliquids
Diesel	Levulinic acid esters
	Levulinic acid dimer esters
	5-HMF (5-hydroxymethylfurfural) esters
	Alkanes
	DME (dimethyl ether)
	Ethanol
	FAEE (fatty acid ethyl ester)
	FAME (fatty acid methyl ester)
	Fischer–Tropsch liquids (from bio-based synthesis gas)
	Deoxygenated and refined primary bioliquids
Kerosene	Fischer–Tropsch liquids (from bio-based synthesis gas)

target of 20% biofuels by 2020. The accumulated worldwide residues from agriculture are estimated to be equivalent to half the crude oil consumption.

In thermochemical conversion of biomass, temperature is a key parameter. At lower temperatures (<300°C) only catalytic processes (e.g., acid catalyzed hydrolysis) are possible. Conversion to a variety of oxygenates such as acids (e.g., levulinic acid), heterocyclic hydrocarbons (furans), and alcohols *via* sugar conversions with promising yields has already been shown to be possible. Lignin is not or hardly decomposed in this temperature regime. In these low temperature processes, most use is made of the composition of biomass by keeping much of the functionality of the sugar building blocks intact. However, pretreatment is required to make the fibers accessible (by, e.g., steam explosion), because native lignocelluloses are inert for hydrolysis at these temperatures. Next to that, the reactions are slow and require often homogeneous catalysts.

In the mid temperature range (300°C < T < 700°C), complete conversion of lignocellulosic biomass is possible. Liquefaction processes (pyrolysis and hydrothermal liquefaction) are significant in this region and yield a multicomponent liquid product containing oxygenate components, gases, and a solid that consists of the remainders of the fiber structure of the feedstock. At higher temperatures (T > 700°C), gasification becomes dominant. Gasification is uncontrolled and in the absence of a catalyst, up to 1300°C, methane is always produced. In the presence of catalysts, steam reforming gasification to syngas (CO + H$_2$) mixtures is possible. Incorporation of a water–gas shift (WGS) reaction (CO + H$_2$O → CO$_2$ + H$_2$) provides flexibility when hydrogen is the required product. Fischer-Tropsch process (FT) conversion of syngas [nCO + 2nH$_2$ → –(CH$_2$)n– + nH$_2$O] allows conversion to liquid fuels such as diesel.

Catalysis will play an important role in the production of biofuels just as catalysis plays a major role in the conversion from fossil feeds to fuels currently.[1,4] Current attention to biofuels (besides ethanol and biodiesel) comes via two important catalytic routes, namely, gasification or liquefaction of the biomass or its components, celluloses or lignin.[5] In the area of liquefaction, most attention has been made to cellulose/sugars conversion to gasoline compatible components via catalytic dehydration, hydrogenation, for example, hydroxyl methyl furan (HMF) and butanol. More details of this can be found in Chapter 8. Other liquefaction processes use the complete lignocellulose as feedstock. Liquefaction of biomass has the following advantages.[1b,6] It allows the decoupling of the locations where biomass is available (e.g., remote, rural areas) and where the bio-based products can be treated at large scale (refineries). Liquefaction allows the use of waste biomass such as timber/forestry wastes (e.g., sawdust) and waste from food production (e.g., rice husk, bagasse). It homogenizes bulky, diverse biomass streams into a liquid that is easier to store, transport, and process. Minerals and metals remain mostly in the solid by-product of the liquefaction process and can thus be recovered at the biomass production site and returned to the soil (closing of the nutrient cycle). As a result, bioliquids contain significantly lower levels of mineral and metal components than the solid biomass feedstock it is produced from, which is beneficial for the subsequent catalytic processing of the oil.

One of the two significant liquefaction processes, hydro thermal upgrading (HTU), coverts all types of lignocellulosic biomass to a biocrude at 300°C–350°C, 120–180 bar, in 5–20 minutes in the presence of water. It produces about 45 wt% biocrude with lower oxygen content than the feedstock biomass.[7] The process was developed in the last decade but has not made much progress. The major hurdles in the development of the HTU process have been (1) cost-effectiveness of the highly concentrated slurries of biomass particles, (2) dealing with fouling and blocking problems caused by the solvent-insoluble hydrophobic product, and (3) product separation. Pyrolysis liquefaction of biomass has, on the other hand, made significant progress in the last years. Most of the earlier work related to thermal pyrolysis in the absence of catalysts. In the last years, application of catalysts in the process for upgrading of the produced bio-oil, *in situ*, during pyrolysis has been of much interest.[8]

In Section 9.2, a noncatalytic, thermal pyrolysis of lignocellulose is outlined. Current issues with the bio-oil made in the process are elaborated. This allows one to understand the need and scope for the development of a pyrolysis process incorporating catalysts. This is outlined later in Sections 9.3.1 and 9.3.2.

9.2   THERMAL PYROLYSIS

One of the promising techniques to generate liquid fuels from lignocellulosic biomass is fast pyrolysis, in which the feedstock is heated rapidly (~100°C/s) in an inert atmosphere (1 bar, N$_2$) to a moderate temperature (400°C–550°C) with short residence time (<2 s) and rapid quenching of the formed vapor. This process has drawn much interest

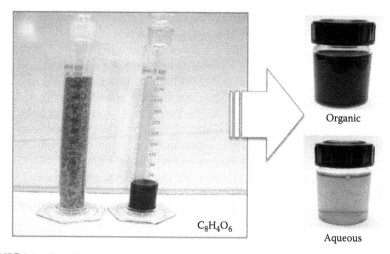

C₈H₄O₆

Organic

Aqueous

FIGURE 9.1   Densification of lignocellulose into bio-oil by pyrolysis. Both the biomass and the bio-oil have the same nominal composition indicated. The bio-oil is a mixture of organic phase and water.

because it produces high yields of a liquid product, named bio-oil, which can contain up to 70% of the energy of the biomass feed.[6,9] The pyrolysis process allows concentrating the biomass into an easier to transport and handle liquid, which is a mixture of organic and aqueous phases (see Figure 9.1). The organic phase, which is the targeted biocrude or bio-oil, is the feedstock for (1) biofuels via dedicated, chemical/catalytic conversions or (2) coprocessing with fossil crude oil in a regular refinery.[10] It contains an aqueous phase where soluble organic components such as acids and alcohols are present. The two phases separate when the water content of the oil is above ~40 wt%.

The literature on biomass fast pyrolysis is quite extensive, and excellent research, technology, and application reviews are available.[1b,6,11] Fast pyrolysis is entering commercial operation; for example, BTG (the Netherlands), ENSYN (Canada), DYNAMOTIVE (Canada), and KIOR (USA) are operating pilot, demonstration, or commercial plants; with typical capacities of one to several tons per hour.[12] Price of pyrolysis oil depends on the feedstock and is reported in the range of 100–170 €/Ton.

A very basic conceptual scheme of the biomass pyrolysis coupled with fuels production is shown in Figure 9.2. This typical process scheme for green fuels via pyrolysis of lignocellulosic biomass is very similar to a current commercial fluid catalytic cracking (FCC) process and provides scope for integration with it. Biomass pyrolysis results in three fractions, bio-oil, gas, and char. In the proposed scheme, solid heat carrier (sand) and catalyst transport char to a regenerator where it is combusted and catalyst regenerated. Heat from regenerator is integrated to the endothermic pyrolysis process similar to FCC scheme in a refinery. The gas stream also contains energy (CO, H_2, CH_4/higher hydrocarbons) and heat can be integrated into the process via combustion.

The characteristics of the so produced bio-oil are compared in Table 9.2 with that of a fuel oil and the biomass from which it is produced.[13] The composition of the bio-oil and the parent biomass from which it is made is rather similar. The biggest

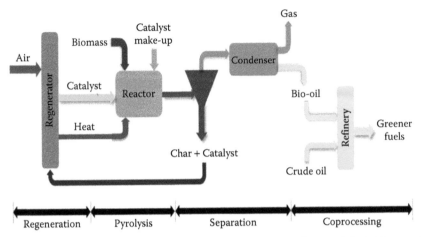

FIGURE 9.2 Conceptual pyrolysis scheme for the conversion of lignocellulosic biomass. In the case of thermal pyrolysis, sand is used to transport char and heat. In the case of catalytic pyrolysis, solid catalyst is mixed with sand. (Courtesy Seshan, K. et al., *Biomass Bioenergy*, 2013. With permission.)

TABLE 9.2
Comparison of Bio-Oil with Biomass and Fossil Fuel-Oil

Materials	C*	H*	O*	Energy Density (MJ/kg)	pH	Water Content*
Biomass	48.74	5.80	45.46	16	–	5.6
Bio-oil	52.11	5.70	42.34	19	2.4	30.5
Fuel-oil	85.30	11.47	1.05	40	5.7	0.1

* Data in mass fraction (%).

technical challenge for such a green refinery is the high oxygen content of the bio-oil, which restricts its applications. First, the heating value is low due to the high oxygen content, biocrude has energy content of 19 MJ/kg, compared to that of a fossil crude of 30 MJ/kg. This already limits its application and requires upgrading. Second, the oil is too acidic and hence corrosive. Typical thermal pyrolysis oils have a pH of ~2.4. Third, the high oxygen content makes the bio-oil polar and prevents miscibility with fossil hydrocarbon fuels. Use of bio-oil as a source of fossil compatible fuels requires upgrading which should address mainly the issue of high oxygen content, which in turn causes the problem characteristics of the oil.

Analysis of pyrolysis bio-oil is complex. A typical analysis via gas chromatography-mass spectrometry (GC-MS) is shown in Figure 9.3. The chromatogram contains a variety of components that are decomposition products of cellulose, hemicelluloses, and lignin.[13]

Since bio-oil is a mixture of hundreds of different organic components, it is reasonable to classify them into different groups based on their chemical functionality.

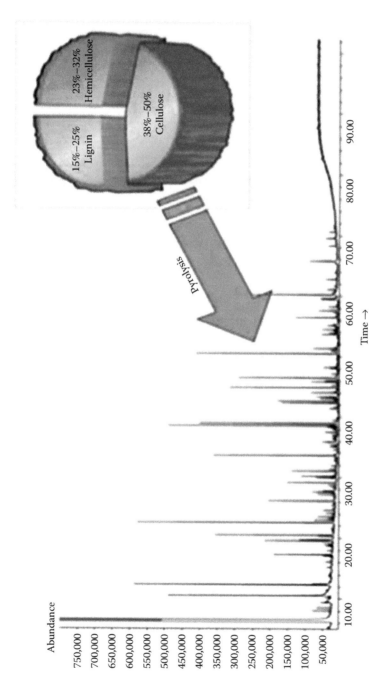

FIGURE 9.3 GC–MS analysis of typical thermal pyrolysis oil shown in Table 9.2.

The rationality of such a classification is that chemical functionalities determine its properties and the molecules in each group have similar characteristics. This also allows judging the quality and at the same time simplifying analysis of bio-oil. In the case of fossil hydrocarbons, a similar method is often used, namely, PIONA (paraffins, iso-paraffins, olefins, naphthenes, and aromatics) analysis.

The classification of the components of bio-oil is shown in Figure 9.4. In this figure, the vertical axis shows the relative proportions of different chemical functional groups in the bio-oil calculated by summing the total ion chromatogram (TIC) area percentages of all the compounds belonging to these groups.[13] This method, commonly used by other authors, generates semiquantitative results.[14] It can be used to compare the concentrations of a certain component in bio-oil obtained under different conditions. The best available GC-MS methods account for only about 80% of the components in the bio-oil. This has the disadvantage that exact catalytic chemistry is difficult to be established. However, relative changes of identifiable components and establishing which component of the biomass they originate from already allows for appreciable progress in catalyst development to upgrade them. As can be seen from Figure 9.4, bio-oil contains a variety of carboxylic acids, for example, formic and acetic acids, mainly formed by the decomposition of cellulose and hemicellulose fractions of biomass.[13] These are the main components that cause acidity of the bio-oil.[13] Phenols that are formed by the decomposition of lignin also contribute to acidity but to a much lesser extent. Hydrocarbons and furans (formed from holocelluloses) are the required components due to their high-energy content.[8]

Carbonyl compounds such as aldehydes and ketones, present in the bio-oil, have a tendency to undergo condensation reactions causing viscosity increase and make the oil unstable.[13] Figure 9.5 shows the gel permeation chromatography (GPC) of the bio-oil revealing the increase in higher molecular weight components after storage at room temperature for 6 months. In the corresponding samples, the decrease in phenol and aldehyde contents and increase in water content (shown in the inset of Figure 9.5) clearly indicate occurrence of aldol-type condensation reactions.

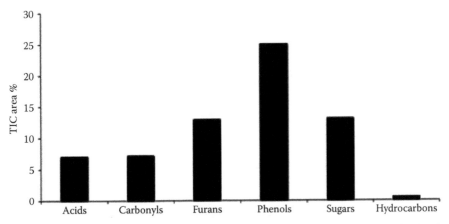

FIGURE 9.4   Relative amounts of components in the bio-oil grouped based on their chemical functionality.

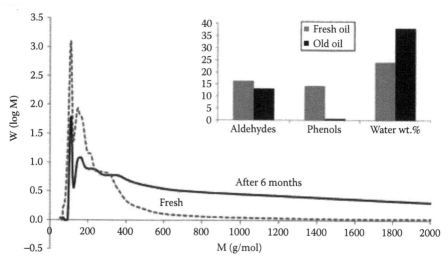

FIGURE 9.5 Gel permeation chromatography of a fresh bio-oil and a sample stored at room temperature for 6 months. The inset shows the changes in aldehydes, phenols, and water contents on storage.

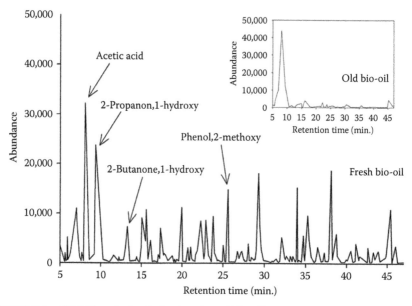

FIGURE 9.6 GC-MS analysis of a fresh and aged bio-oil as shown in Figure 9.5.

Further, these reactive oxygenate, such as aldehydes and ketones, tend to oligomerize readily causing chemical instability. This can be seen in the GC-MS analysis of fresh and aged bio-oil (same samples as in Figure 9.6), which shows that components with carbonyl functional group disappear with time. This instability causes pyrolysis oil to form coke/char when heating it. Another adverse property of pyrolysis oil is the large fraction of heavy components (MW > 1000 g/mol, also present

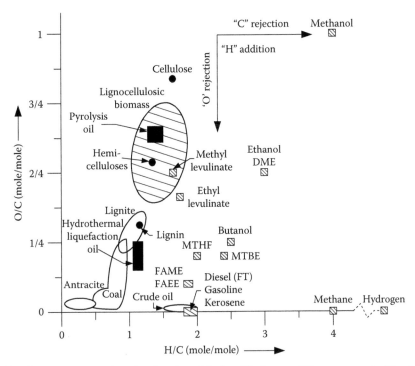

FIGURE 9.7 Van Krevelen classification of fuel and fuel compatible components. The classification is based on the C, H, and O contents.

in fresh bio-oil) that causes problems in downstream conversion units. The alkalis, metals, and heteroatoms like Cl, S, and N still present in pyrolysis oil are susceptible to cause problems in catalytic upgrading steps.

Figure 9.7 classifies the current and future fuel/fuel compatible components based on their composition (C,H,O contents) as in a van Krevelen diagram. In this figure, hydrocarbon fuels are on the x-axis, and the standard upgrading of crude oil in a refinery via coking (carbon removal) or hydro-processing (hydrogen addition) occurs from left to right. In the case of biofuels, they move upward along the y-axis depending on their oxygen content. Lignocelluloses and the bio-oil have high oxygen content as can be seen from Figure 9.7. Thus in addition to carbon rejection and hydrogen addition, as in the case of fossil crude, biocrude upgrading also requires oxygen rejection as shown in Figure 9.7.

In general, chemical/catalytic routes for improving the properties and applicability of bio-oil require decreasing its oxygen content that is deoxygenation. Oxygen removal can be attempted at pre-, *in situ*, or postpyrolysis stages and the options to this, mainly, are deoxygenation,[15] hydrodeoxygenation (HDO),[15b,16] and esterification.[16c,17] Deoxygenation is an easier and promising route[15,16] as it does not require the use of other chemicals, for example, alternative HDO process requires hydrogen that is expensive and not easily available, esterification requires alcohols.

9.3 CATALYTIC PYROLYSIS

9.3.1 NEED AND SCOPE FOR CATALYSTS TO UPGRADE BIO-OIL

A conceptual scheme for the upgrading of bio-oil to fossil fuel compatible components is shown in Figure 9.8. Fossil fuels have a typical composition of $[CH_2]$. Most of the oxygen-containing, non aromatic biofuels can be represented as $[xCH_2]$ plus water. This water content (caused by partial oxidation, e.g., ethanol can be considered as ethylene plus water, elemental composition wise) lowers its energy density compared to hydrocarbon components. For biomass and bio-oil, the H/C ratio is in the region of ~0.5. This is similar to fossil crude oil, the difference being that biomass/bio-oil also contains lots of oxygen in its composition (up to 40 wt%).

In Figure 9.8, as mentioned earlier, refinery upgrading of crude oil occurs along the x-axis via coking or hydrogenation. In order to make a fossil fuel compatible bio-oil, a biorefinery requires removal of carbon and oxygen, that is, decarboxylation and/or decarbonylation, instead of coking. Dehydration via hydrogenation (HDO) is another option to convert the oxygen content to water, an expensive route. Complete deoxygenation even in the absence of hydrogenation is possible ($C_6H_8O_4 \rightarrow 4.6CH_{1.2} + 1.4CO_2 + 1.2H_2O$) and results in an aromatic hydrocarbon mixture (H/C ~1–1.2) due to the low hydrogen content of the starting biomass (H/C ~1.3).[18] In the case of complete deoxygenation, as shown above, the organic yield is about 42% (mass fraction), and this corresponds to 50% energy recovery from the biomass feedstock. Incomplete deoxygenation, that is, retaining some of the oxygen in the organic fraction, will help to enhance the liquid yields; however, this is an option only if the resulting product has properties that are compatible with fossil fuel/fuel additives.

9.3.2 SELECTIVE DEOXYGENATION

In general, deoxygenation of biomass results in elimination of oxygen as carbon oxides, water and, depending on the extent of oxygen removal, a mixture of oxygen

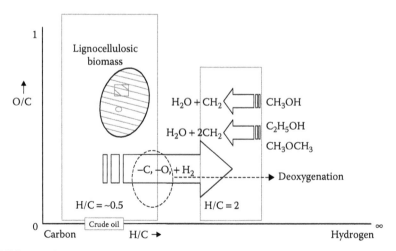

FIGURE 9.8 Conceptual scheme for upgrading biomass/bio-oil to fossil compatible fuels.

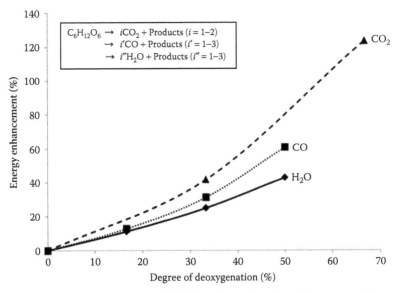

FIGURE 9.9 Energy enhancement of the bio-oil as a function of the extent of deoxygenation for three different routes, namely, decarbonylation, decarboxylation, and dehydration.

containing organic aliphatic/aromatic molecules, such as acids, aldehydes, and alcohols, will be formed. Figure 9.9 shows the influence of three different routes for deoxygenation in the case of a typical sugar monomer, *by* the elimination of (1) CO_2 via decarboxylation, (2) CO via decarbonylation, and (3) H_2O via dehydration. The Figure 9.9 shows that elimination of oxygen by forming carbon dioxide allows the bio-oil to retain maximum energy of the feed biomass. This is because the efficiency of deoxygenation is optimum as only one equivalent of carbon is lost per two equivalents of oxygen. Formation of water decreases the hydrogen content of the deoxygenated product and hence the least desirable. Decarbonylation is a better compromise, if not avoidable, than dehydration to retain maximum energy content of the feed biomass in the deoxygenated product.

This implies that catalysts for pyrolysis in addition to being active for deoxygenation should also allow the selective deoxygenation via formation of CO_2. This is a key design criterion for the development of catalysts for pyrolysis upgrading of lignocelluloses. In order to provide a more fuel compatible product, deoxygenation should also attempt to remove harmful oxygenates such as carboxylic acids (to lower acidity), aldehydes/ketones that cause stability problems, and retaining other oxygenates which have higher energy content.[13,19] Catalyst design should attempt to optimize such selective deoxygenation.

Typically, a heterogeneous solid catalyst would be the optimal choice considering the temperature window of operation (450°C–550°C), tendency for coke char formation and the necessity of frequent regeneration, and the large scale of the process. Incorporation of a solid catalyst in the pyrolysis process will require very efficient solid–solid (biomass-catalyst) contact. Methods such as mechanical mixing, milling, and impregnation of the catalyst precursor on to the biomass have been attempted.

In practice, biomass vaporizes at the pyrolysis temperatures rapidly, and contact of the vapors (1) with a solid catalyst floating around (Fluid catalyst bed) or (2) placed downstream in the flow (fixed bed) are possible options for a process design.

9.3.3   Solid Acid Catalysts

In situ catalytic pyrolysis is considered as one of the options to overcome the problem characteristics of pyrolysis oil mentioned above, the oxygen content being the most important. To start with, catalysts could play a role in the primary decomposition of biomass and in promoting necessary and preventing unwanted reactions of the products. Most importantly, biomass pyrolysis will be carried out predominantly at remote locations and in a distributed manner [1–10 ton/hour]. Thus, the process and the catalysts should be cheap and simple to use.

Deoxygenation reactions are catalyzed by acids and the most studied are solid acids such as zeolites and clays. Atutxa et al.[20] used a conical spouted–bed reactor containing H-MFI (HZSM-5), and Lapas et al.[21] used H-MFI and H-USY (steam stabilized H-FAU) zeolites in a circulating fluid bed to study catalytic pyrolysis (400°C–500°C). They both observed excessive coke formation on the catalyst, and, compared to noncatalytic pyrolysis, a substantial increase in gaseous products (mainly CO_2 and CO) and water and a corresponding decrease in the organic liquid and char yield. The obtained liquid product was less corrosive and more stable than pyrolysis oil. In general, it could be concluded that the slight improvement of the oil is not worthwhile considering the loss in oil yield.

The influence of H-MFI zeolite catalyst on the pyrolysis of biomass based on rice husks in a fixed-bed system has been investigated.[22] It was found that, in the presence of H-MFI, oxygen in the feedstock was removed mostly as H_2O at lower temperatures (<500°C) and as CO/CO_2 at higher temperatures (>550°C). At the higher temperatures, the bio-oils obtained from catalytic experiments contained significantly higher amounts of single ring and polycyclic aromatic hydrocarbons (PAHs) because of deeper deoxygenation. The evaluation of different commercial catalysts,[23] including H-MFI, FCC catalysts, transitional metals (Fe/Cr), and aluminas (γ, α) for biomass pyrolysis, has been carried out in a fixed-bed reactor.[23] It was concluded from this study that H-MFI is a promising catalyst for the selective production of aromatic hydrocarbons, while transitional metal catalysts lead to the selective production of phenol and light phenolics from biomass feedstocks. Deoxygenation of glycerol or sorbitol, as model compounds representing bio-oil, has been reported over several catalysts [H-MFI, γ-Al_2O_3, H-FAU (USY), SiC, and commercial FCC catalyst],[5] and it was concluded that, at high deoxygenation levels, achieved with H-FAU catalyst, aromatic hydrocarbons (fossil fuel compatible), and coke were the major products.

Huber[24] reported pyrolysis tests of cellulose in the presence of H-MFI. After reaction at 600°C in a specialized reactor, aromatic compounds including naphthalene, ethylbenzene, toluene, and benzene were obtained; by-products included coke, H_2O, CO, and CO_2. Low hydrogen content of the biomass implies that while converting it to hydrocarbon fuels, formation of aromatics is the obvious result. The C_6, C_7, and C_8 aromatics can be blended with the gasoline pool. It was proposed that using raw biomass as feedstock should also yield results similar to pure cellulose once the process is optimized.

Our own studies, reported in Figure 9.10, show the influence of H-FAU based catalysts on the pyrolysis of whole biomass.[13] These experiments were carried out with pinewood. Only marginal influence on the chemical composition of the bio-oil was observed. The bio-oil had an energy density of 22 MJ/kg compared to that from a thermal/noncatalytic experiment of 20 MJ/kg.[13] As can be seen from Figure 9.11, the liquid yields decreased in the presence of the acidic zeolite, and as expected, the

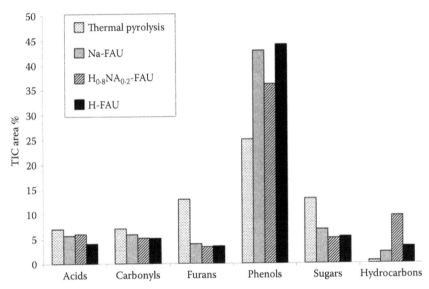

FIGURE 9.10 Influence of zeolite solid acids on the pyrolysis of lignocellulose (pinewood), 500°C, 7 s contact time. Variation in bio-oil components.

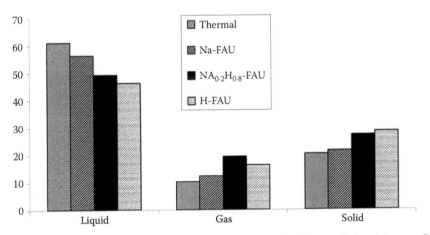

FIGURE 9.11 Influence of zeolite solid acids on the pyrolysis of lignocellulose (pine wood), 500°C, 7 s contact time. Variation in product yields.

Brønsted acidity of the H-FAU caused more cracking and enhanced the yields of gas and char/coke formation.[13] Higher extent of dehydration and water formation was also observed in the presence of H-FAU. Similar results were also observed with H-ZSM5 catalysts.[13] Factors that have to be taken into account while designing catalysts are (1) bulky nature of organic molecules (large molecules up to 2000 g/mol) that escape from the biomass matrix, (2) the need to control the extent of pyrolysis/cracking, and (3) selective scission of bonds, that is, C–C > C–O > C–H to help maximize oxygen removal as CO_2. Thus, texture (pore size, geometry, etc.) and acidity (strength, concentration of acid sites) are the two important parameters for design.

Strong acidity of the zeolites led in general to deep deoxygenation and severe coke formation.[25] To tackle this problem, mesoporous materials with milder acidity such as SBA-15, Al-MSM-41, and Al-MSU-F have been developed.[26] However, the degree of deoxygenation using these catalysts was low compared to the strong Brønsted acid zeolites such as H-FAU and H-MFI.

9.3.4 ALKALI METAL–BASED CATALYSTS

Several researchers have shown that alkali present in the feedstock influences the yields and compositions of the pyrolysis products.[27] For example, it was observed that[28] demineralized corn stover (alkali-free, obtained via washing with dilute nitric acid) resulted in pyrolysis oil that contained ~20 wt% levoglucosan (normally 1–3 wt% anhydrosugar—levoglucosan is present in pyrolysis oil). Alkali metals and alkaline earth metals have attracted attentions as promising catalysts for the upgrading of bio-oil in recent years.[29] Recently, we showed that the presence of sodium ions in the faujasite matrix helped improve the deoxygenation of biomass pyrolysis vapors, reduced the amount of acids, aldehydes/ketones, and enhanced the hydrocarbon content and thus the energy density of the oil.[13] However, the deoxygenation obtained with Na-FAU catalysts was still not optimal. We also found that, with Na_2CO_3, extensive deoxygenation could be achieved during the catalytic pyrolysis of chlorella algae, as indicated by the significant increase in the energy density of bio-oil from 21 MJ/kg (noncatalytic) to 32 MJ/kg (with Na_2CO_3) at 450°C.[9] It has been shown[30] that the incorporation of Na_2CO_3 to white pine during pyrolysis led to a catalytic bio-oil with a significant degree of deoxygenation.

Sooknoi et al.[31] used Cs catalyst in combination with zeolite NaX for deoxygenation of methyl esters and found that Cs plays a crucial role in decarbonylation of methyl esters and when Cs was not present on the zeolite the activity of the catalyst for decarbonylation decreased. Recently, it was shown that supported Na_2CO_3 on γ-Al_2O_3 was more effective on reducing the oxygen content and enhancing the energy content of bio-oil compared to Na_2CO_3.[8] Modification with alkalis, which reduce acidity is an option, especially because alkali metals are also good deoxygenation catalysts.[8]

9.3.5 CASE STUDY: CATALYTIC PYROLYSIS OVER NA-BASED CATALYSTS

In this section, details of pyrolysis experiments with pinewood, carried out in the presence of Na-based catalysts, are taken as a case study to elaborate what can be achieved. Catalysts were placed as fixed bed, downstream to the pyrolysis chamber to

TABLE 9.3
List of Catalysts Used in the Study and Their Characteristics

Sample	Loading (wt%)	Pretreatment Procedures	Surface Area (m²/g)
γ-Al₂O₃	–	Calcined	249
Na₂CO₃	–	Calcined	–ᵃ
Na₂CO₃/γ-Al₂O₃	20	Wet impregnation/ calcined	196

ᵃ Below limit of detection.

allow efficient contact of the biomass vapors with the catalyst. Details of the catalysts are used in this study are reported in Table 9.3. Choice of this system to elaborate is based on the fact that (1) this is one of the most promising systems reported so far, (2) the catalyst is very cheap and easy to prepare, and (3) the results are obtained in the research group of the author and hence details are easily available.

Details of the pyrolysis experiments and the results obtained are shown in Table 9.4. In this table, the total liquid yield (organic + aqueous) of 37 wt% in the catalytic experiment is lower than that of thermal pyrolysis (61 wt%). This is due to (1) the increased loss of carbon via cracking of the pyrolysis vapors on the surface of the catalysts to heterogeneous char and (2) the formation of more gaseous products. Since the catalysts and biomass are physically separated, catalyst is not expected to interfere with the primary decomposition of biomass. Since all pyrolysis

TABLE 9.4
Mass Balance of Catalytic and Noncatalytic Pyrolysis of Biomass

	Yield (wt%)					
	Organic phaseᵃ	Aqueous phaseᵃ	Charᵇ	Cokeᶜ	Gas	Total
Thermal	61ᵈ	–	19	0	13	93
γ-Al₂O₃	7	37	19	9	18	90
Na₂CO₃	13	27	18	17	16	91
20 wt% Na₂CO₃/γ-Al₂O₃	11	26	19	15	23	94

Note: Reaction conditions: $T_{pyrolysis}$ = 500°C, $T_{catalyst\ bed}$ = 500°C, $t_{vapor\ residence}$ = 4 sec, P = 1 bar, $He_{flow\ rate}$ = 70 ml/min.

ᵃ Two different phases of bio-oil which are not miscible in one another, the sum of organic and aqueous phase yields is equal to the total liquid yield.

ᵇ Homogeneous char left in the pyrolysis chamber, r.

ᶜ Hetergeneous char/coke formed on the catalyst.

ᵈ There is no phase separation in thermal pyrolysis. The value shown is the total liquid yield.

experiments were carried out at the same conditions, the yields of char formed in the pyrolysis chamber are the same in all cases. The mass balance shown in Table 9.4 supports this (char yield of 19 wt% for both experiments). The catalysts additionally convert part of the pyrolysis vapor into heterogeneous char (yield of 15 wt%) and more gas (nearly double in yield compared to thermal pyrolysis).

Influence of the catalyst on the deoxygenation of biomass vapors can be inferred from Table 9.4. It can be seen that the formation of CO_2 was favored on the Na-based catalysts. CO yields also increased but to a smaller extent. The formation of water was only slightly enhanced with the catalyst compared to thermal experiment. It is known that both Lewis and Brønsted acid sites exist on the surface of γ-Al_2O_3, which can be responsible for the dehydration of organic compounds in pyrolysis vapors enhancing water formation.[32]

Oxygen content of heterogeneous and homogeneous char formed during pyrolysis is also shown in Table 9.4. Besides the increase in the yields of H_2O and CO_x, the catalyst also influenced the deoxygenation of biomass in the nature of heterogeneous char that was formed. The char in all cases has, within experimental error, identical elemental composition of 78.7 wt% C, 2.8 wt% H, and 18.8 wt% O (ash-free basis). This is also similar to the composition of char obtained from fast pyrolysis as reported in literature,[33] which confirms that the experiments reasonably mimic of actual fast pyrolysis. The catalyst, on the other hand, caused the formation of heterogeneous char, which is different in composition to that of the char formed homogeneously. The oxygen content of the heterogeneous char obtained is 41.5 wt%. The high oxygen content of this heterogeneous char, together with its relatively high yield of 15 wt% (Table 9.4), makes heterogeneous char formation on Na-based catalysts an important deoxygenation route. Significantly, the deoxygenation obtained is very high with only 6.9 wt% [O] from the biomass ending up in bio-oil, the rest distributed in water (43.1%), CO_x (28.6%), heterogeneous char (13.7%), and char (7.7%). Compared to noncatalytic experiment, there is 14 wt% more of [O] from biomass removed via CO_x and 13.7 wt% [O] via heterogeneous char in catalytic upgrading. This shows that deoxygenation via heterogeneous char and gas formation is equally important in achieving high-quality oil.

To summarize, the sodium supported on alumina catalyst showed improved performance compared to noncatalytic test, as shown by (1) only a slight increase in water formation (from 20 wt% to 22 wt%), (2) a significant increase in CO_2 yield (from 5.1 wt% to 12.2 wt%), and (3) a significant deoxygenation via the formation of oxygen-containing heterogeneous char compared to noncatalytic pyrolysis. Results shown in Table 9.3 indicate that the Na_2CO_3/γ-Al_2O_3 is very promising for the deoxygenation of pyrolysis vapors. Oxygen content of the catalytic bio-oil is about 12 wt% compared to 42% in the case of thermal pyrolysis oil. This is a significant level of deoxygenation. Remarkably, the resulting bio-oil (organic phase) has an energy content (37 MJ/kg), which is close to that of traditional fuel oil.[8] This result shows that the 20 wt% Na_2CO_3/γ-Al_2O_3 is a potential catalyst for the upgrading of biomass vapors with reference to energy content.

Among the desired compounds in bio-oil, hydrocarbons are the most valuable fuel component because of their high heating value. It can be seen in Table 9.5 that the

TABLE 9.5
Composition of Bio-Oils Based on Functionalities Analyzed with GC-MS

	Acids	Carbonyls	Furans	Phenols	Sugars	HCs	Others
Thermal	12.0	14.1	12.9	25.0	14.0	0.5	1.7
γ-Al$_2$O$_3$	4.6	22.5	6.8	27.0	0.0	22.0	0.1
Na$_2$CO$_3$	0.0	24.0	5.9	50.3	0.5	2.5	1.4
20 wt% Na$_2$CO$_3$/γ-Al$_2$O$_3$	0.0	40.0	0.4	33.0	0.0	12.4	1.4

catalyst has significantly enhanced the formation of hydrocarbons, which is shown by an increase in the HCs from 0.5% (noncatalytic oil) to 17.8% (catalytic oil). This increase in HC concentration of is correlated to the tremendous increase in energy density of bio-oil, as shown in Table 9.6. Hydrocarbons can be formed during biomass pyrolysis via one of the three possible routes: (1) decarbonylation, decarboxylation, and oligomerization of dehydrated products of sugars, that is, furans into aromatic hydrocarbons; (2) decarboxylation of carboxylic acids into aliphatic hydrocarbons and CO_2; and (3) hydrogenation/hydrogenolysis of phenols into aromatic hydrocarbons.[34] It can be seen from Table 9.5 that both acids and sugars have been completely removed and the content of phenols has increased in the catalytic bio-oil compared to noncatalytic one. This suggests that routes (1) and (2) contribute more than route (3). Moreover, since the aromatic HCs are the dominant among all HCs (Table 9.5), it can be concluded that sugars are the main precursor for HCs in our study.

Pyrolysis of wood results in the formation of carboxylic acids and other acidic components, which make the bio-oil corrosive and negatively affect its applicability as a fuel. It has been shown[19] that, among the various components present in bio-oil, carboxylic acids contribute to the acidity of bio-oil the most (60%–70%). Other components, which contribute to the acidity of bio-oil, are phenols (5%–10%) and sugars (20%). Carboxylic acids and sugars are unwanted components since they greatly contribute to the acidity of bio-oil. As can be seen in Table 9.5, the catalyst

TABLE 9.6
Influence of the Catalysts on the Deoxygenation of Bio-Oil

	CO_2[a]	CO[a]	H_2O[a]	C[b]	H[b]	O[b]	HHV[c]
Thermal	5.1	5.0	21	52	5.7	42	19
γ-Al$_2$O$_3$	8.3	7.8	34	67.8	8.6	22.8	32
Na$_2$CO$_3$	10.4	5.8	24	67.9	8.2	23.3	31
20 wt% Na$_2$CO$_3$/γ-Al$_2$O$_3$	12.2	7.2	24	78.2	8.7	12.3	37

[a] Yield (wt%) based on initial weight of biomass.
[b] Elemental composition (wt%) of the organic phase (except for thermal), dry basis. Oxygen content was calculated by difference.
[c] Higher heating value (MJ/kg) of the organic phase (except for thermal), dry basis.

has completely removed carboxylic acids from bio-oil. The two carboxylic acids originally present in noncatalytic oil, namely, acetic acid and propionic acid with the contents of 6.8% and 5.1%, respectively, were not observed at all in the catalytic oil. The same trend occurs to the sugars in bio-oil. The main sugar component observed in noncatalytic bio-oil is D-allose was also completely removed in catalytic oil.

Figure 9.12 shows the direct relationship between the amount of carboxylic acids present in the bio-oil and its acidity value [total acid number (TAN)]. In the presence of catalyst, carboxylic acid and sugar components in the pyrolysis vapor have been completely removed and hence it can be expected that the bio-oils obtained from this catalyst have lower acidity than noncatalytic oil. In TAN measurement, per gram of catalytic bio-oil needs only 3.8 mg of KOH to titrate compared to 119 mg for noncatalytic oil. The pH measurement correlated well with this result, in which catalytic bio-oil showed to be almost neutral with pH = 6.5, while noncatalytic bio-oil is acidic (pH = 2.6). The oil obtained from this catalyst, therefore, is much less corrosive and safer to transport in metal tanks and pipes.

It was observed in our GC-MS analysis that the mentioned removal of acids (acetic and propionic) was accompanied by the formation of ketones, namely, acetone, 2-butanone, and 3-pentanone, which were absent in the noncatalytic oil. Moreover, it has been shown[35] that acetic acid can be converted into acetone with very high conversion and selectivity (>99%) on a basic catalyst $CeO_2/1\%$ K_2O/TiO_2. Thus, the following mechanism is suggested for the conversion of carboxylic acids on $Na_2CO_3/\gamma\text{-}Al_2O_3$ catalyst.

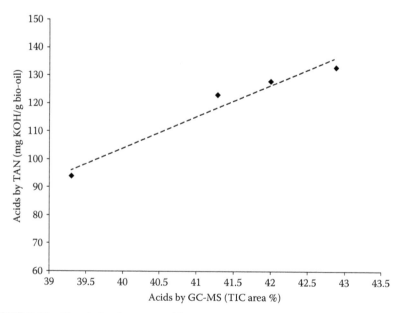

FIGURE 9.12 Correlation between acid components measured by GC-MS and by TAN measurements.

Removal of carboxylic

$$R_1COOH + R_2COOH \rightarrow R_1COR_2 + CO_2 + H_2O$$

acids according to this route is still a problem because (1) the deoxygenation of acids occurs through both decarboxylation and dehydration reactions and (2) the formation of ketones, which are the precursors of the instability of bio-oil as discussed next. In general, presence of ketones and aldehydes cause problems for the stability of the bio-oil by undergoing condensation reactions, which enhance the formation of higher molecular weight components and increase viscosity.[36]

However, this problem is offset to a certain extent by the fact that carboxylic acids can also be decomposed to CO_2 and hydrocarbons (decarboxylation). Figure 9.13 reveals a correlation between reductions in carboxylic acids and the increase in yield of CO_2 in gas stream. Catalyst design should take into account good activity for decarboxylation as this has multiple advantages: (1) it has a strong influence on the acidity of the bio-oil as carboxylic acids contribute strongly to this, (2) the corresponding deoxygenation occurs as CO_2 which is required, and (3) part of the decomposition products, hydrocarbons are high energy fuel components. Recent studies show that pyrolysis in the presence of Cs-based catalysts lowers carbonyl contents of the bio-oil to a large extent and at the same time Cs is a good catalyst for the selective formation of aliphatic hydrocarbons from lignocelluloses. Thus, the promotion of Na-based catalysts with Cs is an attractive option.[37]

Results discussed so far show that the presence of Na_2CO_3/γ-Al_2O_3 during pyrolysis allows extensive deoxygenation and the resulting bio-oil has excellent characteristics to be suitable for the fuel applications. This catalyst could also withstand three cycles of coke combustion regeneration showing its promise for practical applications as in an FCC reactor.[8] In the following section, an analysis of the catalyst and its properties in relation to its performance is discussed.

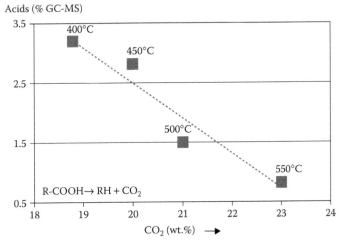

FIGURE 9.13 Correlation between carboxylic acid contents in bio-oils and the corresponding CO_2 formation during pyrolysis. Data are from different catalytic experiments.

9.3.5.1 State of the Catalyst

In the case of alumina, X-ray diffractograms[38] resembles that of the γ-Al$_2$O$_3$. No sharp crystalline peaks corresponding to any of the sodium compounds were observed. It can be concluded from XRD that Na$_2$CO$_3$ is well dispersed as small clusters on γ-Al$_2$O$_3$. EDX elemental mapping also indicated that Na was well dispersed as small clusters (<100 nm) on the alumina matrix. The thermogravimetric analysis (TGAs) of the catalyst and its two components are shown in Figure 9.14. It is known that when heated, the metastable γ-Al$_2$O$_3$ transforms into the thermally stable α-Al$_2$O$_3$ via the formation of two transitional phases δ- and θ-Al$_2$O$_3$.[39] These phase changes are often accompanied by the removal of hydroxyl groups, which explains the decrease in weight of the γ-Al$_2$O$_3$ when heated (Figure 9.14). Moreover, the weight loss of 2.8 wt% of this material up to 800°C completely agrees with that reported in literature about the amount of chemisorbed water removed when a γ-Al$_2$O$_3$ sample, with similar surface area, was heated to 800°C.[32]

Na$_2$CO$_3$, on the other hand, shows to be very stable until 850°C, which follows by a sharp decrease in weight related to the decomposition of the material into Na$_2$O and CO$_2$. The reaction is shown below: Na$_2$CO$_3$ → Na$_2$O + CO$_2$. Based on the stoichiometry of reaction, it is possible to calculate the weight loss to be 41.5 wt% if we assume 100% conversion of Na$_2$CO$_3$. However, it can be seen from Figure 9.14 that the weight loss of Na$_2$CO$_3$ at 1100°C is 62.6 wt%, which is significantly higher than the theoretical maximum weight loss above. This can be explained by the fact that at higher temperature there are three processes occurring almost simultaneously in the Na$_2$CO$_3$ sample, these are decomposition, melting, and vaporization. The vaporization of Na$_2$CO$_3$ is responsible for this extra weight loss. However, these changes in Na$_2$CO$_3$ hardly influence the activity of the catalyst since all catalytic experiments were carried out at 500°C, which is well below the 900°C in the TGA.

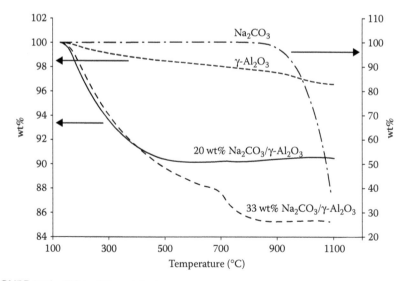

FIGURE 9.14 TGA of the catalysts.

The decomposition of the 20 wt% Na_2CO_3/γ-Al_2O_3 catalyst occurs quite differently. The catalyst showed weight decrease already at 135°C. This weight loss has a faster rate and also is larger than the maximum possible weight loss due to that of the γ-Al_2O_3 present (1.2 wt% dehydration up to 500°C). This is very interesting since the pure Na_2CO_3, as mentioned previously, does not decompose until 850°C. This also suggests that Na_2CO_3 is modified in the presence of γ-Al_2O_3 and is present as a different chemical species in the catalyst. In Figure 9.14, the weight loss for the catalyst up to 500°C, where pyrolysis experiments are carried out, is 9.6 wt%. Taking into account its mass fraction of 80 wt% in the catalyst, the weight loss corresponding to water removal from γ-Al_2O_3 is 1.2 wt%. Thus there is an extra weight loss of 8.4 wt%. Taking into account that Na_2CO_3 loading is 20 wt%, the weight loss corresponding to its decomposition to Na_2O would be 8.3 wt%. An MS was used to analyze the volatiles coming out of the TGA, which indicated both CO_2 and water are released during the weight loss. This suggests that Na is probably present in the catalyst as a hydrated carbonate phase.

The [23]Na (magic angle spinning nuclear magnetic resonance) MAS NMR analysis of the catalyst together with that of Na_2CO_3 is shown in Figure 9.15. The [23]Na spectrum of pure Na_2CO_3 is characterized typically[40] by a sharp peak at 5.5 ppm a broad one around −8.5 ppm and a small peak at −1 ppm. The peaks at −1 and −8.5 are usually associated with Na ions interacting with hydroxyl groups, that is, hydrated species.[41] Interestingly, in the case of Na_2CO_3/γ-Al_2O_3, the two dominating peaks have either almost disappeared (5.5 ppm) or reduced in intensity and became a shoulder (−8.5 ppm). In the meantime, the (relative) intensity of the peak at −1 ppm has significantly increased. NMR analyses of 5–20 wt% Na_2CO_3/γ-Al_2O_3 samples were carried out[41] and the peak around −1 ppm was tentatively assigned to the formation of Na^+ ions coordinating with hydroxyl group of alumina as shown below. It is possible that such an interaction is responsible for the formation of a hydrated Na_2CO_3 species as shown here, which decomposes at lower temperatures.

Interestingly, in the TGA experiments we see weight loss corresponding to that of Na_2CO_3 decomposition to Na_2O. We also see MS signals corresponding to both CO_2 and water during this transition. This also implies that sodium is present in the catalyst as a hydrated Na_2CO_3. The catalytically active species under the pyrolysis temperatures, thus, could be sodium cations coordinated with the hydroxyl groups of alumina in equilibrium with water and carbon dioxide present in the vapor phase. Biomass pyrolysis involves C–C, C–H, and C–O bond scission and acid–base properties of the catalyst plays an important role in this. Interaction of sodium ions with the hydroxyl groups of alumina probably changes the acid–base properties of γ-Al_2O_3. The improved performance of the catalyst may be attributed to this.

The key properties of noncatalytic and catalytic bio-oil and those of fuel oils are compared in Table 9.7. As can be seen, with the 20 wt% Na_2CO_3/γ-Al_2O_3 it is possible to make a bio-oil, which is by far superior to noncatalytic oil and very similar in properties to that of traditional fuel oil. The catalytic bio-oil has low oxygen content (12.3 wt%), almost neutral (pH = 6.5), and high energy density (37 MJ/kg). In order to reduce the oxygen content of this catalytic bio-oil to the same level as in fuel oil, a mild hydrogenation is required. Hydrogenation is also necessary to minimize the amount of carbonyls in this bio-oil as discussed previously, and to increase the yield of bio-oil.

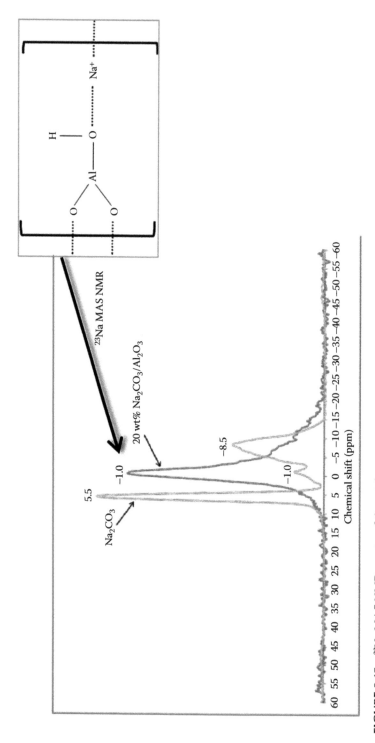

FIGURE 9.15 ^{23}Na MAS NMR spectra of the catalysts.

TABLE 9.7
Comparison of the Bio-Oil Obtained over a 20% $Na_2CO_3/\gamma-Al_2O_3$ via Catalytic Pyrolysis with That of a Noncatalytic Bio-Oil and Fossil Fuel Oil

Characteristic	Noncatalytic Oil	Catalytic Bio-Oil	Fuel Oil
Water content (wt%)	34	3	0.1
C (wt%, dry)	52	78.2	85.3
H (wt%, dry)	5.7	8.7	11.5
O (wt%, dry)	42	12.3	1
HHV (MJ/kg)	19	37	40
pH	2.6	6.5	5.7

The biggest drawback of the catalytic upgrading process is the low yield (9 wt%) of the desired product, the catalytic bio-oil. Because of this low yield, only 21% of the initial energy in biomass was transferred into this catalytic bio-oil, compared to 48% in noncatalytic experiment, even though the energy density and other properties of the catalytic oil are much more suitable for fuel applications. The above difference in energy recovery of bio-oil resulted from the redistribution of the initial biomass energy into: (1) heterogeneous char and (2) aqueous phase of the catalytic liquid. Thus, to overcome this problem and to increase the energy efficiency of the process, it is necessary to extract the energy from the two mentioned side products, for example, by burning the heterogeneous char to generate process heat and reforming the aqueous phase to produce hydrogen. In short, despite the high deoxygenation achieved with Na/Al_2O_3 catalyst, the extremely high amount of heterogeneous char formed on the catalyst not only makes the process less efficient by reducing the yield of bio-oil and the energy recovery, but also poses a huge problem in catalyst handling. Intensive research therefore should be carried out to inhibit heterogeneous char formation on the catalyst and taken into account in the catalyst design before it can be applied in real life. It can be seen from Table 9.4 that the liquid product obtained from our catalytic pyrolysis experiment consists of 22 wt% yield of water, 9 wt% of organic phase (bio-oil), and 6 wt% of aqueous phase (oxygenates). This organic phase contains 12.3 wt% of oxygen. Thus, complete deoxygenation by hydrogenation of the phase will require 1.86 moles of hydrogen per 1 kg of the total liquid. Hydrogen can be generated by reforming of the oxygenates in the aqueous phase of bio-oil.[42] It was shown recently by us that[42] the maximum amount of hydrogen that can be generated from these oxygenates via a combination of steam reforming and WGS reactions is 15.6 moles per 1 kg of the total liquid. These calculations show that the amount of hydrogen obtained from the aqueous phase is more than enough for deoxygenation of the organic phase to achieve similar oxygen content as in fuel oil. Hence, the integration of reforming of the aqueous phase to generate hydrogen to the upgrading of the organic phase shows very promising in an economical point of view.

9.4   CONCLUSIONS

Catalytic primary liquefaction is still in an embryonic stage of development. These processes require cheap and robust catalysts that can cope with severe fouling and poisoning conditions caused by the complex feedstock. Upgrading of solid lignocellulosic biomass, as discussed in this chapter, with cheap sodium-based catalysts, shows that catalytic pyrolysis shows tremendous promise. Catalytic pyrolysis is challenging because of the complexity of the biomass/bio-oil, lack of tools to completely analyze the products of pyrolysis, and high reactivity of the components to char/coke formation. All these factors hinder development of suitable catalysts. However, much can be achieved, just as in the case of fossil fuel development years ago, efficient heterogeneous catalysts showing promise are already a fact and the future for catalytic pyrolysis of lignocellulosic biomass to contribute toward green fuels is promising

ACKNOWLEDGMENTS

I wish to thank three colleagues who have contributed enormously to the work, Dr. I. V. Babich, Dr. S. Nguyen, and Ir. M. Zabeti. I also wish to acknowledge the many discussions I have had on the topic with colleagues as well as collaborators, Dr. P. O'Connor and Professors S. R. A. Kersten, G. Brem, L. Lefferts, and J. A. Lercher. Financial support for the work from (1) BIOeCON, the Netherlands; (2) KIOR, USA; (3) the Dutch Academy of Sciences, NWO, the Netherlands, via the GSPT (Green Sustainable Energy Technologies) and CATCHBIO programs; (4) University of Twente, the Netherlands, via the IMPACT program is gratefully acknowledged.

REFERENCES

1. (a) Huber, G. W.; Iborra, S.; Corma, A., Synthesis of transportation fuels from biomass: Chemistry, catalysis, and engineering. *Chemical Reviews* **2006**, 106, 4044; (b) Kersten, S. R. A.; van Swaaij, W. P. M.; Lefferts, L.; Seshan, K., Options for catalysis in the thermochemical conversion of biomass into fuels. *Catalysis for Renewables* **2007**, 119.
2. Greene, N. N. R. D. C., http://www.nrdc.org/air/energy/biofuels/biofuels.pdf **2004**.
3. Huber, G. W.; Corma, A., Syngeries between bio-oil and refineries for the production of fuels from biomass. *Angewandte Chemie International Edition* **2007**, 46, 7184.
4. Ragauskas, A. J.; Nagy, M.; Kim, D. H.; Eckert, C. A.; Hallett, J. P.; Liotta, C. L., From wood to fuels—Integrating biofuels and pulp production. *Industrial Biotechnology* **2006**, 2, 55.
5. Corma, A.; Huber, G. W.; Sauvanaud, L.; O'Connor, P., Processing biomass-derived oxygenates in the oil refinery: Catalytic cracking (FCC) reaction pathways and role of catalyst. *Journal of Catalysis* **2007**, 247, 307.
6. Bridgwater, A. V.; Peacocke, G. V. C., Fast pyrolysis processes for biomass. *Renewable & Sustainable Energy Reviews* **2000**, 4, 1.
7. Goudnaan, F.; van de Beld, B.; Boerefijn, F. R.; Bos, G. M.; Naber, J. E.; van der Wal, S.; Zeevalkink, J. A., Thermal efficiency of the HTU® process for biomass liquefaction. In *Progress in Thermochemical Biomass Conversion*; Bridgwater, A.V., Ed.; Blackwell Science: Oxford, England, **2001**, 1312–1325.
8. Nguyen, T. S.; Zabeti, M.; Lefferts, L.; Brem, G.; Seshan, K., Conversion of lignocellulosic biomass to green fuel oil over sodium based catalysts. *Bioresource Technology* **2013**, 142, 353–360.

9. Babich, I. V.; van der Hulst, M.; Lefferts, L.; Moulijn, J. A.; O'Connor, P.; Seshan, K., Catalytic pyrolysis of microalgae to high-quality liquid bio-fuels. *Biomass and Bioenergy* **2011**, 35 (7), 3199–3207.

10. Czernik, S.; Bridgwater, A. V., Overview of applications of biomass fast pyrolysis oil. *Energy & Fuels* **2004**, 18 (2), 590–598.

11. Scott, D. S.; Majerski, P.; Piskorz, J.; Radlein, D., A second look at fast pyrolysis of biomass—The RTI process. *Journal of Analytical and Applied Pyrolysis* **1999**, 51 (1/2), 23–37.

12. Meier, D.; van de Beld, B.; Bridgwater, A. V.; Elliott, D. C.; Oasmaa, A.; Preto, F., State-of-the-art of fast pyrolysis in IEA bioenergy member countries. *Renewable & Sustainable Energy Reviews* **2013**, 20, 619–641.

13. Nguyen, T. S.; Zabeti, M.; Lefferts, L.; Brem, G.; Seshan, K., Catalytic upgrading of biomass pyrolysis vapours using faujasite zeolite catalysts. *Biomass and Bioenergy* **2013**, 48, 100–110.

14. Meier, D.; Scholze, B., Fast pyrolysis liquid characteristics. *Biomass Gasification and Pyrolysis: State of the Art and Future Prospects* **1997**, 431–441.

15. (a) Priecel, P.; Capek, L.; Kubicka, D.; Homola, F.; Rysanek, P.; Pouzar, M., The role of alumina support in the deoxygenation of rapeseed oil over NiMo-alumina catalysts. *Catalysis Today* **2011**, 176 (1), 409–412; (b) Stefanidis, S. D.; Kalogiannis, K. G.; Iliopoulou, E. F.; Lappas, A. A.; Pilavachi, P. A., In-situ upgrading of biomass pyrolysis vapors: Catalyst screening on a fixed bed reactor. *Bioresource Technology* **2011**, 102 (17), 8261–8267.

16. (a) Fernandez, M. B.; Sanchez, J. F.; Tonetto, G. M.; Damiani, D. E., Hydrogenation of sunflower oil over different palladium supported catalysts: Activity and selectivity. *Chemical Engineering Journal* **2009**, 155 (3), 941–949; (b) Fisk, C. A.; Morgan, T.; Ji, Y. Y.; Crocker, M.; Crofcheck, C.; Lewis, S. A., Bio-oil upgrading over platinum catalysts using *in situ* generated hydrogen. *Applied Catalysis A: General* **2009**, 358 (2), 150–156; (c) Lohitharn, N.; Shanks, B. H., Upgrading of bio-oil: Effect of light aldehydes on acetic acid removal via esterification. *Catalysis Communications* **2009**, 11 (2), 96–99.

17. Mahfud, F. H.; Melian-Cabrera, I.; Manurung, R.; Heeres, H. J., Biomass to fuels—Upgrading of flash pyrolysis oil by reactive distillation using a high boiling alcohol and acid catalysts. *Process Safety and Environmental Protection* **2007**, 85 (B5), 466–472.

18. Bridgwater, A. V.; Meier, D.; Radlein, D., An overview of fast pyrolysis of biomass. *Organic Geochemistry* **1999**, 30 (12), 1479–1493.

19. Oasmaa, A.; Elliott, D. C.; Korhonen, J., Acidity of biomass fast pyrolysis bio-oils. *Energy & Fuels* **2010**, 24, 6548–6554.

20. Atutxa, A.; Aguado, R.; Gayubo, A. G.; Olazar, M.; Bilbao, J., Kinetic description of the catalytic pyrolysis of biomass in a conical spouted bed reactor. *Energy & Fuels* **2005**, 19 (3), 765–774.

21. Lappas, A. A.; Samolada, M. C.; Iatridis, D. K.; Voutetakis, S. S.; Vasalos, I. A., Biomass pyrolysis in a circulating fluid bed reactor for the production of fuels and chemicals. *Fuel* **2002**, 81 (16), 2087–2095.

22. Williams, P. T.; Nugranad, N., Comparison of products from the pyrolysis and catalytic pyrolysis of rice husks. *Energy* **2000**, 25 (6), 493–513.

23. Samolada, M. C.; Papafotica, A.; Vasalos, I. A., Catalyst evaluation for catalytic biomass pyrolysis. *Energy & Fuels* **2000**, 14 (6), 1161–1167.

24. Ritter, S., A fast track to green gasoline. *Chemical and Engineering News* **2008**, 86 (16), 10.

25. (a) Bridgwater, T., *Fast Pyrolysis of Biomass, A Handbook*, 2008, Vol 2, CPL press, ISBN Nr 1872 6914 71; (b) Carlson, T. R.; Tompsett, G. A.; Conner, W. C.; Huber, G. W., Aromatic production from catalytic fast pyrolysis of biomass-derived feedstocks.

Topics in Catalysis **2009**, 52 (3), 241–252; (c) Thring, R. W.; Katikaneni, S. P. R.; Bakhshi, N. N., The production of gasoline range hydrocarbons from Alcell (R) lignin using HZSM-5 catalyst. *Fuel Processing Technology* **2000**, 62 (1), 17–30; (d) Williams, P. T.; Horne, P. A., The influence of catalyst type on the composition of upgraded biomass pyrolysis oils. *Journal of Analytical and Applied Pyrolysis* **1995**, 31, 39–61.

26. (a) Jackson, M. A.; Compton, D. L.; Boateng, A. A., Screening heterogeneous catalysts for the pyrolysis of lignin. *Journal of Analytical and Applied Pyrolysis* **2009**, 85 (1/2), 226–230; (b) Pattiya, A.; Titiloye, J. O.; Bridgwater, A. V., Fast pyrolysis of cassava rhizome in the presence of catalysts. *Journal of Analytical and Applied Pyrolysis* **2008**, 81 (1), 72–79; (c) Triantafyllidis, K. S.; Komvokis, V. G.; Papapetrou, M. C.; Vasalos, I. A.; Lappas, A. A., Microporous and mesoporous aluminosilicates as catalysts for the cracking of Fischer-Tropsch waxes towards the production of "clean" bio-fuels. From Zeolites to Porous Mof Materials: The 40th Anniversary of International Zeolite Conference, *Proceedings of the 15th International Zeolite Conference* **2007**, 170, 1344–1350.

27. Agblevor, F. A.; Besler, S., Inorganic compounds in biomass feedstocks.1. Effect on the quality of fast pyrolysis oils. *Energy & Fuels* **1996**, 10 (2), 293–298.

28. Patwardhan, P. R.; Satrio, J. A.; Brown, R. C.; Shanks, B. H., Influence of inorganic salts on the primary pyrolysis products of cellulose. *Bioresource Technology* **2010**, 101 (12), 4646–4655.

29. (a) Fahmi, R.; Bridgwater, A. V.; Darvell, L. I.; Jones, J. M.; Yates, N.; Thain, S.; Donnison, I. S., The effect of alkali metals on combustion and pyrolysis of Lolium and Festuca grasses, switchgrass and willow. *Fuel* **2007**, 86 (10/11), 1560–1569; (b) Mullen, C. A.; Boateng, A. A., Chemical composition of bio-oils produced by fast pyrolysis of two energy crops. *Energy & Fuels* **2008**, 22 (3), 2104–2109.

30. O'Connor, P.; Stamires, D.; Daamen, S., Process for the conversion of biomass to liquid fuels and specialty chemicals. US Patent 2012190062 2012.

31. Sooknoi, T.; Danuthai, T.; Lobban, L. L.; Mallinson, R. G.; Resasco, D. E., Deoxygenation of methylesters over CsNaX. *Journal of Catalysis* **2008**, 258 (1), 199–209.

32. Medema, J.; Van Bokhoven, J. J. G. M.; Kuiper, A. E. T., Adsorption of bases on γ-Al$_2$O$_3$. *Journal of Catalysis* **1972**, 25 (2), 238–244.

33. (a) Henrich, E.; Bürkle, S.; Meza-Renken, Z. I.; Rumpel, S., Combustion and gasification kinetics of pyrolysis chars from waste and biomass. *Journal of Analytical and Applied Pyrolysis* **1999**, 49 (1/2), 221–241; (b) Stals, M.; Carleer, R.; Reggers, G.; Schreurs, S.; Yperman, J., Flash pyrolysis of heavy metal contaminated hardwoods from phytoremediation: Characterisation of biomass, pyrolysis oil and char/ash fraction. *Journal of Analytical and Applied Pyrolysis* **2010**, 89 (1), 22–29.

34. Huber, G. W.; Iborra, S.; Corma, A., Synthesis of transportation fuels from biomass: Chemistry, catalysts, and engineering. *Chemical Reviews* **2006**, 106 (9), 4044–4098.

35. Deng, L.; Fu, Y.; Guo, Q., Upgraded acidic components of bio-oil through catalytic ketonic condensation. *Energy Fuels* **2008**, 23 (1), 564–568.

36. Diebold, J., A review of the chemical and physical mechanisms of the storage stability of fast pyrolysis bio-oils; Report No. SR-570-27613; National Renewable Energy Laboratory: Golden, Colorado, January 2000, p. 60.

37. Zabeti, M.; Nguyen, T. S.; Lefferts, L.; Heeres, H. J.; Seshan, K., *In situ* catalytic pyrolysis of lignocellulose using alkali-modified amorphous silica alumina. *Bioresource Technology* **2012**, 118, 374–381.

38. Khaleel, A.; Al-Mansouri, S., Meso-macroporous γ-alumina by template-free sol–gel synthesis: The effect of the solvent and acid catalyst on the microstructure and textural properties. *Colloids and Surfaces A* **2010**, 369 (1–3), 272–280.

39. (a) Zhou, R. S.; Snyder, R. L., Structures and transformation mechanisms of the eta, gamma and theta transition aluminas. *Acta Crystallographica Section B* **1991**, 47, 617–630; (b) Santos, H. D.; Santos, P. D., Pseudomorphic formation of aluminas from

fibrillar pseudoboehmite. *Materials Letters* **1992**, 13 (4/5), 175–179; (c) Bodaghi, M.; Mirhabibi, A. R.; Zolfonun, H.; Tahriri, M.; Karimi, M., Investigation of phase transition of γ-alumina to α-alumina via mechanical milling method. *Phase Transitions* **2008**, 81 (6), 571–580.

40. Jones, A. R.; Winter, R.; Greaves, G. N.; Smith, I. H., ^{23}Na, ^{29}Si, and ^{13}C MAS NMR investigation of glass-forming reactions between Na_2CO_3 and SiO_2. *Journal of Physical Chemistry B* **2005**, 109 (49), 23154–23161.

41. Deng, F.; Du, Y.; Ye, C.; Kong, Y., Adsorption of Na^+ onto γ-alumina studied by solid-state [23]Na and [27]Al nuclear magnetic resonance spectroscopy. *Solid State Nuclear Magnetic Resonance* **1993**, 2 (6), 317–324.

42. de Vlieger, D. J. M., Design of efficient catalysts for gasification of biomass-derived waste streams in hot compressed water. Towards industrial applicability. PhD dissertation, 2013, University of Twente, the Netherlands, ISBN Nr 9789 0365 3492 5.

Recent Trends in the Purification of H$_2$ Streams by Water–Gas Shift and PROX

A. Sepúlveda-Escribano and J. Silvestre-Albero

CONTENTS

10.1 Introduction 281
10.2 Water–Gas Shift 282
    10.2.1 Platinum-Based Catalysts 284
    10.2.2 Gold-Based Catalysts..285
    10.2.3 Non-Noble-Metal–Based Catalysts 287
    10.2.4 Importance of the Catalyst Support...287
10.3 Preferential Oxidation of CO in H$_2$-Rich Streams 288
    10.3.1 Group VIIIB Metal Catalysts 289
        10.3.1.1 Monometallic Catalysts 289
        10.3.1.2 Bimetallic Catalysts...291
        10.3.1.3 Reducible Metal Oxide–Promoted Catalysts.....................291
    10.3.2 Group IB Metal Catalysts...294
    10.3.3 Reaction Mechanism 295
References..296

This chapter reviews the latest trends in the use of catalyst for the purification of hydrogen streams at low temperatures through the water–gas shift (CO + H$_2$O → H$_2$ + CO$_2$) reaction and the preferential oxidation of CO in the presence of H$_2$ (PROX).

10.1 INTRODUCTION

Hydrogen is a promising energy vector to be used to replace or complement the current energy model based on fossil fuels [1]. It is nonpolluting if it is obtained from renewable sources, and it can be easily and efficiently transformed into energy both chemically and electrochemically [2–4]. In this sense, the most promising technologies to obtain energy are based on fuel cells, with the proton exchange membrane fuel cell (PEMFC) standing out.

Hydrogen can be obtained from a great variety of sources, including fossil fuels (natural gas, oil, coal) and renewable sources (water, biomass). The energy needed can be obtained both from fossil fuels and nuclear energy and renewable sources, including sunlight. Depending on the raw materials different methodologies are used. When fossil fuels and biomass are used, involved technologies are mainly reforming (reaction with steam at high temperatures), gasification (reaction with oxygen, but avoiding complete oxidation), and pyrolysis (reaction at high temperature under inert atmosphere). Hydrogen from water is usually obtained through photolysis and electrolysis, obtaining very pure hydrogen streams.

Currently, the most widely used technology to obtain hydrogen is the steam reforming of methane, although many other hydrogen-containing compounds such as C_2–C_4 hydrocarbons [5] and alcohols, mainly methanol and ethanol [6,7], can also be used. One of the main drawbacks of this process is the formation of CO as a by-product, whose relative amount depends on the raw material and the technology used. Thus, the gas stream from the process has to be treated to remove CO, as well as CO_2, in order to obtain as pure as possible hydrogen stream. This step is very important mainly if the produced hydrogen is to be used in low temperature fuel cell such as PEMFC, as CO is a strong poison for its electrodes [1,8–10]. High concentrations of CO can be removed from the reformate streams by the water–gas shift (WGS) reaction, during which it reacts with water to yield CO_2 and H_2. However, the residual concentration of CO after this reaction, around 1000 ppm [10], is still usually too high to feed fuel cells working at low temperatures. It has to be taken into account that fuel cell anodes can typically tolerate up to about 50 ppm CO [8–10] in the feed gas, but the reformate stream usually contains 1–3 vol% CO [11]. One of the best approaches to decrease the residual CO concentration to values that can be tolerated by the fuel cell anode is the preferential oxidation of CO (PROX) in the H_2-containing stream. A PROX catalyst must be able to oxidize CO at low temperatures (usually 80°C–150°C), but without catalyzing the hydrogen oxidation reaction. In this chapter, the recent developments in catalysts for these two reactions, low temperature WGS and PROX will be reviewed. There is a very large amount of literature covering these issues, and it is impossible to mention in this chapter all the important studies that have been carried out, although an attempt will be made to present the main advances and the perspectives in these interesting reactions.

10.2 WATER–GAS SHIFT

Since the middle of the last century, the WGS reaction has been an important step in the industrial production of hydrogen for different processes such as methanol, ammonia, and Fischer–Tropsch syntheses. It has been used for regulating the CO/H_2 ratio of the reformate stream, or even for nearly completely removing poisonous CO, to make it useful for the different applications through the reaction between CO and steam to produce CO_2 and more H_2:

$$CO + H_2O \leftrightarrow CO_2 + H_2 \qquad \Delta H^0_{298\,K} = -41.1\,kJ/mol$$

Because of its exothermic character and the existence of thermodynamic limitations at high temperature, this reaction has been carried out through two stages: the high-temperature shift (623–643 K) with iron-based catalysts [12], and the low temperature shift (473–493 K) with copper-based catalysts [13]. However, when dealing with the purification of hydrogen streams to be used in small-scale operations, such as for mobile or stationary fuel cells, the classical low-temperature copper-based catalysts are not useful, as they show important disadvantages: They need long-term activation processes, they are pyrophoric, they show a bad performance in shut-up/shut-off cycles, and they are intolerant to poisons, water condensation, and oxidation [9,14]. It is thus highly desirable to develop catalysts which are able to perform the reaction at temperatures close to that of the fuel cell working conditions, with high activity and stability.

Several catalyst compositions have been investigated for this reaction, including different metals and supports. Regarding the supports, an excellent behavior has been found when working with partially reducible supports, such as CeO_2 or TiO_2 [15,16]. When metal nanoparticles are supported on these oxides, their redox properties are enhanced in such a way that the temperature at which their surface is reduced is decreased and the extent of reduction is increased. This is very important, as the proposed mechanisms involve the participation of the support's surface in the reaction. Thus, on the one hand, the redox or "regenerative" mechanism [17–19] involves a first step with the chemisorption of CO on the surface of the metal particles, and its further oxidation by oxygen supplied at the metal–oxide interface by the support. This generates an oxygen vacancy on the surface of the support, which is capped by water yielding H_2.

$$CO + * \rightarrow CO*$$

$$H_2O + M\text{-}O \rightarrow H_2 + M\text{-}O_2$$

$$M\text{-}O_2 + CO \rightarrow CO_2 + M$$

(M: metal cation in the support; * active metal site)

On the other hand, the so-called associative mechanism [20–23] proposes that the reaction occurs through the interaction of adsorbed CO with active OH groups on the support's surface, yielding carbon-containing surface intermediates such as formates, carboxyls, carbonates, and/or bicarbonates, which subsequently decompose to yield CO_2 and H_2, whereas the surface of the support is reoxidized.

$$H_2O + 2* \rightarrow HO* + H*$$

$$CO + * \rightarrow CO*$$

$$CO* + HO* \rightarrow *O\text{-}CH = O$$

$$*O\text{-}CH = O \rightarrow CO_2 + H*$$

$$2\,H* \rightarrow H_2 + 2*$$

Thus, the surface redox properties of the support are of paramount importance, both in terms of extension (large surface areas are needed, which are able to form large metal–oxide surface interface sites) and in terms of reducibility. In addition, the active metal should to be able to activate CO on its surface and be able to interact with the oxide support. In this sense it has been claimed that, for ceria supported catalysts, the active species for the reaction are atomically dispersed metals, [Au-O$_x$]-Ce or [Pt-O$_x$]-Ce species, strongly bound to the support, whereas supported metal nanoparticles are mere spectator species [15,24].

The most used metals in the development of low temperature WGS reaction are platinum and gold; however, due to their lack of availability and high price, also base metals such as nickel and copper have been studied.

10.2.1 PLATINUM-BASED CATALYSTS

Platinum is considered as a promising catalyst for the low temperature WGS reaction, and it has been the object of many studies which use different metal oxides as supports, mainly CeO$_2$ [15,25–28] and TiO$_2$ [29–31], as well as mixed oxides [27,32].

As aforementioned, it is commonly accepted that the WGS reaction takes place at the interface between the metal particle and the surface of the oxide. Thus, the importance and the nature of the interfacial sites involved have been the subject of many studies by using different supports and experimental techniques.

Aranifard et al. have recently used density functional theory and microkinetic modeling to study the mechanism of the WGS reaction at the three-phase boundary of Pt supported on a CeO$_2$ (111) surface [33]. They found that the interface sites were 2–3 orders of magnitude more active than Pt(111) and stepped Pt surface sites, and thus these sites determined the overall catalytic activity. It is thus of paramount importance to prepare catalysts in which the amount of these interfacial sites are maximized, and an important approach would be to decrease the particle size of the metal nanoparticles.

Different attempts have been made to modify the properties of interfacial sites and to enhance the reducibility of the oxide support. The most common has been to modify the ceria support by the addition of a second metal to form mixed oxides on different compositions. Thus, Kalamaras et al. have recently prepared La^{3+}-doped CeO$_2$ supports for small (1.0–1.2 nm) platinum nanoparticles [32]. They concluded that the WGS reaction followed both the "redox" and "associative" mechanism on these catalysts, although the prevalence of each of them depended on the support composition. Furthermore, the catalytic activity was higher for Pt/Ce$_{0.8}$La$_{0.2}$O$_2$ than for platinum supported on either pure CeO$_2$ or pure La$_2$O$_3$, and it was explained as due to a larger extent of the reactive zone and the higher reactivity of active sites. In fact, formation of a mixed oxide enhanced the formation of labile oxygen and its surface mobility.

The effect of the composition of CeO$_2$–ZrO$_2$ mixed oxides on the catalytic behavior of platinum and gold was studied by Boaro et al. [34]. First, they obtained better performance for Au catalysts. For platinum, they did not find any significant correlation between the WGS activity and the support composition. These authors concluded that the redox and structural properties of the support, which can be modified by Zr addition in this case, play a secondary role in the reaction, and it is the nature of the metal–support interface which is more important; this can be tailored by the choice of the metal precursor and the synthesis procedures. However, Duarte de Farias et al. found a beneficial effect of the addition of Zr^{4+} to CeO$_2$ in such a way that a higher activity was obtained for Pt supported on Ce$_{75}$Zr$_{25}$, Ce$_{50}$Zr$_{50}$, and Ce$_{60}$Zr$_{40}$, but activity decreased for the $Ce_{25}Zr_{75}$ support. In fact, the activity of $Pt/Ce_{50}Zr_{50}$ was 50% higher than that of Pt/CeO_2. It was concluded from this study that the main factor controlling the catalytic performance was the chemical composition of the oxide support, whereas its reducibility affects neither the activity nor the stability of the catalysts [35].

The use of CeO_2–TiO_2 mixed oxides as supports for Pt nanoparticles has also shown promising results in low temperature WGS [27,36]. The effect of the Ce/Ti ratio in these materials has been recently reported [31]. Platinum nanoparticles of 1.2–2.0 nm size were supported on $Ce_{1-x}Ti_xO_2$ ($x = 0$, 0.2, 0.5, 0.8, and 1.0) supports, and their catalytic behavior was studied in the WGS reaction at 200°C–350°C. The best results were obtained with the $Pt/Ce_{0.8}Ti_{0.2}O_2$ catalysts, which showed a CO conversion at 250°C to be 2.5 times larger than that of Pt/TiO_2 and 1.9 times larger than that of Pt/CeO_2. The results were explained on the basis of structural stability under reaction conditions, moderate acidity and basicity, and better reducibility at lower temperatures. Thus, the chemical composition of the support has a great influence on the catalytic performance of supported platinum.

Although partially reducible oxides have been largely studied as supports for Pt in WGS catalyst, it has been also shown that nonreducible supports such as silica or activated carbon can also be used when alkali and alkaline earth cations are used as promoters [37–40]. It has been proposed that the active site is a partially oxidized Pt-alkali-$O_x(OH)_y$ species on which both the CO adsorption and the H_2O activation take place, and that the presence of the alkali is needed to stabilize these species [38]. This approach has been used to prepare SiO_2-encapsulated Na-promoted Pt catalysts in which the water dissociation and the hydroxyl regeneration steps take place on the alkali metal promoted Pt-OH_x active sites; furthermore, the core–shell configuration of these catalysts favors its stability under reaction conditions (350°C) [39]. It has been even proved that a metal oxide support is not necessary for obtaining active platinum catalyst. Thus, in a very recent study, Zugic et al. have shown that platinum supported on multiwalled carbon nanotubes and promoted by sodium ions is active in WGS under ideal feed stream composition (only CO and H_2O) at temperatures lower than 300°C, while carbon-supported Pt catalysts have been reported to be inactive in this reaction [41].

Activated carbon has also been used as a support for Pt but, in this case, promoted by ceria. The aim was to obtain and stabilize small ceria crystallites, with a large surface area, which could provide a large amount of interfacial Pt-CeO_2 sites and simultaneously reduce the amount of the ceria used in the catalyst formulation as compared to a bulk ceria support [42]. The effect of the solvent of the platinum precursor, $[Pt(NH_3)_4](NO_3)_2$, was also assessed. The best results were obtained with the Pt-40wt%CeO_2 catalyst prepared with an aqueous solution of the platinum precursor, and it was attributed to the better interaction between platinum and ceria crystallites in this catalyst, as assessed by temperature-programmed reduction experiments. It showed a better behavior than the classical Pt/CeO_2 catalyst, whereas platinum supported on activated carbon was shown to be inactive.

10.2.2 GOLD-BASED CATALYSTS

Since Haruta's discovery of the exceptional catalytic properties of small gold nanoparticles for some reactions [43], there has been an increasing interest during the last decade in the development of gold-based catalysts for the low temperature WGS reaction [44–46]. Although different supports have been used, mainly partially reducible metal oxides, superior performance has been obtained for ceria-supported gold [47–51], although there are also interesting studies on gold supported on TiO_2, Fe_2O_3, and mixed oxides.

One of the first research groups reporting interesting results on gold-based catalysts for low temperature WGS was Andreeva et al. They first studied $Au/\alpha\text{-}Fe_2O_3$ catalysts, for which they found a very high activity at low temperature, which they attributed to a specific interaction between gold and the support [52]. An associative mechanism was proposed for this system by which water dissociated on the small gold nanoparticles, and the active hydroxyl groups formed were transferred by spillover to adjacent sites of the support. The support participates in the formation and decomposition of carbon-containing species through a redox cycle between Fe^{2+} and Fe^{3+} [53]. They also studied different preparation routes and found that the best results were obtained when gold hydroxide was deposited by precipitation of freshly prepared iron hydroxide [54]. Some years later, this team started working on the Au/CeO_2 system. They investigated the role on the preparation route by using two different approaches for the deposition–precipitation method, that is, by using calcined CeO_2 or freshly prepared $Ce(OH)_3$ as support [55], and they found that much smaller gold particles were obtained with CeO_2 and, thus a better catalytic performance.

For the Au/CeO_2 system, it has been argued that the active phases for the low temperature WGS reaction can be both well-dispersed Au nanoparticles and gold oxides, which would be stabilized by interaction with ceria through its surface oxygen vacancies [15,56]. Thus, the preparation route is of paramount importance, as it determines not only the size of the gold nanoparticles, but also the size and crystallinity of the ceria support which, in turn, determine its reducibility and the extent of its interaction with supported gold [57]. Furthermore, it is well accepted that the oxide support also plays an important role in the reaction pathway, favoring the water dissociation step and providing active sites at the interface at which the reaction between CO and hydroxyl groups can take place [58–60].

As stated earlier, one of the main roles of ceria in Au/CeO_2 catalysts is to disperse and stabilize the Au nanoparticles, which is achieved through the interaction of the metal with the surface oxygen vacancies. In this sense, small ceria crystallites showing a large surface area are desired. Thus, it has been shown that ceria nanoparticles with a size between 3 and 4 nm and a large number of surface vacancies showed enhanced activity for CO oxidation [61]. However, catalyst deactivation is observed in some cases, and it is mainly assigned to the collapse or restructuring of the gold surface–ceria interface as a consequence of temperature and/or components of the reactive gas stream. In this sense, Ta et al. [57] used atomic resolution environmental transmission electron microscopy (ETEM) to observe the behavior of Au/CeO_2 catalysts under conditions close to those of the WGS reaction, and they observed a restructuration of the size and shape of the gold nanoparticles as a consequence of the reaction conditions. They concluded that this undesired effect could be hindered by increasing the gold–ceria interaction by, for instance, thermal treatments ~673 K.

Similarly to what has been discussed earlier for Pt-based catalysts, the effect of the composition of the ceria-based support has also been studied for gold catalysts through doping the ceria support with small amounts of other metals. In this sense, Tabakova et al. prepared Ce–Fe mixed oxides of different compositions by co-precipitation with urea, and used them as supports for gold catalysts [51]. They observed a detrimental effect of the presence of iron in the support on the activity for WGS at temperatures ranging from 400 to 600 K, which was assigned to differences in gold particle sizes depending on the support composition and to the lower concentration of oxygen vacancies and Ce^{3+} cations in the mixed oxides. More recently, Vindigni et al. prepared mixed CeO_2–ZrO_2 oxides and prepared gold catalysts by deposition–precipitation [62]. They obtained the following trend for catalytic activity, AuCe50Zr50 > AuCe80Zr20 > AuCe, which was explained on the basis of the effect of the presence of zirconia in the structural characteristic of the supports, on their acid–base properties, and in differences in gold particle size for the different supports. Ceria doping with iron and manganese has also been shown to provide active catalysts if the proper preparation route is used [63], its main effect being the increase of the Ce^{3+} content on the support and, in this way, in the number of oxygen vacancies.

10.2.3 NON-NOBLE-METAL–BASED CATALYSTS

The high prices and low availability of noble metals such as platinum and gold have encouraged the research toward the use of other less expensive metals; among them, best results have been obtained with nickel and copper [17,21,64,65].

In spite of their good performance, one of the main drawbacks of nickel catalysts in WGS is that they are also active for methanation, thus consuming some hydrogen [66]; therefore, research is being carried out to try to minimize this undesired reaction. Thus, it has been reported that methanation can be limited on massive nickel powder by addition of potassium [67]; at the same time, activity for WGS is enhanced and the catalyst stability increased as a consequence of limited carbon deposition.

Alumina-supported Ni, Cu, and bimetallic Cu–Ni nanoparticles have been prepared by Lin et al. from metal colloids [68,69]. They obtained very interesting results with core–shell Cu–Ni particles consisting of a Cu core and a Ni shell, which showed similar catalytic activity as the pure Ni catalyst, but with lower methanation ability.

The nickel/ceria system has been the center of several studies [21,64,70]. Interestingly, a strong metal–support interaction effect has been shown that modifies the structural properties of the nickel particles with reduction at high temperature under hydrogen and thus its catalytic behavior. Limited methanation was observed in catalysts prepared from reduction of ceria–nickel mixed oxides [70].

10.2.4 IMPORTANCE OF THE CATALYST SUPPORT

From the examples discussed in the previous sections, it is clear that ceria is one of the main constituents of active low temperature WGS catalysts, and that their properties affecting the catalytic behavior are the object of numerous studies. In addition to its surface properties (acid–base and redox), its structural properties can also affect the catalytic behavior, both from the point of view of its interaction with active metals and by its direct participation in the reaction pathway.

The role of the ceria shape and morphology on its behavior as a catalyst in WGS has been investigated by Agarwal et al. [71]. They prepared rods, cubes, and also used commercial octahedral ceria nanoparticles and carried out the WGS reaction with a CO/H_2O reaction feed ratio of 1:3 at 350°C and atmospheric pressure. Their results showed that the WGS activity normalized per square meter was identical for ceria octahedral and rods, but cubes were much more active. They also observed, by Fourier transformed infrared (FTIR) analysis of adsorbed CO and OH groups, that the surface structure was similar for octahedra and rods, and different from that of the cubes. Transmission electron microscopy (TEM) analysis allowed to conclude that the similar reactivity of octahedra and rods was due to the fact that both exposed {111} surfaces, whereas the ceria cubes exposed {100} surface, which seem to be more active under the reaction conditions used. However, theoretical calculations have shown that the formation energy of anion vacancies in different CeO_2 crystal surfaces follows the order {110} < {100} < {111} [72], thus indicating that ceria nanorods would be the most adequate supports. In fact, other authors have shown that the use of nanorods, which are rich in oxygen vacancies, enhanced the gold activity in WGS [47,73] and also in CO oxidation [74]. These authors prepared nanorods, nanocubes, and nanopolyhedra by hydrothermal methods and introduced gold nanoparticles by deposition/precipitation. They noticed a different degree of interaction between the ceria support and gold nanoparticles, which was especially weak for ceria nanocubes. Consistent with these observations, the WGS reactivity decreased in the order rods > polyhedral >> cubes. These results have been also confirmed by other studies, in which rod-shaped ceria supports produced much higher activity in CO oxidation than spherical nanoparticles, and this was attributed to their easiness in the generation of oxygen vacancies [75].

In a recent paper, Vindigni et al. [76] prepared two CeO_2 supports by different methods, the homogeneous precipitation with urea from an aqueous $(NH_4)_2Ce(NO_3)_6$ solution (urea gelation co-precipitation, UGC) and the precipitation with K_2CO_3 from an aqueous solution of $Ce(NO)_3\cdot6H_2O$. They used these supports to prepare Au/CeO_2 catalysts, and they observed a better catalytic behavior for that using the UGC support, in spite of the fact that both catalysts contained the same amount of gold and a similar amount of exposed Au sites. From the characterization of the supports, they obtained that the UGC method showed a higher surface area and smaller particle size. Furthermore, this synthetic route produced a more defective ceria support with an enhanced reactivity of the Ce^{4+} sites toward the formation of Ce^{3+} under reductive treatments. The authors claimed that this higher reactivity could favor the water dissociation step and in this way enhance the overall WGS activity.

10.3 PREFERENTIAL OXIDATION OF CO IN H_2-RICH STREAMS

As described in previous sections, hydrogen produced by auto-thermal reforming of hydrocarbons or alcohols and followed by the WGS reaction still possesses large amounts of CO (around 1%) that must be removed before being suitable to be used in fuel cell applications, mainly the polymer-electrolyte membrane fuel cell (PEMFC) [77]. Among the different technologies to reduce the amount of CO to trace levels below 10 ppm from hydrogen-rich streams, Pd-membrane separation [78], catalytic methanation [79], and preferential oxidation of CO (PROX) [80–82] are the most widely investigated, PROX reaction being the most promising one.

The preferential oxidation of CO is based on two competitive reactions:

$$CO + \frac{1}{2}O_2 \rightarrow CO_2 \qquad \Delta H_{298} = -283 \, kJ/mol$$

$$H_2 + \frac{1}{2}O_2 \rightarrow H_2O \qquad \Delta H_{298} = -242 \, kJ/mol$$

Thermodynamically both reactions are very similar; the main challenge being the design of a catalyst able to selectively oxidize CO to CO_2 avoiding undesirable consumption/oxidation of H_2 to H_2O. Furthermore, taking into account that the PROX reaction will be placed between the WGS unit (working at ~200°C) and the PEMFC (working at ~80°C), the designed catalyst must be able to work in a wide temperature range keeping a good catalytic activity and selectivity toward CO_2 [83]. Current catalysts for the PROX reaction can be classified into group VIIIB metal catalysts (Pt, Ru, Ir, and Rh) and group IB metal catalysts (Cu-, Ag-, and Au-based catalysts). Besides the nature of the metal nanoparticle, the support is also a crucial component in the design of the catalyst being classified as "inert" supports, such as Al_2O_3, SiO_2, and carbon, or "active" supports, such as CeO_2 and TiO_2. In the next sections we will briefly summarize the main achievements and drawbacks of these formulations, with special emphasis on the platinum group metal catalysts.

10.3.1 GROUP VIIIB METAL CATALYSTS

10.3.1.1 Monometallic Catalysts

Platinum group catalysts (mainly Pt and Rh) are perhaps the earlier studied systems for the PROX reaction. These systems are characterized by a good catalytic performance in the temperature range of 60°C–150°C together with an acceptable selectivity to CO_2, mainly when prepared in the presence of a promoter (reducible metal oxide support). When used in the presence of an "inert" support, these systems lack acceptable activity at low reaction temperatures.

Monometallic Pt nanoparticles supported on alumina, zeolites, and carbon materials have been widely investigated [84–93]. In general, un-promoted Pt catalysts exhibit a noticeable PROX activity above 150°C–200°C (close to 100% conversion) with selectivity to CO_2 usually above 50%. Interestingly, the catalytic performance on inert supports not only depends on the reaction conditions (O_2 concentration, gas composition, etc.), but also on the surface chemistry of the support, the nature of the metal precursor, and the nanoparticle size. Using Pt catalysts supported on multi-walled carbon nanotubes, Jardim et al. reported that the light-off temperature shifts to lower values after an increase in the calcination temperature (see Figure 10.1) [93]. Apparently, large platinum nanoparticles exhibit an improved performance in the conversion of CO although associated with a slight decrease in the selectivity.

As described earlier, catalytic performance can be modulated by modifying the feed composition. In general, increasing the O_2 concentration in the gas inlet gives rise to an improvement in the CO conversion although associated with a decrease in the selectivity due to the competing oxidation of H_2 to H_2O [84,87,94]. Furthermore, the incorporation of H_2O and/or CO_2 to simulate real reaction conditions strongly

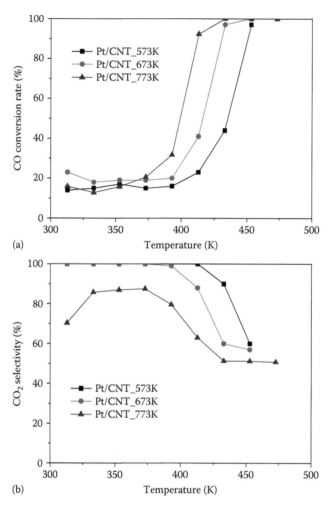

FIGURE 10.1   Effect of the thermal treatment temperature on the (a) catalytic activity and (b) selectivity to CO_2 for 1 wt% Pt/MWCNT catalysts in the PROX reaction at different reaction temperatures. (Reprinted from Jardim, E. O. et al., *Appl. Catal. B: Environ.,* 113–114, 72, 2012. With permission.)

modifies the catalytic activity and selectivity. Manasilp et al. reported a 10-fold increase in the CO conversion after H_2O incorporation on Pt/Al_2O_3 catalysts while the conversion drastically dropped after CO_2 incorporation [84].

Another important catalyst for PROX reaction concerns supported monometallic Rh and Ru nanoparticles [95]. In a similar manner to Pt nanoparticles, catalytic performance of Ru and Rh catalysts is highly sensitive to the preparation conditions, that is, metal precursor, reducing agent, and pretreatment conditions [95–101]. Chin et al. reported a detrimental effect of Cl− coming from the metal precursor ($RuCl_3$) compared to the same catalyst but prepared using a nitrate derivative precursor [$Ru(NO)(NO_3)_3$] [96]. The lower CO conversion in the chlorinated catalyst was attributed to the blocking of the active sites by Cl^- or by Cl^- induced structural rearrangement. Comparing different metal contents, Kim et al. observed that the 5 wt% Ru/Al_2O_3 catalyst with a low chemisorption capacity for CO and O_2 was a promising catalyst in terms of activity and selectivity (CO conversion of 8.3% and 61% selectivity at 60°C) [98]. Using $Ru/\gamma-Al_2O_3$ catalysts, Han et al. reported a close correlation between the CO conversion at low temperature and the amount of Ru^0 in the catalyst surface, as revealed by x-ray photoelectron spectroscopy (XPS) [99]. Apparently, the reaction at low temperature is controlled by O_2 activation on reduced Ru nanoparticles, with two different reactions taking place on these nanoparticles: (1) CO methanation with hydrogen and (2) CO oxidation with oxygen.

Catalytic studies under realistic conditions (in the presence of H_2O and CO_2) using Rh nanoparticles supported on different zeolites (type-A zeolites) have shown that the zeolite composition (nature of the counterions), the pore size, and the crystal structure are important parameters defining the PROX activity versus the undesired methanation of CO_2 [100,101]. The appropriate design of the catalyst composition and reaction conditions (e.g., oxygen deficient conditions) allows the achievement of a 100% CO conversion in the temperature range of 80°C–120°C, with an acceptable selectivity to CO_2 (above 35%).

In summary, catalytic results show that un-promoted platinum group metal catalysts exhibit a good performance in the PROX reaction at high temperatures (above 150°C), ruthenium being the most promising one. However, the use of a second metal (bimetallic systems) or the use of "active" supports is mandatory to improve the performance of these systems at low temperature in the PROX reaction.

10.3.1.2 Bimetallic Catalysts

Among the different compositions investigated, Pt–Ru bimetallic catalysts supported on different supports (alumina, silica, and mordenite) have been proposed as promising catalysts for the PROX reaction with an improved behavior at low temperatures compared to the monometallic ones [102–104]. Independently of the support (either Al_2O_3 or SiO_2), Pt–Ru catalysts exhibit a CO conversion close to 100% in the temperature range of 100°C–150°C with a selectivity to CO_2 above 50% [102,103]. Similar catalysts supported on mordenite showed an excellent CO conversion (around 90%) with selectivity to CO_2 close to 90% at 150°C [104].

Besides the Pt–Ru formulation, other bimetallic compositions (e.g., Pt–Au, Pt–Pd) have also been investigated in the PROX reaction. Nakman et al. found that Pt–Au systems supported on zeolite A exhibit a downshift in the light-off temperature by 50°C compared to the monometallic Pt catalyst [105]. Pd-based catalysts modified by the incorporation of trace amounts of Pt (Pd/Pt ratio 7) and supported on CeO_2 were proposed by Parinyaswan et al. as excellent catalyst with a CO conversion close to 100% in the temperature range of 90°C–110°C [106]. Using more complex synthesis methods, Eichhorn et al. developed metal nanoparticles core–shell catalysts (M–Pt) with an excellent behavior in terms of catalytic activity and selectivity in the PROX reaction [107]. In general, bimetallic catalysts exhibit improved stability in the presence of H_2O and/or CO_2 compared to the monometallic catalysts.

10.3.1.3 Reducible Metal Oxide–Promoted Catalysts

The most widely investigated reducible metal oxides for the PROX reaction are iron oxide and cerium oxide. The incorporation of these oxides either as a promoter or as a support gives rise to an additional oxygen supply at the metal–support interface (active lattice oxygen from the support), thus giving rise to a completely different catalytic behavior on the PROX reaction. Qiao et al. reported a promising behavior for Pt nanoparticles supported on FeO_x (CO conversion above 95% at 80°C) compared to conventional catalysts, this improvement being 2–3 times larger when achieving highly dispersed nanoparticles (Pt_1/FeO_x catalysts) [108]. Similarly, Watanabe et al. reported Pt–Fe catalysts supported on mordenite with a superior catalytic performance to conventional Pt/Al_2O_3 and Pt/mordenite catalysts [109]. Pt-promoted FeO_x/mordenite catalysts exhibited a 100% CO conversion in the temperature range of 80°C–150°C. According to Fu et al., the excellent catalytic behavior after iron incorporation is associated with the formation of coordinately unsaturated ferrous sites at the metal–support interface able to activate dioxygen species [110]. The promoter effect of Fe_2O_3 has been also extended to bimetallic formulation such as Pt–Cu with a similar promotion of the catalytic activity at low temperatures due to the synergetic effect of copper and iron oxide in supplying active oxygen to the Pt–Cu alloy nanoparticles [111].

As described earlier, cerium oxide–based noble metal catalysts have been also widely investigated in the PROX reaction. Pt/CeO_2 and $Pt/Ce_xZr_{(1-x)}O_2$ catalysts have shown an improved behavior at low temperatures (60°C–100°C) compared to unpromoted Pt catalysts [112–121]. Partially reducible cerium oxide acts as an oxygen buffer, thus supplying active oxygen at the noble metal–support interface. As it can be observed in Figure 10.2, CeO_2-based platinum catalysts are also highly sensitive to the presence of chloride species, mainly after a reduction treatment at high temperature, due to the formation of CeOCl species that inhibit active lattice oxygen availability [120]. As it can be observed in Figure 10.2, the formation of these CeOCl species is favored after a reduction treatment at high temperature (500°C).

Besides the nature of the metal precursor (chlorine-based precursors), the morphology of the support is also an important parameter to take into account. Gao et al. reported an important effect of the support morphology (CeO_2 cubes, CeO_2 rods, and CeO_2 octahedra) on the redox properties and catalytic behavior of Pt-supported nanoparticles in the PROX reaction [121]. Jardim et al. reported a superior performance in terms of catalytic activity and selectivity for Pt/CeO_2 catalysts deposited on multiwalled carbon nanotubes (MWCNTs) compared to the conventional Pt/CeO_2 catalyst [93]. Despite the promoting effect of cerium oxide, these systems exhibit a low selectivity to CO_2 (approximately 40%) due to the concomitant consumption of H_2.

Other promoters as Sn, Co, Cu, and Ni have also been evaluated in the PROX reaction for Pt-supported catalysts. The incorporation of these secondary species gives rise to the formation of intermetallic compounds (e.g., Pt_3Co) with a completely different catalytic behavior compared to the conventional Pt catalyst [122]. Kugai et al. reported a high O_2 conversion and a high CO_2 selectivity for $Pt–Cu/CeO_2$ catalysts [123]. The presence of partially reduced $Pt–CuO_x$ nanoparticles due to the presence of strong metal–support interactions was proposed as the active sites for the observed behavior. The promoting effect of Co was also studied on Pt catalysts [124–126]. Among the different Pt–Co catalysts reported in the literature, Pt–Co/YSZ was proposed as the most promising one being able to reduce the CO concentration below 10 ppm in the temperature range of 110°C–150°C, even in the presence

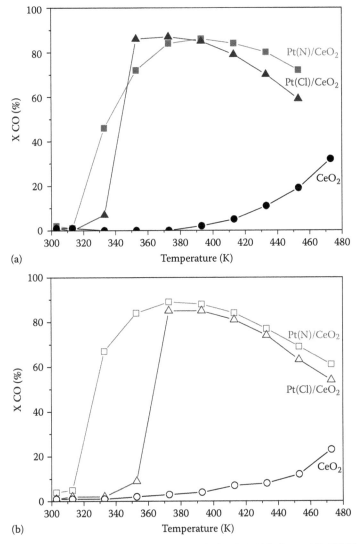

FIGURE 10.2 CO oxidation light-off curves for CeO_2, $Pt(N)/CeO_2$, and $Pt(Cl)/CeO_2$ catalysts, reduced at 523 K (a) and 773 K (b).

of CO_2 and H_2O [126]. Similarly, Pt–Ni catalysts supported in different supports such as carbon nanotubes or alumina were reported as active catalysts for the low temperature preferential oxidation of CO [127,128]. Ko et al. reported a superior performance for co-impregnated Pt-Ni/γ-Al_2O_3 catalysts compared to similar systems prepared by sequential impregnation [127]. Interestingly, these catalysts exhibit the high CO conversion even in the presence of 2% H_2O and 20% CO_2 over a wide temperature range. Besides Pt nanoparticles, other noble metal–based catalysts (mainly Ir) promoted using reducible oxides have been described in the literature for the PROX reaction. Recent studies described in the literature have shown that Ir catalysts supported in CeO_2, TiO_2, or $Ce_xZr_{(1-x)}O_2$ exhibit a promising catalytic activity with a high CO conversion at temperatures around 80°C (~70%) [112,129–132]. In a similar manner to Pt nanoparticles, the presence of chlorine from the metal precursor (H_2IrCl_6) gives rise to a poor catalytic activity due to the formation of Ce–O–Cl species, thus inhibiting oxygen surface mobility [131]. Combining two different preparation methods (deposited and embedded Ir), Lin et al. developed a dual-bed reactor, one using Ir-in-CeO_2 (top) and one using Ir-on-CeO_2 (bottom) [133]. Experimental results show that the dual bed affords a high CO conversion over a wide temperature range (80°C–200°C) together with a good selectivity to CO_2 (over 70%).

10.3.2 GROUP IB METAL CATALYSTS

Among the group IB metal, Au has been the most widely investigated in the PROX reaction. Although initially gold was considered to be inactive in the oxidation of CO, Haruta et al. found that Au nanoparticles below a critical diameter exhibit a surprisingly high catalytic activity [134]. Since this pioneering work by Haruta et al., numerous studies have been devoted to the dispersion of finely dispersed gold nanoparticles on suitable supports [135]. Similarly to group VIIIB metal catalysts, Au catalysts exhibit a poor catalytic activity at low temperatures in the presence of inert supports [136–140]. Quinet et al. found that Au/TiO_2 catalysts are highly active at low temperatures while Au/Al_2O_3 catalysts are not [136]. The catalytic activity of Au catalysts depends among others on the particle size (small nanoparticles ~2 nm are highly active) and the presence of hydrogen (hydrogen promotes CO oxidation at low temperature). Incorporation of a second metal (bimetallic systems) to Au catalysts has been suggested to improve the catalytic activity at low temperature several orders of magnitude when using inert supports. AuSr, AuCu, or AuAg nanoparticles deposited on inert supports such as silica or MgO have shown superior catalytic properties (improved CO conversion and superior selectivity) in the PROX reaction when compared to their constituent elements [137–140].

Another important approach to improve the catalytic activity of Au nanoparticles concerns the use of reducible oxides. The presence of the oxide allows not only the stabilization of a smaller particle size due to the presence of strong metal–support interactions but also the stabilization of ionic Au species (Au^{3+}/Au^+), which are proposed as active sites for the PROX reaction [65–67,141–143]. Interestingly, the catalytic activity of Au/CeO_2 catalysts can be further improved by incorporation of doping cations on the metal oxide network. The groups of Odriozola and coworkers and Bocuzzi and coworkers found that the incorporation of Eu, Zr, Zn, Sm, or Fe doping cations promotes the formation of oxygen vacancies on CeO_2, thus favoring Au dispersion due to the presence of specific interactions between the oxygen vacancies and the Au nanoparticles [144–146]. The improved catalytic behavior compared to the conventional Au/CeO_2 catalyst was attributed to the synergy between Au-Ce-M (M = doping cation), regardless of the nature of the modifier, that is, the higher the reducibility improvement of the CeO_2 support the higher the CO conversion. Furthermore, the presence of the doping cations improves the resistance of these catalysts toward deactivation by CO_2 compared to the Au/CeO_2 catalyst [147].

In spite of the excellent catalytic behavior of gold catalysts at low temperature, their poor catalytic behavior at high temperature, due to the competing H_2 oxidation, their high deactivation at long operation terms, as well as their sensitivity to the preparation conditions makes supported Au catalysts unattractive for a practical application in the PROX reaction.

Besides Au catalysts, in the last few years there has been a growing interest for other group IB metal catalysts, mainly CuO_x/CeO_2 and $Au/CuO_x/CeO_2$ [148–150]. Copper-based catalysts have been proposed as a substitution for the expensive noble metal catalysts for PROX reaction since they show a better catalytic performance. Liu et al. related the catalytic performance to the stabilization of Cu^+ species in the catalyst surface due to the presence of strong copper–cerium oxide interactions [148]. The reaction path was assumed to follow a redox mechanism, involving changes in the oxidation state for both copper (Cu^{2+}/Cu^+) and cerium (Ce^{4+}/Ce^{3+}), the interface sites being the active sites for the reaction. In general, these catalysts exhibit an improved catalytic behavior at low temperatures (high CO conversion) together with a high selectivity to CO_2. Luo et al. reported an improvement in the CO conversion (100% conversion at 80°C) for Cu/CeO_2 catalysts prepared using a surfactant templated method compared to Cu/CeO_2 catalysts prepared by conventional impregnation [151]. The improved behavior was associated with the presence of high-surface area nanosized CuO–CeO_2, that is, there was a correlation between the available surface area and the CO conversion. Besides the total surface area of the support and the nature of the surface copper species, the catalytic activity is also defined by the catalyst formulation. Mai et al. proposed the $Ce_{0.80}Cu_{0.20}O_2$ nanocomposite catalyst as the most promising formulation in terms of catalytic activity and stability [152]. Despite the good catalytic performance, copper-based catalysts are highly sensible to the presence of CO_2 and H_2O which exert a high inhibiting effect [153].

10.3.3 REACTION MECHANISM

The reaction mechanism for the PROX reaction highly depends on the nature of the catalyst used, either un-promoted or promoted noble metals. For un-promoted noble metal–based catalysts the reaction follows the competitive Langmuir–Hinshelwood mechanism where CO, O_2, and H_2 compete for the same reaction site, that is, the noble metal nanoparticle [154–157]. At low reaction temperatures, CO is strongly adsorbed on the metal nanoparticle, thus explaining the low catalytic activity of these systems in the PROX reaction at low temperature as described earlier. Higher temperatures are required to desorb the CO covering the metal nanoparticles, thus allowing O_2 and H_2 chemisorption. Therefore, CO desorption and O_2 adsorption are proposed as the rate-determining steps according to the following equation:

$$r = A \exp\left(\frac{-E_a}{RT}\right) P_{CO}^\alpha P_{O2}^\beta$$

where:
E_a is the activation energy
P_{co} and P_{O2} are the gas partial pressure for both reactants
α and β are the reaction order in CO and O2, respectively

Following these premises, low temperature CO conversion on noble-metal nanoparticles requires the weakening of the CO–noble metal bond to allow O_2 to access the active sites. Among the different approaches to weaken the CO–noble metal bond, the most promising ones involve the incorporation of a second more electropositive metal (bimetallic systems described earlier), the modification of the crystalline phase in the support or the modification of the particle size. As described in Figure 10.1, large platinum nanoparticles are more active due to the preferential growing of extended Pt(111) terraces where CO chemisorption is weaker compared to other crystallographic orientations [93,158]. Unfortunately, large nanoparticles also favor H_2 chemisorption and oxidation to H_2O, with the subsequent decrease in the selectivity. Although the Langmuir–Hinshelwood mechanism is the most widely accepted on noble metal un-promoted catalysts (mainly group VIIIB), there is some controversy concerning Au catalysts. In gold systems the reaction mechanism is more complex. While CO has been accepted to adsorb on Au nanoparticles, several classes of active sites have been suggested for oxygen activation: interfacial metal–support sites, cationic gold sites, and low-coordination gold atoms [159]. Several reaction intermediates adsorbed on gold and bicarbonates, carboxylates, and hydroxycarbonyls adsorbed on gold or at the particle–support interface have been proposed [135].

In the case of noble metal catalysts promoted by metal oxides, the reaction mechanism is more complex. The most widely accepted mechanism is the noncompetitive Langmuir–Hinshelwood mechanism where CO adsorbs on the metal nanoparticle and reacts with oxygen coming from the support at the metal–support interface [160]. However, the partially reducible nature of some oxide supports allows the presence of other reaction mechanisms. Among them, the most widely accepted is the Mars–van Krevelen mechanism in which surface oxygen lattice participates directly in the CO oxidation reaction without the participation of the noble metal nanoparticle, thus leaving oxygen vacancies [161]. Oxygen vacancies can be recovered from O_2 from the gas phase along the reaction due to the electron-rich environment created on the vicinity of these vacancies. However, temporal analysis of products on Au nanoparticles supported on different metal oxides (Al_2O_3, ZnO, ZrO_2, TiO_2) showed that in the presence of noble metal nanoparticles, a noble metal–assisted Mars–van Krevelen mechanism is also possible [162,163]. According to these studies, noble metal nanoparticles assist surface lattice oxygen removal and recovery along the catalytic reaction.

REFERENCES

1. A. Dermirbas, *Bio-hydrogen for future engine fuel demands*. Springer-Verlag London Ltd. London (2009).
2. J.A. Turner, *Science* 285 (1999) 687.
3. K. Liu, C. Song, V. Subramani, *Hydrogen and syngas production and purification technologies*. Wiley, Canada (2010).
4. G.W. Huber, S. Iborra, A. Corma, *Chem. Rev.* 106 (2006) 4044.
5. P.K. Cheekatamarla, C.M. Fimmerty, *J. Power Sources* 160 (2006) 490.
6. D.R. Palo, R.A. Dagle, J.D. Holladay, *Chem. Rev.* 107 (2007) 3992.
7. G.A. Deluga, J.R. Salge, L.D. Schmidt, X. Verykios, *Science* 303 (2004) 993.

8. R.M. Navarro, M.A. Peña, J.L.G. Fierro, *Chem. Rev.* 107 (2007) 3952.
9. J.M. Zalc, D.G. Löffler, *J. Power Sources* 111 (2002) 58.
10. R. Farrauto, S. Hwang, L. Shore, W. Ruettinger, J. Lampert, T. Giroux, Y. Liu, O. Ilinich, *Annu. Rev. Mater. Res.* 33 (2003) 1.
11. D.L. Trimm, Z.I. Önsan, *Catal. Rev. Sci. Eng.* 43 (2001) 31.
12. D.S. Newsome, *Catal. Rev. Sci. Eng.* 21 (1980) 275.
13. F. Huber, H. Meland, M. Roning, H. Venvik, A. Holmen, *Top. Catal.* 45 (2007) 101–104.
14. C. Ratnasamy, J.P. Wagner, *Catal. Rev.-Sci. Eng.* 51 (2009) 325.
15. Q. Fu, H. Saltsburg, M. Flytzani-Stephanopoulos, *Science* 301 (2003) 935.
16. Q. Fu, S. Kudriavtseva, H. Saltsburg, M. Flytzani-Stephanopoulos, *Chem. Eng. J.* 93 (2003) 41.
17. Y. Li, Q. Fu, M. Flytzani-Stephanopoulos, *Appl. Catal. B: Environ.* 27 (2000) 179.
18. C.M. Kalamaras, P. Panagiotopoulou, D.J. Kondarides, A.M. Efstathiou, *J. Catal.* 264 (2009) 117.
19. R.J. Gorte, S. Zhao, *Catal. Today* 104 (2005) 18.
20. T. Shido, Y. Iwasawa, *J. Catal.* 141 (1993) 71.
21. G. Jacobs, E. Chenu, P.M. Patterson, L. Williams, D. Sparks, G. Thomas, B.H. Davis, *Appl. Catal. A: Gen.* 258 (2004) 203.
22. E. Chenu, G. Jacobs, A.C. Crawford, R.A. Keogh, P.M. Patterson, D.E. Sparks, B.H. Davis, *Appl. Catal. B: Environ.* 59 (2005) 45.
23. C.M. Kalamaras, D. Dionysiou, A.M. Efstathiou, *ACS Catalysis* 2 (2012) 2729.
24. W. Deng, A.I. Frenkel, R. Si, M. Flitzany-Stephanopoulos, *J. Phys. Chem. C* 112 (2008) 12834.
25. C.M. Kalamaras, S. Americanou, A.M. Efstathiou, *J. Catal.* 279 (2011) 287.
26. S. Ricote, G. Jacobs, M. Milling, Y. Ji, P.M. Paterson, B.H. Davis, *Appl. Catal. A: Gen.* 303 (2006) 35.
27. J.B. Park, J. Graciani, J. Evans, D. Stacchiola, S.D. Senanayake, L. Barrio, P. Liu, J.F. Sanz, J. Hrbek, J.A. Rodriguez, *J. Am. Chem. Soc.* 132 (2010) 356.
28. P. Panagiotopoulou, D.I. Kondarides, *Catal. Today* 112 (2006) 49.
29. K.G. Azzam, I.V. Babich, K. Seshan, L. Lefferts, *Appl. Catal. B: Environ.* 80 (2008) 129.
30. X. Zhu, M. Shen, L.L. Lobban, R.G. Mallinson, *J. Catal.* 278 (2011) 123.
31. K.C. Petallidou, K. Polychronopoulou, S. Boghosian, S. García-Rodríguez, A.M. Efstathiou. *J. Phys. Chem. C* 117 (2013) 25467.
32. C.M. Kalamaras, K.C. Petallidou, A.M. Efstathiou, *Appl. Catal. B: Environ.* 136–137 (2013) 225.
33. S. Aranifard, S.C. Ammal, A. Heyden, *J. Catal.* 309 (2014) 314.
34. M. Boaro, A. Vicario, J. Llorca, C. de Leitenburg, G. Dolcetti, *Appl. Catal. B: Environ.* 88 (2009) 272.
35. A.M. Duarte de Farias, D. Nguyen-Thanh, M.A. Fraga, *Appl. Catal. B: Environ.* 93 (2010) 250.
36. I.D. Gonzalez, R.M. Navarro, M.C. Alvarez-Galan, F. Rosa, J.L.G. Fierro, *Catal. Commun.* 9 (2008) 1759.
37. Y. Amenomiya, P. Pleizier, *J. Catal.* 76 (1982) 345.
38. Y. Zhai, D. Pierre, R. Si, W. Deng, P. Ferrin, A.U. Nilekar, G. Peng et al., *Science* 329 (2010) 1633.
39. Y. Wang, Y. Zhai, D. Pierre, M. Flytzani-Stephanopoulos, *Appl. Catal. B: Environ.* 127 (2012) 342.
40. J.H. Pazmiño, M. Shekhar, W.D. Williams, M.C. Akatay, J.T. Miller, W.N. Delgass, F.H. Ribeiro, *J. Catal.* 286 (2012) 279.
41. B. Zugic, D.C. Bell, M. Flytzani-Stephanopoulos, *Appl. Catal. B: Environ.* 144 (2014) 243.
42. R. Buitrago, J. Ruiz-Martínez, J. Silvestre-Albero, A. Sepúlveda-Escribano, F. Rodríguez-Reinoso, *Catal. Today* 180 (2012) 19.

43. M. Haruta, N. Yamada, T. Kobayashi, S.J. Iijima, *J. Catal.* 115 (1989) 301.
44. F. Bocuzzi, A. Chiorino, M. Manzoli, D. Andreeva, T. Tabakova, L. Ilieva, V. Iadakiev, *Catal. Today* 75 (2002) 169.
45. J. Lin, N. Ta, W. Song, E. Zhan, W. Shen. *Gold Bulletin* 42 (2009) 48.
46. J. Li, J. Chen, W. Song, J. Liu, W. Shen, *Appl. Catal. A: Gen.* 334 (2008) 321.
47. N. Yi, R. Si, H. Saltsburg, M. Flytzani-Stephanopoulos, *Energy Environ. Sci.* 3 (2010) 831.
48. G. Jacobs, S. Ricote, P.M. Patterson, U.M. Graham, A. Dozier, S. Khalid, E. Rhodus, B.H. Davis, *Appl. Catal. A: Gen.* 292 (2005) 229.
49. H. Sakurai, T. Akita, S. Tsubota, M. Kiuchi, M. Haruta, *Appl. Catal. A: Gen.* 291 (2005) 179.
50. B.S. Caglayan, A.E. Aksoylu, *Catal. Commun.* 12 (2011) 1206.
51. T. Tabakova, M. Manzoli, D. Paneva, F. Boccuzi, V. Idakiev, I. Mitov, *Appl. Catal. B: Environ.* 101 (2011) 266.
52. D. Andreeva, V. Idakiev, T. Tabakova, A. Andreev, *J. Catal.* 158 (1996) 354.
53. D. Andreeva, V. Idakiev, T. Tabakova, A. Andreev, R. Giovanoli, *Appl. Catal. A: Gen.* 134 (1996) 275.
54. D. Andreeva, T. Tabakova, V. Idakiev, P. Christov, R. Giovanoli, *Appl. Catal. A: Gen.* 169 (1998) 9.
55. T. Tabakova, F. Bocuzzi, M. Manzoli, J.W. Sobczak, V. Idakiev, D. Andreeva, *Appl. Catal. B: Environ.* 19 (2004) 73.
56. R. Burch, *Phys. Chem. Chem. Phys.* 8 (2006) 5483.
57. N. Ta, J.Y. Liu, S. Chenna, P.A. Crozier, Y. Li, A.L. Chen, W.J. Shen, *J. Am. Chem. Soc.* 134 (2012) 20585.
58. P. Liu, J.A. Rodriguez, *J. Chem. Phys.* 126 (2007) 164705.
59. J.A. Rodriguez, *Catal. Today* 160 (2011) 160.
60. M. Shekhar, J. Wang, W.-S. Lee, W.D. Williams, S.M. Kim, E.A. Stach, J.T. Miller, W.N. Delgass, F.H. Ribeiro, *J. Am. Chem. Soc.* 134 (2012) 4700.
61. S. Carretin, P. Concepción, A. Corma, J.M.L. Nieto, V.F. Puntes, *Angew. Chem. Int. Ed.* 43 (2004) 2538.
62. F. Vindigni, M. Manzoli, T. Tabakova, V. Idakiev, F. Bocuzzi, A. Chiorino, *Appl. Catal. B: Environ.* 125 (2012) 507.
63. T. Tabakova, L. Ilieva, I. Ivanov, R. Zanella, J.W. Sobczak, W. Lisowski, Z. Kaszkur, D. Andreeva, *Appl. Catal. B: Environ.* 136–137 (2013) 70.
64. S. Hilaire, X. Wang, T. Luo, R.J. Gorte, J. Wagner, *Appl. Catal. A: Gen.* 215 (2001) 271.
65. N. Schumacher, A. Boisen, S. Dahl, A.A. Gokhale, S. Kandoi, L.C. Grabow, J.A. Dumesic, M. Mavrikakis, I. Chorkendorff, *J. Catal.* 229 (2005) 265.
66. S.H. Kim, S.W. Nam, H.I. Lee, *Appl. Catal. B: Environ.* 81 (2008) 97.
67. K.-R. Hwang, C.-B. lee, J.-S. Park, *J. Power Sources* 196 (2011) 1349.
68. J.-H. Lin, V.V. Guliants, *ChemCatChem* 4 (2012) 1611.
69. J.-H. Lin, V.V. Guliants, *Appl. Catal. A: Gen.* 445–446 (2012) 187.
70. L. Barrio, A. Kubacka, G. Zhou, M. Estrella, A. Martínez-Arias, J.C. Hanson, M. Fernández-García, J.A. Rodríguez, *J. Phys. Chem. C* 114 (2010) 12689.
71. S. Agarwal, L. Lefferts, B.L. Mojet, D.A.J.M. Lighthart, E.J.M. Hensen, D.R.G. Mitchell, W.J. Erasmus et al., *ChemSusChem* 6 (2013) 1898.
72. T.X.T. Sayle, S.C. Parker, D.C. Sayle, *Phys. Chem. Chem. Phys.* 7 (2005) 2936.
73. R. Si, M. Flytzani-Stephanopoulos, *Angew. Chem. Int. Ed.* 47 (2008) 2884.
74. Y. Lee, G. He, A.J. Akey, R. Si, M. Flytzani-Stephanopoulos, *J. Am. Chem. Soc.* 133 (2011) 12952.
75. Z.M. Tana, J. Li, H. Li, W. Shen, *Catal. Today* 148 (2009) 179.
76. F. Vindigni, M. Manzoli, T. Tabakova, V. Idakiev, F. Bocuzzi, A. Chiorino, *Phys. Chem. Chem. Phys.* 15 (2013) 13400.
77. I.H. Son, M. Shamsuzzoha, A.M. Lane, *J. Catal.* 210 (2002) 460.
78. S. Tosti, *Int. J. Hydrogen Energy* 35 (2010) 12650.

79. Q.H. Liu, L.W. Liao, X.H. Zhou, G.Q. Yin, *Adv. Mater. Res.* 236–238 (2011) 829.
80. T.V. Choudhary, D.W. Goodman, *Catal. Today* 77 (2002) 65.
81. E.D. Park, D. Lee, H.C. Lee, *Catal. Today* 139 (2009) 280.
82. K. Liu, A. Wang, T. Zhang, *ACS Catalysis* 2 (2012) 1165.
83. C.E. Thomas, B.D. James, F.D. Lomax, I.F. Kuhn, *Int. J. Hydrogen Energy* 25 (2000) 551.
84. A. Manasilp, E. Gulari, *Appl. Catal. B: Environ.* 37 (2002) 17.
85. G. Avgouropoulos, T. Ionnides, Ch. Papadopoulou, J. Batista, S. Hocevar, H. Matralis, *Catal. Today* 75 (2002) 157.
86. S. Ren, X. Hong, *Fuel Process. Technol.* 88 (2007) 383.
87. M. Watanabe, H. Uchida, H. Igarashi, M. Suzuki, *Chem. Lett.* 24 (1995) 21.
88. I. Rosso, C. Galletti, G. Saracco, E. Garrone, V. Specchia, *Appl. Catal. B: Environ.* 48 (2004) 195.
89. R. Andorf, W. Maunz, C. Plog, T. Stengel, US Patent 5955395, 1999.
90. V. Sebastian, S. Irusta, R. Mallada, J. Santamaria, *Appl. Catal. A: Gen.* 366 (2009) 242.
91. P.V. Snytnikov, V.A. Sobyanin, V.D. Belyaev, P.G. Tsyrulnikov, N.B. Shitova, D.A. Shlyapin, *Appl. Catal. A: Gen.* 239 (2003) 149.
92. K. Tanaka, M. Shou, H. Zhang, Y. Yuan, T. Hagiwara, A. Fukuoka, J. Nakamura, D. Lu, *Catal. Lett.* 126 (2008) 89.
93. E.O. Jardim, M. Gonçalves, S. Rico-Francés, A. Sepúlveda-Escribano, J. Silvestre-Albero, *Appl. Catal. B: Environ.* 113–114 (2012) 72.
94. M. Kotobuki, A. Watanabe, H. Uchida, H. Yamashita, M. Watanabe, *Chem. Lett.* (2005) 866.
95. S.H. Oh, R.M. Sinkevitch, *J. Catal.* 142 (1993) 254.
96. S.Y. Chin, O.S. Alexeev, M.D. Amiridis, *Appl. Catal. A: Gen.* 286 (2005) 157.
97. M. Echigo, T. Tabata, *Appl. Catal. A: Gen.* 251 (2003) 157.
98. Y.H. Kim, E.D. Park, H.C. Lee, D. Lee, K.H. Lee, *Catal. Today* 146 (2009) 253.
99. Y.-F. Han, M. Kinne, R.J. Behm, *Appl. Catal. B: Environ.* 52 (2004) 123.
100. C. Galletti, S. Specchia, G. Saracco, V. Specchia, *Ind. Eng. Chem. Res.* 47 (2008) 5304.
101. C. Galletti, S. Fiorot, S. Specchia, G. Saracco, V. Specchia, *Top. Catal.* 45 (2007) 15.
102. S.H. Lee, J. Han, K.-W. Lee, J. Korean, *Chem. Eng.* 19 (2002) 431.
103. S.Y. Chin, O.S. Alexeev, M.D. Amiridis, *J. Catal.* 243 (2006) 329.
104. H. Igarashi, H. Uchida, M. Watanabe, *Chem. Lett.* (2000) 1262.
105. P. Naknam, A. Luengnaruemitchai, S. Wongkasemjit, S. Osuwan, *J. Power Sources* 165 (2007) 353.
106. A. Parinyaswan, S. Pongstabodee, A. Luengnaruemitchai, *Int. J. Hydrogen Energy* 31 (2006) 1942.
107. A.U. Nilekar, S. Alayoglu, B. Eichhorn, M. Mavrikakis, *J. Am. Chem. Soc.* 132 (2010) 7418.
108. B.T. Qiao, A.Q. Wang, X.F. Yang, L.F. Allard, Z. Jiang, Y.T. Cui, J.Y. Liu, J. Li, T. Zhang, *Nat. Chem.* 3 (2011) 634.
109. M. Watanabe, H. Uchida, K. Ohkubo, H. Igarashi, *Appl. Catal. B: Environ.* 46 (2003) 595.
110. Q. Fu, W.X. Li, Y.X. Yao, H.Y. Liu, H.-Y. Su, D. Ma, X.-K. Gu et al., *Science* 328 (2010) 1141.
111. J. Kugai, R. Kitagawa, S. Seino, T. Nakagawa, Y. Ohkubo, H. Nitani, H. Daimon, T.A. Yamamoto, *Appl. Catal. A: Gen.* 406 (2011) 43.
112. F. Mariño, C. Descorne, D. Duprez, *Appl. Catal. B: Environ.* 54 (2004) 59.
113. O. Pozdnyakova, D. Teschner, A. Wootsch, J. Kröhnert, B. Steinhauer, H. Sauer, L. Toth et al., *J. Catal.* 237 (2006) 1.
114. J.L. Ayastuy, A. Gil-Rodríguez, M.P. González-Marcos, M.A. Gutiérrez-Ortiz, *Int. J. Hydrogen Energy* 31 (2006) 2231.

115. D. Teschner, A. Wootsch, O. Pozdnyakova-Tellinger, J. Kröhnert, E.M. Vass, M. Hävecker, S. Zafeiratos et al., *J. Catal.* 249 (2007) 318.

116. O. Pozdnyakova-Tellinger, D. Teschner, J. Kröhnert, P.C. Jentoft, A. Knop-Gericke, R. Schlögl, A. Wootsch, *J. Phys. Chem. C* 111 (2007) 5426.

117. H.–S. Roh, H.S. Potdar, K.-W. Jun, S.Y. Han, J.-W. Kim, *Catal. Lett.* 93 (2004) 203.

118. J.L. Ayastuy, M.P. González-Marcos, A. Gil-Rodríguez, J.R. González-Velasco, M.A. Gutiérrez-Ortiz, *Catal. Today* 116 (2006) 391.

119. A. Wootsch, C. Descorme, D. Duprez, *J. Catal.* 225 (2004) 259.

120. E.O. Jardim, S. Rico-Francés, F. Coloma, J.A. Anderson, J. Silvestre-Albero, A. Sepúlveda-Escribano, *Topics in Catal.* (Submitted).

121. Y. Gao, W. Wang, S. Chang, W. Huang, *ChemCatChem* 5 (2013) 3610.

122. H. Xu, Q. Fu, X. Guo, X. Bao, *ChemCatChem* 4 (2012) 1645.

123. J. Kugai, T. Moriya, S. Seino, T. Nakagawa, Y. Ohkubo, H. Nitani, T. Akita, Y. Mizukoshi, T.A. Yamamoto, *Chem. Eng. J.* 223 (2013) 347.

124. W.S. Epling, P.K. Cheekatamarla, A.M. Lane, *Chem. Eng. J.* 93 (2003) 61.

125. P.V. Snytnikov, K.V. Yusenko, S.V. Korenev, Y.V. Shubin, V.A. Sobyanin, *Kinet. Catal.* 48 (2007) 276.

126. E.Y. Ko, E.D. Park, H.C. Lee, D. Lee, S. Kim, *Angew. Chem. Int. Ed.* 46 (2007) 734.

127. E.-Y. Ko, E.D. Park, K.W. Seo, H.C. Lee, D. Lee, S. Kim, *Catal. Today* 116 (2006) 377.

128. R.T. Mu, Q. Fu, H. Xu, H. Zhang, Y.Y. Huang, Z. Jiang, S. Zhang, D.L. Tan, X.H. Bao, *J. Am. Chem. Soc.* 133 (2011) 1978.

129. M. Okumura, N. Masuyama, E. Konishi, S. Ichikawa, T. Akita, *J. Catal.* 208 (2002) 485.

130. Y.Q. Huang, A.Q. Wang, X.D. Wang, T. Zhang, *Int. J. Hydrogen Energy* 32 (2007) 3880.

131. Y.Q. Huang, A.Q. Wang, L. Li, X.D. Wang, T. Zhang, *Catal. Commun.* 11 (2010) 1090.

132. J. Lin, Y.Q. Huang, L. Li, A.Q. Wang, W.S. Zhang, X.D. Wang, T. Zhang, *Catal. Today* 180 (2012) 155.

133. J. Lin, Y.Q. Huang, L. Li, B.T. Qiao, X.D. Wang, A.Q. Wang, T. Zhang, *Chem. Eng. J.* 168 (2011) 822.

134. M. Haruta, T. Kobayashi, S. Iijima, F. Delannay, in: M.J. Phillips, M. Ternan (Eds.), *Proc. Int. Congr. Catal.* 9th, vol. 3, 1988, p. 1206F.

135. S. Ivanova, V. Pitchon, C. Petit, V. Caps, *ChemCatChem* 2 (2010) 556.

136. E. Quinet, L. Piccolo, F. Morfin, P. Avenier, F. Diehl, V. Caps, J.-L. Rousset, *J. Catal.* 268 (2009) 384.

137. A.Q. Wang, J.H. Liu, S.D. Lin, T.S. Lin, C.Y. Mou, *J. Catal.* 233 (2005) 186.

138. H. Häkkinen, S. Abbet, A. Sanchez, U. Heiz, U. Landman, *Angew. Chem. Int. Ed.* 42 (2003) 1297.

139. J.C. Bauer, D. Mullins, M. Li, Z. Wu, E.A. Payzant, S.H. Overbury, S. Dai, *Phys. Chem. Chem. Phys.* 13 (2011) 2571.

140. T. Deronzier, F. Morfin, M. Lomello, J.-L. Rousset, *J. Catal.* 311 (2014) 221.

141. L.F. Liotta, G. Di Carlo, G. Pantaleo, A.M. Venezia, *Catal. Today* 158 (2010) 56.

142. X.Y. Liu, A. Wang, T. Zhang, C.-Y. Mou, *Nano Today* 8 (2013) 403.

143. T. Takei, T. Akita, I. Nakamura, T. Fujitani, M. Okumura, J. Huang, T. Ishida, M. Haruta, *Adv. Catal.* 55 (2012) 1.

144. W.Y. Hernández, F. Romero-Sarria, M.A. Centeno, J.A. Odriozola, *J. Phys. Chem. C* 114 (2010) 10857.

145. O.H. Laguna, F. Romero Sarria, M.A. Centeno, J.A. Odriozola, *J. Catal.* 276 (2010) 360.

146. M. Manzoli, G. Avgouropoulos, T. Tabakova, J. Papavasiliou, T. Ioannides, F. Bocuzzi, *Catal. Today* 138 (2008) 239.

147. T. Tabakova, G. Avgouropoulos, J. Papavasiliou, M. Manzoli, F. Bocuzzi, K. Tenchev, F. Vindigni, T. Ioannides, *Appl. Catal. B: Environ.* 101 (2011) 256.

148. W. Liu, M. Flytzani-Stephanopoulos, *J. Catal.* 153 (1995) 317.

149. A. Di Benedetto, G. Landi, L. Lisi, G. Russo, *Appl. Catal. B: Environ.* 142–143 (2013) 169.
150. O.H. Laguna, W.Y. Hernández, G. Arzamendi, L.M. Gandía, M.A. Centeno, J.A. Odriozola, *Fuel* 118 (2014) 176.
151. M.-F. Luo, J.-M. Ma, J.-Q. Lu, Y.-P. Song, Y.-J. Wang, *J. Catal.* 246 (2007) 52.
152. H. Mai, D. Zhang, L. Shi, T. Yan, H. Li, *Appl. Surf. Sci.* 257 (2011) 7551.
153. J.L. Ayastuy, A. Gurbani, M.P. Gonzalez-Marcos, M.A. Gutierrez-Ortiz, *Int. J. Hydrogen Energy.* 35 (2010) 1232.
154. Y.-F. Han, M.J. Kahlich, M. Kinne, R.J. Behm, *Phys. Chem. Chem. Phys.* 4 (2004) 389.
155. M.J. Kahlich, H.A. Gasteiger, R.J. Behm, *J. Catal.* 171 (1997) 93.
156. A. Sirijaruphan, J.G. Jr. Goodwin, R.W. Rice, *J. Catal.* 227 (2004) 547.
157. J. Xu, X.-C. Xu, L. Ouyang, X.-J. Yang, W. Mao, J.J. Su, Y.-F. Han, *J. Catal.* 287 (2012) 114.
158. S. Mukerjee, J. Mcbreen, *J. Electroanal. Chem.* 448 (1998) 163.
159. A. Cho, *Science* 299 (2003) 1684.
160. M.M. Schubert, M.J. Kahlich, G. Feldmeyer, M. Hüttner, S. Hackenberg, H.A. Gasteiger, R.J. Behm, *Phys. Chem. Chem. Phys.* 3 (2001) 1123.
161. X.S. Liu, O. Korotkikh, R. Farrauto, *Appl. Catal. A: Gen.* 226 (2002) 293.
162. D. Widmann, Y. Liu, F. Schüth, R.J. Behm, *J. Catal.* 276 (2010) 292.
163. D. Widmann, R.J. Behm, *Angew. Chem. Int. Ed.* 50 (2011) 10241.

Index

Note: Locators followed by *f* and *t* denote figures and tables in the text

A

Alkali metal-based catalysts, 266
Anderson–Schulz–Flory model, 100, 151, 182
Aqueous-phase reforming (APR), 194, 228, 238
Autothermal reforming (ATR), 194

B

Badische Anilin and Soda Fabrick (BASF), 149
Becquerel effect, 3
Bimetallic catalysts, 291
Biofuels, 214–248, 253
    advantages, 218
    biomass composition, 221–226
    biomass platforms, 226–247
    biorefinery concept, 217–218
    feedstock harvest, 219–220
    pretreatment of biomass, 220–221
    world map of crops, 219*f*
Biomass composition, 221–226
    ash, 225–226
    energy storage molecules
        saccharide, 221
        starch, 222
        sucrose, 222
    lignocellulosic materials, 222–223
        cellulose, 223–224
        hemicellulose, 224
        lignin, 224–225
Biomass platforms, 226–247
    polysaccharides, 235–247
    vegetable oils/animal fats, 226–228
    via bio-oils, 228–235
Biomass pretreatment, 220–221
Biomass-to-liquid via Fischer–Tropsch
    (BTL-FT) synthesis, 149–150
Bio-oil applications, 235
Biorefineries, 215*f*, 217–218
Bi-reforming process, 95
Bowties, 226
"Brute-force" approach, 2, 39
Butler–Volmer model, 21

C

Carbon dioxide (CO$_2$) hydrogenation, 94–96
    carbon monoxide, 96–98
    formic acid, 103–104
    higher alcohols, 103
    higher hydrocarbons, 99–100
    methane, 98–99
    methanol and dimethyl ether, 100–103
    perspective and challenges, 104–105
Carbon dioxide (CO$_2$) to fuels, 93–119
Carbon monoxide, 96–98
Carbon Recycling International (CRI), 101
Cellulose, 223–224
Cellulose hydrolysis, 238–240
Chemical energy transmission system
    (CETS), 129
Chemical vapor deposition (CVD), 171
Cobalt, 25
Cobalt acetate, 155
Cobalt catalyst reduction, 163
Cocatalysts, 29
Combustion and reforming reactions (CRR)
    mechanism, 130
Computational fluid dynamics (CFD), 137
Contact potential differences (CPD), 138
Continuous stirred tank reactor (CSTR), 158
Copper-based catalysts, 97
Core/shell approach, 17
Crude glycerol stream, 207

D

Deep defects, 15
Density functional theory (DFT), 139,
    186, 199
Direct formic acid fuel cells (DFAFCs),
    103–104
Direct partial oxidation (DPO) mechanism, 130
Dopants, 15
Double layer hydroxide (LDH) structure,
    114–115
Dry reforming of methane (DRM) catalysts,
    128, 130
Dye-sensitized solar cell (DSSC), 13
Dye-sensitized water splitting, 42–44

E

Ecalene™ mixed alcohol process, 187
Electron donor–acceptor linked dyad, 63, 66–70
Electron spin resonance (ESR) study, 107
Energy Independence and Security Act (EISA),
    2007, 164

Enzymatic hydrolysis, 240
Ethanol
    bimetallic Cu–Co-based catalysts, 182
    history and current situation, 177–178
    industrial-grade, 177
    Mo-based catalysts, 180
    modified Fischer–Tropsch synthesis catalysts,
        181–182
    modified methanol synthesis catalysts,
        180–181
    noble metal-based catalysts, 179–180
    production from synthesis gas, 186–188
Ethanol steam reforming (ESR), 194, 203–207
Ethylene diamine tetraacetic acid (EDTA), 64
Extended X-ray absorption fine structure
        (EXAFS), 156, 172

F

Face-centered cubic (fcc) structure, 86
Fatty acid methyl esters (FAME), 217
Femtosecond laser excitation, 68
Fischer–Tropsch liquids (FTLs), 217
Fischer–Tropsch (FT) synthesis, 95, 99, 147–165
    critical features of reactors, 151–152, 152t
    prime deactivation pathways, 165
Fixed-bed reactors, 151
Fluidized-bed reactors, 151
Fluidized catalytic cracking (FCC) process, 226
Formic acid, 95, 103–104
Förster resonant energy transfer (FRET), 14
Furfural route, 240–242

G

Gas hour space velocity (GHSV), 134
Gas-to-liquids (GTL) facility, 150
Gel permeation chromatography (GPC), 259, 260f
Glycerol steam reforming (GSR), 195, 207–208
Gold-based catalysts, 285–287
Green diesel, 227
Group IB metal catalysts, 294–295
Group VIIIB metal catalysts
    bimetallic catalysts, 291
    monometallic catalysts, 289–291
    reducible metal oxide–promoted catalysts,
        291–294

H

Haldor Topsøe TIGAS® process, 176
Hemicellulose, 224
    hydrolysis, 237–238
Hexagonal close-packed (hcp) structure, 86
High-resolution transmission electron
        microscopy (HRTEM), 233f
"Holy Grail" of chemistry, 2, 44–45

Hydrocarbon fuels, 93–94
Hydrogen, 281
    purification, 281–296

I

Incident photon-to-current efficiency
        (IPCE), 16
Incipient wetness impregnation (IWI), 160
Indium tin oxide (ITO), 40
Institut Français du Pétrole (IFP), 182
Intermolecular electron transfer
    electron donor–acceptor linked dyad, 66–70
    electron donor to photoexcited
        photosensitizer, 70–75
Isoparaffin-rich diesel, 227
Isosorbide route, 244–246

L

Langmuir–Hinshelwood mechanism, 295–296
Layered metal oxide semiconductors
        (LMOS), 43
Le Châtelier's principle, 173
Levulinic acid route, 242–244
Lignin, 224–225
    hydrogenolysis, 246–247
Lignocellulosic biomass, 214
    catalytic pyrolysis, 253–276
        alkali metal–based catalysts, 266
        need and scope, 262
        over Na-based catalysts, 266–275
        selective deoxygenation, 262–264
        solid acid catalysts, 264–266
        thermal pyrolysis, 255–261
Low-temperature FT (LTFT) process, 152
    catalyst preparation, 154–156
    catalyst pretreatment, 163
    metal dispersion effect, 153–154
    noble metal promoters effect, 156–163
    support effect, 152–153
Lurgi MegaMethanol® process, 176, 176f
Lurgi–Octamix process, 187

M

Marcus theory, 66
Maritime feedstocks, 220
Metastable phase, 88
Metgas, 95
Methane, 98–99, 123–125
    activation processes, 125, 126f
    coupling of, 137–139
    direct nonoxidative conversion, 136–137
    dry reforming, 128
    to methanol, 139–141
    negative value, 124

oxidative coupling, 127–136
  reforming of, 125–136
Methanol, 100–103
  catalysts, 170–173
  history and current situation, 169–170
  industrial synthesis, 175–177
  steam reforming, 195–196
    Cu-based catalysts, 196–197
    Pd-based catalysts, 197–203
  thermodynamics and mechanistic
      considerations, 173–175
Methanol-to-gasoline (MTG) process, 95
Methanol-to-olefin (MTO) process, 95, 101
Monometallic catalysts, 289–291
Mott–Schottky analysis, 7

N

Nanocatalysis, 123
Nanoparticles, 77–88
Nanostructuring techniques, 18
Natural gas, 123–124
Nicotinamide adenine dinucleotide phosphate
    (NADP$^+$), 63–65
Ni nanoparticles, 84–88
Noble metal-based catalysts, 132–133
Nocera's catalyst, 5
Non-noble-metal–based catalysts, 287
Nuclear magnetic resonance (NMR), 186

O

Organosolv process, 247*f*
Oxidative coupling of methane (OCM),
    127–136, 138
Oxygen-evolving complex (OEC), 63
Oxygen storage/transport capacity (OSC),
    129–130

P

Partial oxidation (POX), 194
Partial oxidation of methane (POM), 125, 132*f*
Phosphoric acid fuel cells (PAFCs), 139
Photoactive semiconductor materials, 9–10
Photoanode–photocathode cells, 8
Photocatalytic conversion of CO_2 to fuel
  alternative solid semiconductors, 112–115
  phosphides and sulfides, 115–116
  photocatalytic systems based on TiO_2,
      105–112
  photoreduction of CO_2, 116–119
Photocatalytic hydrogen evolution systems,
    63–89
  electron-transfer behavior, 65–76
  metal nanoparticles on, 77–88
  overall cycle, 64*f*

Photocatalytic water splitting, 27–31
  photoanode materials, 32–34
  photocathode materials, 34–35
Photoelectrochemical tandem cells, 10
Photoelectrochemical water splitting, 31–35
Photoelectrode–photovoltaic hybrid cells, 8
Photosensitizer, 63–64
Photosynthesis, 105, 117
Photosystem I (PS I), 64
Photosystem II (PS II), 64
Photovoltaic-based water splitting, 38–42
Plasmonic-enhanced water splitting, 14
Plasmon resonance energy transfer (PRET), 14
Platinum-based catalysts, 140, 284–285
Polymer electrolyte fuel cells (PEFCs), 139
Polysaccharide hydrolysis, 235–236
Potassium promoters, 97
Potential determining ions, 6
Protonation route, 116
Proton exchange membrane fuel cell (PEMFC),
    193, 281
Pt nanoparticles, 77–84

R

Rapeseed methyl ester (RME), 217
Rate-determining step (RDS), 174
Reducible metal oxide–promoted catalysts,
    291–294
Reverse Boudouard reaction, 108
Reverse water–gas shift (RWGS) reaction, 96
Rhodium, 179
Ru nanoparticles, 80–84

S

Sabatier reaction, 98–99, 110
Saccharide, 221
  hydrolysis, 237
Sacrificial agents, 12
Scanning tunneling microscopy (STM), 160
Second-generation bioethanol, 203
Semiconductor-based solar water splitting, 5–7
Semiconductor–liquid junctions (SCLJs), 5–7
Semiconductor materials
  bulk modifications to enhance light
      harvesting, 14–16
  nanostructuring, 16–19
  surface modifications to enhance light
      harvesting, 13–16
Semiconductor photoelectrodes, 6–7
Simulated countercurrent moving bed
    chromatographic reactor
    (SCMCR), 141
Single-walled carbon nanotubes (SWNTs), 130
Sintering (aging), 165
Slurry-bed reactors, 151

Solar water splitting, 2
    historical background on, 3–5
    main approaches toward, 7–8
    materials requirements and current trends,
        8–27
        modification and nanostructuring, 12–19
        semiconductor materials, 8–12
        water oxidation and reduction catalysis,
            19–27
    principles of semiconductor-based, 5–7
    progress on semiconductor-based
        dye-sensitized water splitting, 42–44
        photocatalytic water splitting, 27–31
        photoelectrochemical water splitting,
            31–35
        photovoltaic-based water splitting, 38–42
        tandem photoelectrochemical systems,
            35–38
Sol–gel technique, 128
Solid acid catalysts, 264–266
South African Coal Oil and Gas Corporation
    (SASOL), 150
Spatial structuring of nanosized materials, 18
SSITKA (steady-state isotopic-transient kinetic
    analysis), 156
Starch, 222
    hydrolysis, 237
Steady-state isotopic-transient kinetic analysis
    (SSITKA), 156
Steam reforming, 193–208
    of monofunctional alcohols
        ethanol steam reforming, 203–207
        methanol steam reforming, 195–203
    of polyols, 207–208
Steam reforming of methane (SRM), 125–128
STM (scanning tunneling microscopy), 160
Strong metal–support interactions (SMSI), 153, 171
Sucrose, 222
    hydrolysis, 236–237
Sunlight-to-fuel energy conversion, 105
Surface science, 199
Syngas, 95, 100, 125, 127–128, 130, 134–136,
    147, 149–150, 164
    to methanol and ethanol, 169–188
Synthetic fuels, 148
Synthetic natural gas (SNG), 95

T

Tafel slope, 21–22
Tandem cell, 35
Tandem photoelectrochemical systems,
    35–38
Thermal pyrolysis, 255–261
Thermo gravimetric analysis (TGA), 158
    of catalyst, 272–273
Third-generation biofuels, 214
TiO_2 photocatalytic performance, 110
Titanates, 114
Transmission electron microscopy (TEM),
    172, 288
Triglyceride esters, 226
Turn over frequency (TOF), 21, 141
Turn over number (TON), 21, 141, 156

U

Urea gelation co-precipitation (UGC), 288

W

Water–gas shift (WGS) reaction, 97, 174, 195,
    255, 282–284
    gold-based catalysts, 285–287
    importance of catalyst support, 287–288
    non-noble-metal–based catalysts, 287
    platinum-based catalysts, 284–285
Water oxidation catalyst (WOC), 13
Water reduction catalyst (WRC), 19–24
Water-splitting tandem cells, 8
Weight hours space velocity (WHSV), 137
Wooden-based biomass, 229–234
World gas scenario, 124f

X

X-ray photoelectron spectroscopy (XPS),
    159, 199
XRD (X-ray diffraction) measurement, 156

Z

Z-scheme concept, 30